INSTRUCTOR'S RESOURCE MANUAL

GEOSYSTEMS

An Introduction to Physical Geography

INSTRUCTOR'S RESOURCE MANUAL

Cecelia Hudelson
Foothill College

Third Edition

GEOSYSTEMS

An Introduction to Physical Geography

Robert W. Christopherson

PRENTICE HALL Upper Saddle River, NJ 07458

Production Editor: *Carole Suraci*
Supplement Editor: *Wendy Rivers*
Special Projects Manager: *Barbara A. Murray*
Production Coordinator: *Benjamin Smith*

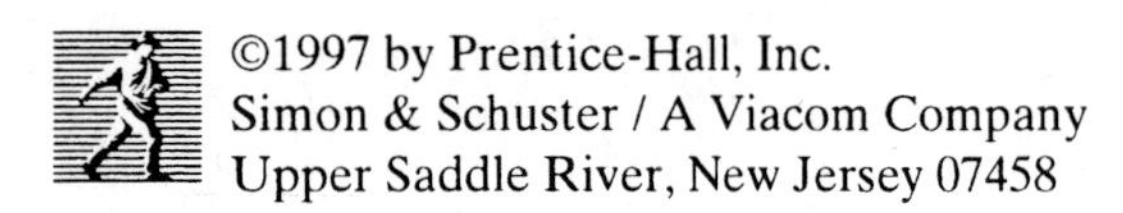

Simon & Schuster / A Viacom Company
Upper Saddle River, New Jersey 07458

Printed in the United States of America

10 9 8 7 6 5 4 3 2 1

ISBN 0-13-505348-X

Prentice-Hall International (UK) Limited, *London*
Prentice-Hall of Australia Pty. Limited, *Sydney*
Prentice-Hall Canada Inc., *Toronto*
Prentice-Hall Hispanoamericana, S.A., *Mexico*
Prentice-Hall of India Private Limited, *New Delhi*
Prentice-Hall of Japan, Inc., *Tokyo*
Simon & Schuster Asia Pte. Ltd., *Singapore*
Editora Prentice-Hall do Brasil, Ltda., *Rio de Janeiro*

Contents

Introduction

Outline

New Beginning
The *GEOSYSTEMS* Package
Geographic Awareness
Format for this Manual
A Tour of GEOSYSTEMS
GEOSYSTEMS Newsletter Update
Teaching Essentials
  Teaching Media
  Grading and Course Structure Suggestions
Final Thoughts about Teaching
  The Teacher as a Personality
  The Teacher as Intellect
  The Teacher as Actor
A Ready Reference
  Table of Contents
  Focus Study List
"Geography I. D."

New Beginning

Welcome to physical geography! Thus begins the third edition of *GEOSYSTEMS–AN INTRODUCTION TO PHYSICAL GEOGRAPHY*. The success of the first and second editions throughout the United States and Canada is owed to the teachers and students who have embraced our approach to physical geography. The third edition is therefore dedicated to you the teachers and the students of the new edition. "*To all the students and teachers of Earth, our home planet, and its sustainable future.*"

The dynamics of physical geography–an important Earth system science–is well illustrated by the June 1991 eruption of Mount Pinatubo in the Philippines. The effects of this century's most powerful eruption is systematically woven throughout the third edition. The injection of 15 to 20 million tons of ash and sulfuric acid mist into the stratosphere produced colorful twilight afterglows worldwide. The increase in atmospheric albedo produced a worldwide average temperature decrease of over 0.5°C (0.9F°) and temperatures decreased about half this amount in the Southern Hemisphere. This evidently lowered temperatures in what would have been one of the warmest years in history. These volcanic aerosols increased shortwave reflectance of the atmosphere by 4.3 W/m^2, whereas longwave absorption increased only by 1.8 W/m^2, producing a net radiation reduction of 2.5 W/m^2. Such measurements are recorded by Earth Radiation Budget (ERB) sensors aboard several satellites. Examine Figure 6-1 for a series of AVHRR images of these aerosols and a visible-wavelength image of the eruption. The general circulation model (GCM) operated by the Goddard Institute for Space Studies (GISS) accurately predicted this temporary cooling effect. Volcanic processes and their relationship to plate tectonics are covered in Chapters 11 and 12, paleoclimatic effects in Chapter 17, and correlation of this eruption to aspects of the nuclear winter hypothesis in Chapter 10. Physical geographers possess the spatial analysis tools to understand the global implications of such an event.

Features of the Third Edition
You will notice many new and rerendered figures and revised text sections in the third edition. Every chapter of the third edition is thoroughly revised, updated, and expanded, with new or improved maps, photographs, art, and tables. There is expanded coverage of global change, climate, the Great Lakes, weather models, and Canada. The many new figures in this edition are fully integrated within the text discussions. Here are a few of the features of the third edition.

- Our widely praised cartography program is updated to reflect the rapid pace of change in political boundaries and physical systems. Of all the maps in the text, 40% are new to this edition. Locator maps accompany most remote sensing images and photographs. Also, numerous topographic maps, many new to this edition, are used to illustrate key features of the landscape.

- Photographs from every continent are included in our 375 photos; 60% are new to this edition. Many photos are integrated with art so you can compare the concept shown in the art with a representative scene from the real world. For example: the four seasons on land and in orbit (Figure 2-21 and 2-22), limestone landscapes and caverns (Figure 13-18), Earth's major deserts (Figure 17-9 and 17-13), and the Great Lakes shorelines (Focus Study 19-1).

- New, expanded coverage of the Great Lakes: lake-effect snow (Chapter 8), formation of the Great Lakes (Chapter 17), and a new focus study on the lakes and their ecosystems (Chapter 20).

- Continued coverage of Canadian physical geography includes a text, figures, and maps of periglacial landscapes (a new focus study in Chapter 17) and Canadian soils (a new Appendix C). Canadian data on a variety of subjects are portrayed on 25 different maps in combination with the United States-physical geography does not stop at the United States-Canadian border!

- Twenty-one "Focus Study" essays, some completely revised and several new to this edition, provide additional explanation of key topics as diverse as the stratospheric ozone predicament, wind power, forecasting the near-record 1995 hurricane season, the El Nino phenomenon, an environmental approach to shoreline planning, the Mount St. Helens eruption, the status of the Colorado River, and the loss of biodiversity.

- Seventy "News Reports" relate topics of special interest. For example: careers in GIS, jet streams and airlines flights times, how one culture harvests fog, the new UV Index, using Earthshine to study global energy budgets, the disappearing Nile Delta, humans eat clay, how sea turtles read Earth's magnetic field, and drilling for oil in the rain forest.

- The *Geosystems Home Page* provides on-line resources for each chapter on the World Wide Web. You will find review exercises, specific updates for items in the chapter, suggested readings, and links to interesting related pathways on the Internet. A click on the Tables of Contents link and selection of a chapter launches you into this new dimension of physical geography.

The *GEOSYSTEMS* Package

To enhance your use of *GEOSYSTEMS* and the teaching of physical geography a full package of supplements and teaching aids are available. In addition to the text please make sure you have received the following.

- *Student Study Guide* (ISBN: 013-573080-5): includes figures from the text. This supplement can be packaged with the text or sold separately at bookstores. The organizational format for each chapter of the study guide is as follows:
  - Chapter Overview and Learning Objectives
  - Glossary Review
  - Learning Activities and Exercises

– Sample Self-test (Answers appear in
the back of the study guide.)

• *Applied Physical Geography: Geosystems in the Laboratory* (ISBN: 013-505405-2): a new laboratory manual that includes 12 comprehensive lab exercise chapters. The chapters are further broken down into a total of 77 convenient steps. A complete manual of solutions and answers for these lab exercises is included in the back of this instructor's manual.

• Overhead Transparencies set (ISBN: 013-505355-2): 105 full-color overhead transparencies of key figures in the text plus 4 of the principal tables. Most of the figures featured are enlarged for better projection.

• A 35-mm slide set (ISBN: 013-505363-3): 60 important pictures and images from the text.

• Test Bank (ISBN: 013-505330-7): over 2672 revised questions and answers to accompany the text. Computer versions available for Mac (013-505389-7) and IBM (013-505371-4).

• *The Macmillan Geodisc* (ISBN: 0-02-322474-6): an interactive laser disk fea-turing 1500 still images, over 50 minutes of motion video, and six minutes of full-color animations. This interactive laser disk is accompanied by an instructor's manual and complete bar-code system.

• Internet Support by Duane Griffin and James E. Burt, University of Wisconsin-Madison, and Robert Christopherson. Our Geosystems Home Page
(http://www.prenhall.com/geosystm) gives you on-line review exercises, opportunities to delve deeper into subjects out on the Net, follow-up answers to specific items in the text (for example, the forecast outcome of the 1996 and 1997 Atlantic hurricane seasons), and links to a wealth of interesting sites that relate to each chapter.

Geographic Awareness

Undergraduate physical geography is an introduction to geography and as such is an opportunity to teach general education students and perhaps interest them in our discipline–a potential harvest of teaching, academic, and practicing geographers for the future. At my college, American River College, our classes are filled with freshman and sophomore students completing their core curriculum courses before transferring to upper-division work. Physical geography is an introduction, and perhaps their only exposure, to science and scientific methodology. Because of the nature of physical geography, a diverse and dynamic approach seems most appropriate. The following ideas are simply one teacher's opinion and perceptions.

Popular news sources have carried many articles concerning the geographic illiteracy of students and the general public. And, strangely enough, the subject of global geographic education became politicized in the 1980s with a former Secretary of Education actually warning the nation against eliminating such illiteracy! Unbelievably, educators were told by this former official that students would "suspend judgment and get wishy" if they knew their geography and history better! (For example, see: *FOCUS*, summer 1990, Vol. 40, No. 2, p. 26, "What is the Essence of Geographic Illiteracy?" by Kit Salter; *Newsweek*, 9/1/86, p. 67, "The Forgotten Subject"; *Newsweek*, 8/8/88, "Lost on the Planet"; *Christian Science Monitor*, 5/1/89, p. 7, "Schools Fight Geographic Illiteracy with National Contest"; *Sacramento Bee* [in the library reference system "Newsbank"], 4/6/89, Section F, p. 1, "Where in the World?")

I suggest that part of a lecture period be devoted to a discussion of geography and the importance of knowing the "where" of physical and cultural things. This is why I begin Chapter 1 with a section defining the geographic approach and our scientific discipline. The five themes authored by the Association of American Geographers and the National Council for Geographic Education are featured. The course can be dedicated to placing each student consciously onto the world stage and in their campus and community–at least in terms of physical systems.

In the Preface of the new *Applied Physical Geography: Geosystems in the Laboratory* manual I ask the students to complete their "Geography I. D." The one-page assignment is a personal application of the five themes to thoroughly locate and place the students in the environment. I include this one-page geography lesson as the last page of this introduction. Let me know what you think.

Public Broadcasting System stations aired a 10-part series entitled, "Spaceship Earth–A Global Geography," produced by Antelope Films. Perhaps check to see if your campus has a site license for this outstanding look at physical and cultural/human geography as tools with which to analyze the status of our planet.

The Association of American Geographers and the National Council for Geographic Education are both active in programs that improve public geographic literacy and upgrade geographic education. (See Appendix A in the text for addresses of related organizations.) The National Geographic Society, assisted by many geographers and educators nationwide, has set up over 60 alliances to promote geographic education. You may want to consult your local alliance and acquaint your students with their programs. These alliances produce newsletters and information useful for teaching. (Contact: the Geography Education Program, Educational Media Division, National Geographic Society, 17th and M Streets, N.W., Washington, D.C. 20036 for more information about state alliances, availability of materials for geographic education, and to receive their excellent newsletter called *Update*.)

By federal mandate, we now have an official "Geography Awareness Week" each November. Also, the National Geographic Society conducts a National Geography Bee beginning at the state level each April and ending with national finals held in Washington D.C. in late May hosted by Alex Trebek from *Jeopardy!* In 1993 they sponsored a first-ever International Geography Olympiad in London that was broadcast on PBS. Geography is now specified as one of five key core curricula for national emphasis, along with history, mathematics, science, and English. As a discipline, geography is witnessing a strong resurgence as the 1990s progress.

Format for this Manual

In this resource manual, each text chapter is supported from several perspectives. You will find each chapter of this manual organized in the following manner.

Overview: A brief introductory statement summarizing the content of the chapter and my overall goals for the text presentation. Included are references to specific coverage in the *Student Study Guide* and the *Applied Physical Geography* lab manual.

New to the Third Edition: A review of new material, figures, and revisions that you will find in each chapter of this new edition.

Key Learning Concepts: A list of the learning objectives addressed in each chapter.

Expanded Outline Discussion: I use the first, second, and third order headings directly from each chapter and briefly discuss related background, additional sources, (including web sites) added examples, tangential material, suggestions for discussion, or topical materials. Not all text headings are discussed.

Glossary Review: Conveniently grouped together in alphabetical order are all the **boldfaced** terms and concepts that appear in the chapter. In the student study guide, these glossary terms appear with fill-in spaces for the students to record definitions in their own words. The second edition features 700 boldfaced entries in the glossary.

Annotated Chapter Review Questions: The review questions at the end of each chapter are specifically answered with wording from the text.

Overhead Transparencies: As an adopter of *GEOSYSTEMS* you have received a packet of 100 high-quality, full-color overhead transparencies of figures and key tables from the text. The overheads selected for each chapter are listed with their captions.

A Tour of *GEOSYSTEMS*

Earth's physical geography systems are complex, interwoven threads of energy, air, water, weather, climates, landforms, soils, plants, animals, and the physical Earth itself. *GEOSYSTEMS* is thoroughly up-to-date, containing the latest information about the status of Earth's physical systems as viewed through the *spatial analysis* approach. *GEOSYSTEMS* is an introductory text in which relevant human-environmental themes are integrated with the principle core topics usually taught in a course of physical geography. The text is conveniently organized into four parts that logically group chapters with related content. After completion of Chapter 1, "Essentials of Geography," these four parts may be taught in the integrated text sequence or in any order convenient to the instructor.

GEOSYSTEMS is organized around process systems and is essentially non-mathematical, with text material organized in the same direction as the flow of energy and matter, or arranged in a manner consistent with time and the flow of events. Particular attention was invested in the rewriting and revision of *GEOSYSTEMS* to achieve readability and clarity appropriate to the student in university- and college-level courses. The student will benefit from thorough review questions, varied and generally available suggested readings, a comprehensive glossary, and full-color high-production values as evidenced by more than 550 figures. Some of the text features include:

- Inside the front cover is a new image of "The Living Earth" from Earth Imaging and Spaceshots, Inc. The satellite-based computer illustration is created from thousands of satellite images gathered by Advanced Very High Resolution Radiometer (AVHRR) sensors aboard several NOAA satellites. The text in Chapter 20 asks students to compare the global terrestrial biome map in Figure 20-4 with this illustration. "The Living Earth" features natural colors typical of local summer under cloudless conditions. Ocean-floor features are derived from other remotely sensed bathymetric data.
- After Chapter 1, which presents the essentials of physical geography, including a discussion of geography, systems analysis, latitude and longitude, time, and cartography; a logical arrangement follows through the next 20 chapters which are organized into four parts.
- *GEOSYSTEMS* avoids the typical approach of tacked-on relevancy in isolated boxes or inserts, or the omission of such material entirely. Care is taken to discuss applied topics with the core material where appropriate; in this way the student can relate basic subject matter to the real world.
- Twenty-one "Focus Study" inserts, many with figures, highlight with greater depth key topics related to text material. The emphasis varies among the focus studies from traditional (Chapters 1, 10, and 17) to topical subjects (Chapters 3, 8, 12,

15, 18, and 19), with some handling specific tools as in the focus studies in Chapters 2 and 5. See the complete list of focus studies for the text at the end of this section.

- The text and all figures use metric/English measurement equivalencies. A complete and expanded set of measurement conversions is conveniently presented in an easy to use arrangement inside the back cover of the textbook. For more information about metric conversion materials and programs contact: the Metric Program Office, National Institute of Standards and Technology, U.S. Department of Commerce, Gaithersburg, MD 20899. A useful report was prepared by Barry N. Taylor, ed., *Interpretation of the SI for the United States and Metric Conversion Policy for Federal Agencies*, NIST Special Publication 814, NIST, Department of Commerce, October 1991. And, nongovernmen-tal organizations: the U.S. Metric Association, 818-368-7443; HRD Press, 1-800-822-2801; Marlin Industrial Division, 1-800-344-5901. The conversion law is Public Law 100-418, 100th Congress, H.R. 4848, 8/23/88, the "Metric Conversion Act" that amended the original 12/23/75 act.
- Appendix A contains addresses of important geographic and environmental organizations and agencies, general reference information, and sources of geographic resources.
- Appendix B describes topographic maps, and includes a hypothetical quadrangle.
- Appendix C explains the Canadian System of Soil Classification, including a section of review questions to aid student comprehension.
- GEOSYSTEMS ends with a capstone chapter, "Earth, Humans, and the New Millennium," which is unique in physical geography. *GEOSYSTEMS* attempts to look holistically at Earth and the world in this concluding treatment. This chapter can be used to facilitate class discussions concerning the impact of humans on Earth systems, as well as the important role played by geographic analysis and physical geography in particular. A focus study detailing the 1992 Earth Summit is presented in Chapter 21.
- You will notice the inclusion of the latest relevant scientific information throughout the text. This helps the student see physical geography as an important Earth system science linked to other sciences in this era of expanded data gathering, computer modeling, and discovery. Geography is a dynamic science not a static one!
- An up-to-date treatment of climate change, including global warming, is integrated throughout the text. The student is shown how climate change specifically relates to many aspects of physical geography. Note that the topic does not appear as one of the focus studies; rather, it is treated within the text. Although it is politically controversial there is a degree of scientific consensus relative to global warming, with this difference clearly explained. The four reports from the Intergovernmental Panel on Climate Change (1990 and 1992) provide an important resource for these sections.
- In key chapters (climate, arid landscapes, soils, biomes), large, integrative tables are presented to help synthesize content. If you choose, the rest of the chapter can be studied for deeper analysis of the subject. This approach adds to the adaptability of the text for different levels of presentation and for quarter, trimester, or semester length classes. In fact, Table 20-1 in Chapter 20 can be referenced throughout the term, as the total Earth system is organized within the various terrestrial biomes.
- The Glossary in *GEOSYSTEMS* is prepared with text-related definitions and explanations for 700 terms and

concepts. Over 1800 terms and concepts appear in the relational index.

GEOSYSTEMS Update Newsletter

As an adopter you will receive a *Geosystems Update* Newsletter. The purpose is to share the latest summaries of scientific information related to the content of the text. Such things as general solar, weather, earthquake, and volcanic activities; astronomical happenings, and environmental events and traumas, among other items, are included. Also, included are ideas for classroom activities, addresses and career information for student use, relevant professional meetings and activities, and additional resources for the teaching of physical geography.

Teaching Essentials

I do not know if there is a correct way to teach or if any individual has discovered the "right" approach. This is part of teaching's mystery and intrigue. The variation of practices among teachers is as subjective and variable as the individuals themselves. As more experience and maturity is gained in the individual personality, so too does the definition of the teaching craft evolve within that individual. The teacher is constantly redefining the inner "ultimate potential teacher" concept. So, to pin down specific rules for teaching is impossible, although some generalizations are appropriate.

I think it is time that we all accept the status of teaching as a true profession and that practitioners are true professionals. No matter how hard we stress academic geography and published research, the flow of new geographers into the discipline and the enrollments so vital to funding and administrative internecine campus battles are both keyed to our ability to attract people to our classes. This enrollment in turn generates interest from the institution. Think for a moment of the master teacher who started your inner-fire.

Dr. Susan Hanson, past-president of the Association of American Geographers, summarized the importance of teaching and teaching methodologies in her column in the January 1991, *AAG Newsletter* (Vol. 26, No. 1, p. 1):

> ...an important message for academic geographers: good teaching is essential to the renewal and vitality of our discipline, and now is the time to capture students' imagination and idealism by conveying the conviction that geographic insights can empower people to make a difference....With students now clearly wanting to contribute to solving problems like urban poverty, environmental pollution, uneven development, land degradation, and resource depletion, we have the opportunity to demonstrate the strengths of the geographic tradition in tackling problems. The key is effective teaching.
>
> It is time to see that the strength of our discipline lies not only in pursuing new knowledge about the world but also in infecting others with an eagerness to pursue new geographic knowledge on their own. It is time to openly acknowledge and cherish and learn from the good teachers among us. It is time to reflect closely upon, and to share with colleagues, what happens on those days when our teaching "clicks" and we leave the classroom with that glow that comes from knowing students have caught the fire of our enthusiasm. It is time to think anew about how people learn, and about how we might tune our teaching to accommodate the diversity of learning styles in our classrooms. It is time to take advantage of new technologies that facilitate learning.

Teaching Media. The student of today is visually motivated to a great degree, and therefore responsive to media. If you are fortunate enough to have a dedicated geography classroom, then a plan can be worked out among instructors for making it part of the media for teaching. Possible media resources may include the classroom walls and ceiling (for display of maps, charts, satellite images, an events bulletin board, and posters), overhead projector, slide projector with remote control at the screen, video monitor and playback with remote, cassette tape player, shortwave radio, chalkboard, and a teaching "tool box."

• **Classroom**. I place posters, photographs, images, and maps that relate to topics under discussion in different parts of the room so that I can leave the front and teach from other locations in the room. For example, a Mount Saint Helens poster, map projections chart, image of Hurricane Gilbert, relative age of the ocean crust and ocean floor maps, weather map symbols, and world physiographic map are mounted around the room. I have placed a mounted poster of the Andromeda Galaxy in the back of the room so that I can walk the 2 million light years to get to it during that discussion. The National Geographic Society's "The World" mural map (120" x 72") is reasonably priced and suitable for permanent installation. If your materials must be transported, it might be convenient to place those items that are not on rollers on large pieces of cardboard. (Appliance and automotive dealers throw away such cardboard. Various acetate coverings are available from suppliers.)

I have two screens placed at angles in each front corner–one for overheads and the other for slides. In this way both screens and the chalkboard can be used simultaneously. The video monitor is installed in the front-center of the room above the chalkboard, with the VCR installed in a modified cabinet drawer on the side of the room. In other rooms a mobile cart is used, a situation probably similar to that on most campuses.

I had the room retrofitted with some switches in the front that turn the lights and the slide projector on and off. I recommend a note-taking light fixture in the back of the room, so that media can be shown with enough light present for students to take notes. This back light is also on a switch so the room can be completely darkened when desired. I am sure there are many possible variations, depending on the room and your specific needs.

• **Overhead projector**. With the availability of thermal films and machines, it is possible to make overheads from any source. As dictated by room size, the overhead can be used in concert with the chalkboard. I do not recommend its exclusive use in rooms with under 100 chairs because an inherent visual disconnection between the teacher and the student is associated with its chronic use. In larger rooms, however, nearly continuous overhead use is necessary.

• **Slide projector**. Slides are an essential resource in physical geography, for this is a course studying the physical elements of Earth's environment. Pictures from many sources, including personal photos, bring the world into the classroom. Any camera with a 50 mm-macro lens can be used to take slides for class from various sources. If a copy stand is not available, then simply use daylight film and set up outdoors with diffused natural lighting–I have made many slides on the patio. Or, using tungsten film, you may make slides on the dining room table or your desk with two 75-watt floodlights in simple reflectors. Use slower films (ASA-64, daylight, or KPA-40, tungsten) for finer grain and clarity. (It is important to check copyright laws with all copied materials. Be sure they are not used for profit and that they are exclusively for classroom purposes.)

Install in your room, or carry with you, a remote cord and projector controller so that you can stand at the screen or even in the picture as the slide is explained. The "voice in the shadows" and "this is a..." type of presentation is dead in the dark. Use slides the same way you would tutor a student with a figure in the text, pointing things out with your hands.

• **Video monitor and playback**. I think any media can be abused through overuse, and this includes videotapes of science shows especially when such tapes are used with the teacher out of the picture. I believe that the use of video in a manner similar to the use of slides, like single-concept pictures in motion, is preferable. Simple editing to produce succinct pieces is easy with two VCRs placed side-by-side, using an SP mode to produce a better second-generation image. The remote allows you to freeze-frame and reverse for repeated showing. An excellent use of this is for weather satellite loops or coverage of environmental events such as earthquakes, floods, hurricanes, etc. Public domain NASA videos are shown on many local cable access channels. I sometimes have these orbital images of Earth running as the students come into class. Check with your local Public Broadcasting System station for more information on availability of videography or with your local cable company for NASA coverage. Your campus probably has a site license for many of these shows. Additionally, C-SPAN allows you to tape key congressional debates and votes, pertinent lectures, hearings, and national conferences that deal with environmental topics associated with physical geography. These video segments can work effectively to augment lectures and to bring a point home.

If you have access to a video camera, you may want to tape some lecture segments at a location in the field. The outdoors are certainly usable in class, for example, to incorporate interviews with park rangers and technicians and personal on-camera explanations.

• **Interactive laser disk.** *GEOSYSTEMS* is supported by a new interactive laser disk–*The Macmillan Geodisc*. An instructor's manual is provided with the disk, complete with bar codes for all the photographs, illustrations, motion video, and color animations. Most disk players provide access to numbered frames either through a numeric pad or a bar-code reader. This flexibility allows you to customize your presentation with a variety of media resources. As an example, let's say you are discussing the global impacts of the Mount Pinatubo eruption and with a single bar-code sweep, you call up a four-minute video segment from the BBC that brings the sights and sounds of the actual eruption to the classroom!

• **Cassette tape player.** National Public Radio news shows have featured segments on geographic education and physical geography suitable for classroom use. When introducing certain topics, the use of music or the sound of waves, rain, waterfalls, and thunder, etc. to accompany slides is a nice feature that varies the student's sensory experience, possibly evoking a feeling about the meaning or beauty of a subject.

• **Shortwave radio.** As mentioned in Chapter 1 of this manual, a shortwave can be used to play Coordinated Universal Time, solar forecasts, and related foreign broadcasts for classes.

• **Chalkboard.** I resorted years ago to buying my own high-quality colored chalk (I recommend HygaColor–not too hard nor too soft, and easily erasable). Chalk color can be suggestive of the subject at hand and can prove valuable in giving the student some visual variation. White chalk that is too hard and writes too light is deadly; this is why I try to use at least eight colors. Our texts are in color, reality is in color, so why not our chalk?!

• **Teaching tool box.** To avoid the before-class scramble for colored chalk, overhead pens, stapler, tissues, paper clips, throat lozenges, I have a small reinforced box (shoe-box size) and simply take this "tool box" with me to lecture. No matter what shape the room was left in by the last instructor or janitor, the tool box has you ready to go on arrival.

Grading and Course Structure Suggestions. All of us develop our own grading system and course structure and then we generally never discuss our system with anyone. It seems as though this becomes the big, sacred mystery! The following in no way is meant as a definitive statement, but rather as just some suggestions–for I am sure we all have good ideas.

• **Course syllabus.** This should include a clearly stated presentation of catalog description, grading policies, goals, test make-up policy, office location, office hours, mail box, office phone number, and e-mail address. (I have found that an answering machine on the office phone is a great investment, increasing student contact easily by a third. E-mail increases student access considerably. Make sure you check it frequently, students using e-mail often expect quick feedback.)

• **Grading.** I structure the class assignments and tests based on 1000 points. I give three equally-weighted exams and four exercises, representing 75 percent and 25 percent of the grade respectively. Relative to testing, please see the separate test bank that I have prepared to accompany *GEOSYSTEMS.* (And, of course, a computerized grade book is invaluable to assist with such multiple-grading inputs.) Exercises may take many forms but can include a library exercise, a weather observation exercise, climographs and a sample station exercise, an article analysis exercise, many versions of mapping exercises, etc. These more subjective aspects of the course have proved to be invaluable in drawing the class together and making physical geography more interesting. Drop me a note if you would like me to send you a copy of any of these exercises, and please feel free to send me examples of your equally valid lecture-class exercises. The following is just a sampling of several ideas.

• **Library exercise.** I have always been amazed at how little exposure the students have had to the library--"that big building in the middle of campus, don't you know?!" Yet facility with the library is key to their academic success. I work with the librarians to refine my exercise and they seem quite pleased as the geography students show up each semester. The students, after-the-fact, are constantly referring to the impact the exercise had on them. The items in the four-page handout include work with computer card files, computerized magazine and journal summaries, computerized newspaper summary system, periodicals, pamphlet files, and reference collections. And I end it with a request that they get a library card and a list of the library's operating hours–seemingly obvious tasks. Granted, much of this should have been handled in high school and in college English courses, but my experience is that such an exercise exposes the student to new resources–strongly nurturing their college experience.

•**Weather observation exercise.** This is a simple handout for seven days of basic weather observations using the standardized weather station symbols published by the National Weather Service as shown in text Table 8-1 (p. 203). Included are temperature, pressure, state of the sky, weather type, wind direction and speed, and cloud type. Students need to locate a source for the weather data: the cable Weather Channel, a weather phone number, or instruments on campus. I ask them to turn it in whenever the seven days are done, which drives some of them nuts because they want to know exactly when it is due.

• **Climographs and sample station exercise.** Using the basic climograph form as it appears in Chapter 10 and the climatic data in Appendix A, I have students graph any six stations of their choice spread through a variety of climates, then describe the Köppen classification and determine the range of air pressures between January and July, as well as the dominant soil order and the appropriate terrestrial biome designation from maps in Chapters 6, 18, and 20. This requires that they locate their cities in an atlas and transfer that location to the maps in the text. The second part of the exercise asks them to select one of their six cities and write a brief geographic analysis of the place and its regional setting; this can include aspects of both physical and human geography.

• **Article analysis exercise.** The purpose of this semester-long exercise is to find and analyze articles that appear in newspapers, magazines, journals, or any other media that relate to the geography course. Students are asked to prepare a couple of pages of analysis for each item. In addition to printed sources, they may also utilize the Public Broadcasting System (PBS), National Public Radio (NPR), and C-SPAN and such programs as NOVA, the MacNeil-Lehrer News Hour, National Geographic, Nature, the Morning Report, and All Things Considered. This is assigned the first week of class and is not due until the last week of the term, therefore providing an opportunity to do a couple of minutes of opening warm-up in almost every class using applied topics to "hook" their interest. In this way, they can see how much of the course content and related issues and controversies are actually occurring out there in "the real world." Also, this allows you to point out items in professional journals and related literature, and it encourages students to possibly upgrade their research papers for other classes with geographic content.

Final Thoughts about Teaching

Obviously, those of you who are veteran teachers may have had many of these thoughts in your own experience and you probably could have written this section yourself. With this in mind, I offer the following. The attribute that is the sustaining force behind teaching is an applied common sense, a feel for the appropriateness of a practice. No matter what skills teachers develop, if they do not see, hear, and feel what is happening in the class, then the potential of the moment is defeated and the class will drag painfully through the term. A teacher must develop the ability to imagine the feelings, sensations, and thinking that is going on in the student. This can only be done through being with them and not being isolated and detached. This is the *empathy force.*

I like to refer to the "eyebrow index" as an indicator of what is happening in the student. The constant sampling of all the eyes in the room directs me in my presentation. Frowns and furrows trigger a repeat variation of the material just presented, followed by another sequence of questioning–show, discuss, apply–show, discuss, apply. Admittedly, when you are teaching in a large classroom, such empathetic eye contact is difficult; perhaps if you can move up the aisles it will facilitate seeing the faces. Time does not allow for the resolution of all the negative eyebrows, but movement in the majority signals that new material can be presented. Since students' facial expressions are somewhat unconscious, they are surprised and thrilled when the teacher perceives their need for help. To intensify the eyebrow indicator, I refer back to students who previously asked questions or made statements. And if I remember, I try to work such referrals in somehow, even if it is later in the semester. It has impact on students to see that what they say and do is remembered. This also improves the quality of questions, for they know their questions are taken seriously.

Many characteristics can be listed as applicable to the teacher. I have chosen to simplify this consideration to three idealized categories. Each of these can be thought of as broad paradigms of behavior and attributes, variable in their mix and intensity within each individual.

The Teacher as a Personality. Each of us has our own individual personality and projects a persona or social mask to others, hopefully one that is appropriate for the social role we play in society. When in front of a class, every teacher generates a certain teacher persona that faces out to a roomful of student personas. I believe that the thinner these necessary masks become, the greater the energy that is generated for the teacher-student interaction. For some, merely stepping out from behind the podium or lectern might raise the pulse rate. Sharing an important personal anecdote to heighten a point might quicken your own breathing and heart rate. Experiment with this in your own situation and feel the increase in nervous energy. The challenge is to translate this energy into positive fuel for the presentation.

Allowing your primary person to shine through your mask and learning how to channel the energy thus derived is an important aspect of successful teaching. The student can see a real personality in all its glory–perfections and flaws and the works. There are few thrills that compare with that rush of goose-bumps one gets while explaining some majestic point about Earth's physical geography to receptive faces–especially when it's for the umpteenth time after umpteen years of teaching! It's all right to feel emotion and instill emotion over these dramatic environmental subjects. After all, this is a dramatic and beautiful planet!

The Teacher as Intellect. Intelligence is an especially important trait of a teacher. Critical to the craft of teaching is this power to know and understand, for the teacher must comprehend information specific to the study at hand. You may have experienced insights and new learning yourself while right in the middle of lecturing on an "old" topic. These flashes are always interesting and certainly keep one humble, because you never know when the lightning is going to strike! To have up-to-date material, constantly in revision, is critical to teaching, and the freshness that is derived is one of the rewards of continuing learning. The effective teaching of information is different from researching and publishing that information.

Teachers of physical and cultural geography must be eclectic, that is, able to draw from many disciplines as well as from their own field. The ability to synthesize requires patterned seeing and relational thinking, if notes for the classroom discussion are to be dynamic and full of energy.

The Teacher as Actor. Teaching is a performing art and therefore involves a performance. Voice is very important–enunciation, modulation, and projection. Direct your voice outwards, not into the lectern. And remember, what seem like exaggerated gestures and voice patterns will seem like normal speech in the back of the room, whereas a normal voice from the lecturer will come across as flat and sterile. You might want to practice: using a recorder, tape your voice for a minute or two, reading, lecturing, and speaking; then, with that feedback, adjust accordingly. Perhaps you could have a lecture-discussion videotaped for review. (Note the nervous energy when you go to push the record button–tap into that!)

Importantly, because the goal is communication and learning, the performance should not be for its own sake but be a means to the educational objectives in operation: it is a catalytic vehicle. With such an exciting discipline as physical geography, we must watch out for excess, however; "Gee, that teacher is entertaining but we don't learn much," is probably not the response we're looking for.

There is constant pressure in the classroom for substance in content and presentation, acting almost like positive feedback in a geographic system. Classroom teaching seems capable of absorbing any increased energy input—seemingly without limit! The possibilities are infinite, and with Earth as the subject, the challenges are exhilarating.

Please accept this *Instructor's Resource Manual* as a dialog from one teacher to another. If you have comments, criticisms, or further information, please drop me a note–such feedback and conversation is always welcome.

Write to:

Robert W. Christopherson,
American River College,
4700 College Oak Drive,
Sacramento, California 95841

Or e-mail:
Bobobbe@aol.com

For Ready Reference

The table of contents and complete focus study list are printed here for your convenience.

Table of Contents

Focus Study List

9-2:	Ogallala Aquifer Overdraft
Chapter 10-10-1:	The El Niño Phenomenon
Chapter 12-12-1:	The 1980 Eruption of Mount Saint Helens
Chapter 13-13-1:	Vaiont Reservoir Landslide Disaster
Chapter 14-14-1:	Floodplain Strategies
Chapter 15-15-1:	The Colorado River: A System Out of Balance
Chapter 16-16-1:	An Environmental Approach to Shoreline Planning
Chapter 17-17-1:	Periglacial Landscapes
Chapter 18-18-1:	Selenium Concentration in Western Soils
Chapter 19-19-1:	The Great Lakes
Chapter 20-20-1:	Biodiversity and Biosphere Reserves
Chapter 21-21-1:	Earth Summit 1992

The "Geography I. D." Assignment:

(* Derived from *Applied Physical Geography: Geosystems in the Laboratory* by Robert W. Christopherson, © 1994 MacMilan College Publishing Company.)

GEOGRAPHY I.D.

* **To begin:** complete your personal *Geography I. D.* Use the maps in a physical geography text, an atlas, college catalog, and additional library materials if needed. Your laboratory instructor will help you find additional source materials for data pertaining to the campus.

On each map find the information requested noting the January and July temperatures indicated by the isotherms (the small scale of these maps will permit only a general determination), January and July pressures indicated by isobars, annual precipitation indicated by isohyets, climatic region, landform class, soil order, and ideal terrestrial biome. Record the information from these maps in the spaces provided. The completed page will give you a relevant geographic profile of your immediate environment. As you progress through your physical geography class and physical geography lab the full meaning of these descriptions will unfold. This page might be one you will want to keep for future reference.

GEOGRAPHY I. D.

NAME:______________________________ LAB SECTION:______________________

HOME TOWN:______________________ LATITUDE:____ LONGITUDE:______________

COLLEGE/UNIVERSITY:____

CITY/TOWN:____________________________ COUNTY (PARRISH):__________________

LOCATION: STANDARD TIME ZONE:____.

LATITUDE:_____________________ LONGITUDE:__________________________

ELEVATION (include location of measurement):___________________________

PLACE (tangible and intangible aspects that make this place unique):__________

____.

REGION (aspects of unity shared with the surrounding area):__________________

____.

POPULATION:

CITY: ____________________________________. METROPOLITAN STATISTICAL

AREA (CMSA, PMSA, if applicable):_____________________________________.

ENVIRONMENTAL DATA:

JANUARY AVG. TEMPERATURE:______ JULY AVG. TEMPERATURE:________________

JANUARY AVG. PRESSURE (mb): _____ JULY AVG. PRESSURE:____________________

AVERAGE ANNUAL PRECIPITATION (cm/in.):____________________________

AVG. ANN. POTENTIAL EVAPOTRANSPIRATION (if available; cm/in.):__________

CLIMATE REGION (Köppen Symbol and name description):____________________

____.

TOPOGRAPHIC REGION OR STRUCTURAL REGION:_______________________________

____.

BIOME (terrestrial ecosystems description): ______________________________

____.

1

Essentials of Geography

Overview

Our study of physical geography begins with the essentials of this important discipline. The "Essentials of Geography" chapter contains the basic tools for the student to use in studying the content of physical geography. After completion of this chapter you should feel free to either follow the integrated sequence of the text or treat the four parts of the text in any order that fits your teaching approach.

Most students have differing notions as to what geography is and what geographers do. Some think they are going to memorize the capitals of the states, while others think that they are embarking on a search for lands and peoples, or that they will be learning things for success with a popular "Jeopardy!" category. This uncertainty as to the nature of geography provides an excellent opportunity to review the state of affairs relative to geographic awareness. For information about geography as a discipline and career possibilities contact the Association of American Geographers (AAG's address is in Appendix B of the text) and ask for their pamphlets, "Careers in Geography" by Salvatore J. Natoli (1990), "Geography: Today's Career for Tomorrow," and "Why Geography?"

In my own physical and cultural geography classes I find that many students are unable to name all the states and provinces or identify major countries on outline maps. When we add the complexity of the spatial aspects of Earth's physical systems, i.e., atmospheric energy budgets, temperatures, wind patterns, weather systems, plate tectonics, earthquake and volcano locations and causes, global ecosystem diversity, and terrestrial biomes, the confusion even among the best informed is great and there in lies our challenge!

These are certainly dramatic times, with an expanding global concern for the environment moving to the forefront of the political and scientific agenda. For the first 12 days of June 1992 the United Nations Conference on Environment and Development (UNCED) held an international gathering of scientists, political leaders, and citizens in Rio de Janeiro. A key portion of the conference dealt with issues central to physical geography and our human-environment theme: preservation of biological diversity, climate change and global warming, stratospheric ozone depletion, transboundary air pollution, deforestation, soil erosion and loss, desertification, permanent drought, fresh water resources, and resources from the ocean and coastlines. In addition to these environmental topics, issues relating to human societies also were addressed. Full international attention improves as negative feedback from Earth's physical environment continues to draw the world together.

The *Student Study Guide* presents 20 "Learning Objectives" to guide the student in reading the chapter. The *Applied Physical Geography* lab manual has one exercise with seven steps that involve many aspects of this chapter.

New to the Third Edition

(Note: This section highlights major changes, new features, and additions in the third edition. This does not describe all the rewrite and recast of the text.)

1. A list of key learning concepts begins the chapter.

2. A new introduction opens the third edition. Numerous locations and environmental events are referred to, and offer you an opportunity to work with a classroom map and introduce physical geography to your students. These examples demonstrate the significance of geographic events to the national economy, international politics and future resource reserves.

3. A detailed description of **Earth systems science**. This will enable students to recognize the centrality of Earth studies in creating an integrative world view.

4. Figure 1.3 illustrates the car as open system, allowing the students to connect to systems which they operate on a daily basis.

5. Use of examples, such as Mount Pinatubo, in a manner which will enhance the student's ability to see connections between spheres as well as chapters. In the first chapter, Mount Pinatubo is used as an example of dynamic equilibrium - its fluctuating activity spewing volcanic ash into the atmosphere causing increased precipitation levels on a global basis.

6. Removal of the topographic mapping section. Now discussed more fully in Appendix B.

7. A new graph, Fig. 1-15, illustrates great circles and small circles.

8. Figure 1-22 portrays Goode's homolosine map. Sections of the map which are comprised of homolosine and sinusoidal projections are labeled to show students how map projections can be combined to reduce distortion.

9. The causes of map distortion are presented in a bullet format.

10. An entirely new section which describes the process of remote sensing and its applicability in geographic information systems (GIS). Included in this section are new graphs which delineate the process of remote sensing (Figure 1-25) and illustrate the amount of information which a GIS data plane contains (Figure 1-27). Also included in this section is a satellite image of the Kliuchevskoi volcano on the Kamchatka Peninsula as captured by the Space Shuttle Endeavour, September 1994.

11. A new section has been added to the third edition of *GEOSYSTEMS*. Throughout the chapters there are News Reports which contain short articles concerning material in this chapter that is news worthy. For chapter one, there are two reports. News Report #1 is titled, "Careers in GIS", and discusses the use of GIS in geographical analysis and career opportunities which have developed for GIS degree holders. News Report #2, "GPS: A Personal Locator", examines geographical positioning systems (GPS) and addresses the commercial use of satellite images to increase transport, urban and environmental planning.

12. A section called, "In the Next Chapter", has been added to the end of each chapter. This section attempts to create constructs for greater comphrension of the material and guide the student to recognize connections between each chapter. The focus of this chapter's section is to direct the student's comprehension of open and closed systems towards Earth-Sun relationships addressed in Chapter 2, i.e., the flow of energy which enters the atmosphere (open system) and stimulates oceanic and atmospheric circulation patterns (closed system).

Key Learning Concepts

The following learning objectives help guide the student's reading and comprehension efforts. The operative word is in italics. These are included in each chapter of *Geosystems* and the *Geosystems Workbook*. The student is told: "After reading the chapter you should be able to:

1. *Define* geography and physical geography in particular.

2. *Describe* systems analysis, open and closed systems, feedback information, and systems operations, and *relate* these concepts to Earth systems.

3. *Explain* Earth's reference grid: latitude and longitude, plus latitudinal geographic zones and time.

4. *Define* cartography and mapping basics: map scale and map projections.

5. *Describe* remote sensing and *explain* geographic information systems (GIS) methodology as a tool used in geographic analysis."

Expanded Outline Discussion

The following headings (boldfaced) match some of the first, second, and third order headings in Chapter 1. The narrative under each heading contains information, sources, and anecdotal facts relating to portions of the chapter. Not all text headings are discussed.

THE SCIENCE OF GEOGRAPHY

Geography (from *geo*, "Earth," and *graphein*, "to write") is the science that studies the interdependence among geographic areas, natural systems, society, and cultural activities *over space*. The term *spatial* refers to the nature and character of physical space; to measurements, relations, locations, and the distribution of things. Everyone is a geographer at one time or another; our mobility is great in the modern era and most of us cover some distance each day. Most students have already traveled more than the circumference of Earth as measured on an automobile odometer.

Geography, rather than being governed in a body of knowledge (as are most other disciplines), is instead governed by a method: spatial analysis. Geography is an approach, a way of looking at things, and by its very nature is eclectic, integrating aspects of many disciplines through the spatial bias to form a coherent whole.

Relative to the five geographic themes, an excellent map poster titled, "Maps, the Landscape, and Fundamental Themes in Geography" is available from the National Geographic Society (17th and M Streets N.W., Washington, DC 20036; 1-800-638-4077), the Association of American Geographers (1710 16th Street N.W., Washington, D.C. 20009), and the National Council for Geographic Education, (Indiana University of Pennsylvania, Indiana, Pennsylvania 15705).

The Geographic Continuum

GEOSYSTEMS consistently uses *Earth* as a proper noun and may be the first geography text to do so; and without the article *the* in front of it. The lower-case treatment can be viewed as a subtle, subconscious denial of Earth and illustrates further the physical-human dichotomy. This point can be made by asking students why Earth was ever referred to as a common noun, unlike any of the other planets and despite specific rules of grammar.

Since the students may be taking courses in some of the other disciplines listed in Figure 1-2, this might be an opportunity to give examples of our geographic approach, e.g., compare a historical approach and a geographical approach to the study of settlement in the Midwest and the Great Plains.

As an example, geographers focus on the spatial problems presented by the environment and attempt to reconstruct conditions representative of the second quarter of the 19th century: great distances, groundwater too deep for hand-dug wells, lack of energy for pumping, lack of fencing materials, the existence of densely matted sod that caused plowing difficulties, and mobile and adaptable native populations. All these conditions represented spatial problems for analysis and later resolution, i.e., railroads, oil-drilling techniques for water wells, wind energy, barbed wire, John Deere's self-scouring steel plow, and, for the unfortunate Native Americans, the Colt six-shooter.

EARTH SYSTEMS CONCEPTS

We seem to possess a basic aptitude for perceiving distinct systems smaller than ourselves, but we are quite slow in seeing those systems that function on a scale larger than us: neighborhood, city, state and province, country, international regions, the whole Earth system, the solar system, the Milky Way Galaxy, our local group, and the universe.

Ask the students to think about their place within these vast Earth systems as they walk to their next class. What is our place in this dynamic biosphere (discussion to follow) operating all about us? How might we maximize our position? Optimize our sustainability?

Systems Theory

Systems analysis has moved to the forefront as a method for understanding operational behavior in many disciplines, including geography. *GEOSYSTEMS* is not a rigorous systems analysis; rather, it is an essentially non-mathematical analysis of systems. There are so many systems about us that teaching material is readily at hand–at least for open systems. Discuss with students the existence of any possible closed systems, or aspects of open systems that behave as if closed.

Use course grades as an example of negative feedback since such feedback can affect the student's effort and level of operation; or use the possible connection between greenhouse gases and global warming with increasing temperature as an example of a feedback to which we may or may not respond, or the Los Angeles earthquake of 1993 as negative feedback for society to implement improvements in hazard perception, or the impact of further elimination of natural biomes and the loss of plant and animal diversity.

Positive feedback is exemplified by photosynthesis, for oxygen is transparent to incoming insolation that drives the process and therefore a key by-product does not inhibit system operations. Or in an economic sense, the concept of gross domestic product (G.D.P.) as presently conceived exemplifies positive feedback since negative impacts are not calculated as negative–all growth is regarded as positive regardless of consequences. Economist Herman Daly referred to this positive feedback economic system as "hyper-terminal growthmania."

Earth's Energy Equilibrium: An Open System

Part One of *GEOSYSTEMS* (Chapters 2 through 6) is organized along energy pathways from the Sun across space to the top of the atmosphere, varying with seasons, cascading down through the atmosphere to Earth's surface, producing patterns of temperatures, pressures, and global winds. This discussion sets up energy patterns for later energy budgeting topics in the atmosphere and at the surface.

Earth's Physical Matter: A (Nearly) Closed System

Indicative of Earth's closed material system is the modern urban landfill (dump). Thousands of them nationwide are nearing capacity or are already full as the disposal crisis grows. The 3100 tons of garbage on the wandering barge *Mobro* a few years ago illustrates a simple but important concept: One system's output becomes another system's input. Shipping it out of the community (out of sight) that generated it does not resolve the problem. Unfortunately, instead of being subjected to modern recovery techniques, the resources on the barge were eventually incinerated in Brooklyn, rendering 400 tons of ash, which was buried in another landfill dump.

The grassroots popularity of resource recycling in many communities generally is not supported by the national government. The virgin-material industries lobby heavily and effectively to protect primary resource exploitation and extraction. Continued public pressure and the increasing expense of extraction and imports will eventually get the national system into a recycling mode!

The idea that there is no practical way to bring resources into Earth's systems (air, water, material resources) presents a basis for a recycling discussion. Does your campus operate a recycling center? How long has it run? Is recycling sponsored by your university or college? a student club? a fraternity or sorority? Or, is it operated by a community-based group? Can material resources be taken there? What kind of closed-system feedback information do students see on a daily basis?

Earth's Four "Spheres"

A quick review of the solar system reveals that Earth is unique relative to all four of these sphere models, each indicative of one of the four "parts" that form the basis of organization in *GEOSYSTEMS*: atmosphere (Part 1), hydrosphere (Part 2), lithosphere (Part 3), and biosphere (Part 4).

Biosphere

Chapter 19 discusses the abiotic and biotic aspects of the biosphere. The interactions and flows among the abiotic spheres produce Earth's dynamic biosphere. An interesting article appeared in *Nature* (365, no. 6448, October 1993, pp. 715-21) titled "A Search for Life on Earth from the Galileo Spacecraft." Instead of other planets, the sensors aboard this modern interplanetary spacecraft were turned on Earth during a fly-by on its way to Jupiter. This article presents what *Galileo* detected that suggests life and conscious intelligence on the third planet from the Sun: abundant gaseous oxygen, a widely distributed surface pigment with absorption in the red part of the visible spectrum, atmospheric methane levels, and the presence of narrow-band, pulsed, amplitude-modulated radio transmissions. The premise and content of this article can form the stimulus for a great classroom discussion. (See the two *Galileo* images in Chapter 6, Figures 6-10 and 6-13, and note the captions.)

Related News Report: Mention of the ***Biosphere 2*** experiment/adventure operating near Oracle, Arizona is appropriate here. A new science of biospherics is unfolding in the Arizona desert, 64 km (40 mi) north of Tucson. A 1.3 hectare (3.15 acre) glass and steel habitat containing seven complete ecosystems of matched soils, plants, and animals, including the most intensive agricultural plot on Earth, and a living ocean with corals and marine life is in operation.

Scientists and engineers have actually constructed a 180-meter long (600-foot) scale-model of Earth's tropical terrestrial and oceanic biomes: rain forest, savanna, scrub and thorn forest, desert, ocean, including 500 coral colonies, and salt and fresh water marshes, including 3800 species of plants and animals. The habitat was only entered and exited through an airlock during the year before it was sealed.

Following five years of research and development and four years of construction of this largest-ever sealed habitat, the "biospherians" entered September 26, 1991. As the four women and four men (ages 29 to 69; 2 with advanced degrees, 5 with bachelors degrees, and 1 technician) struggled inside their closed material system at a subsistence level, controversy and a cynical media swirled around in the outside world. Meanwhile the biospherians assisted scientists on the outside with more than 60 on-going research projects inside the biosphere. Their entire diet was grown on an intensive agricultural plot of only a half acre of soil–80% during Mission 1 and 20% planted and grown by the 8 biospherians in the months before final closure. Biosphere 2 completed its first two-year mission–Mission 1–on September 26, 1993, when eight people emerged on "Re-Entry Day"– a re-entry into Biosphere 1 and Earth's environment. Transition 1 continued until March 6, 1994, when Mission 2 officially began with a new international crew of seven.

We were able to visit the facility the week before the 8 people entered on their two-year mission and again we were there on the day of re-entry back into Biosphere 1. The biospherians emerged thin and healthy (loss of an average of 16% of body weight, low-blood cholesterol levels of 125–an average decrease of 35%, lowered blood glucose, and lowered blood pressure). Their diet was low-fat (less than 10% of calories), nutrient-dense, and high fiber. All 8 had the same pale skin tone for they had not experienced ultraviolet light for two years because it is filtered-out by the space frame. I will always remember that dawn in the chill of the morning desert air as over 3000 cheering people greeted the 8 explorers of this high-tech inner world. They each reacted strongly to their first breath of nontropical air and normal oxygen content for the first time in 2 years.

Biosphere 2 is so tightly sealed that it allows scientists to track the oxygen and carbon dioxide mix–these are at 20.946% and 0.036% respectively in the air we breathe. Sensors inside the enclosure can measure momentary increases in CO_2 during the passage of a cloud overhead (reduced photosynthesis)! As Mission 1 progressed oxygen levels began to drop; however, CO_2 levels did not increase as expected. If there was increased respiration going on, CO_2 levels should increase. Scientists wanted to know where the missing CO_2 went. Oxygen concentrations continued to drop to a low of 14.2% by December 1992. To compound this problem southern Arizona was experiencing record cloudiness caused by a persistent El Niño in the Pacific (see focus study in Chapter 10) that produced lowered photosynthetic rates.

This turn of events placed the biospherians in unknown biomedical territory–to have people at normal atmospheric pressure for their altitude yet at only 14.2% oxygen. They experienced various symptoms of high-altitude sickness. If the mission was to continue and if scientists were to work inside during the first transition later in 1993, the oxygen levels had to be increased.

In consultation with scientists from Lamont-Doherty Earth Observatory and several other institutions the decision was made to inject a controlled amount of oxygen to bring levels up to 19% to continue the experiment. Thus, this biospheric experiment had reached new ground.

According to a report in the January 18, 1994, EOS, microorganisms in the organically rich soil were consuming oxygen faster than photosynthetic replacement. The missing excess CO_2 thus produced was reacting with concrete surfaces inside the containment, forming calcium carbonate, and effectively masking the oxygen sink in the soils. This is an on-going dynamic situation of unknown outcome based on empirical measurements, assumptions and hypotheses, and real-time operations– an interesting scientific experiment.

Dr. Eugene P. Odum, Institute of Ecology, University of Georgia, wrote in

Science (*Science*, 260, no. 5110, May 14, 1993, p. 879):

> **The [Biosphere 2] experiment is not reductionist, discipline-oriented science, but a new, more holistic level of ecosystem science....Biosphere 2 is as much a human experiment as a scientific one....When one considers that nothing on the scale of Biosphere 2 has been attempted before and how little we really know about how our Biosphere 1 works, a measure of success will have been achieved if the biospherians come out alive and healthy this fall after the 2-year isolation. Certainly the experiment will have improved our understanding of human-biosphere interrelations and helped answer the question of how much natural environment must be preserved for life support.**

The scientific director of research for Biosphere 2 is Dr. John Corliss, the developer of the deep-sea diving Alvin and former Scripps Institute and NASA/Goddard scientist. His presence has helped the enterprise recover from initial skepticism in academia and a decidedly negative media.

My view is that this construction in Arizona is a mission, much like a space mission, where some scientific experiments are conducted even though the missions are not purely scientific. I do think that much has been learned already about closed-system habitats and biomedical responses in humans–the scientific implications are intriguing.

The construction is described as remarkable in an issue of *Architecture*, "Desert Shield" (May 1991, pp. 77-81+); its sheer size and concept really takes one's breath away. The structure has won many architectural and design awards.

Whatever you think of the venture, it does provide a useful teaching metaphor for Spaceship Earth, or Biosphere I, for in a sense we are biospherians dependent completely on natural systems to provide the substances necessary for life. We have a fixed resource base, air, water, and food production systems just as they do. The difference is one of scale and complexity. More importantly, Biosphere 2 provides a focal point for improving our planetary awareness. There is a new book written by two of the biospherians: Abigail Alling and Mark Nelson, *Life Under Glass–The Inside Story of Biosphere 2*, Oracle, AZ: The Biosphere Press, 1993. And a cookbook of the nutrient dense, high fiber, low-fat diet they grew and ate inside the containment: Sally Silverstone, *Eating In–From the Field to the Kitchen in Biosphere 2*, Oracle, AZ: The Biosphere Press, 1993.

A SPHERICAL PLANET

This is an exciting topic for a classroom discussion and presentation. What would Pythagoras (ca. 580-500 B.C.) have noticed that would have led him to conclude Earth's sphericity? Four possible examples are noted in the text.

Eratosthenes' brilliant measurement is discussed in Focus Study 1-1. In my research, I found several versions as to how he determined the linear distance between Alexandria and Syene, e.g., soldiers pacing off the distance, or paid workers using set lengths of rope to measure the distance segment-by-segment, or the travel time required by camels in caravan. The camel story is interesting and I included this in the description in the second edition.

The 16 percent error in his measurement is attributable to the fact that Syene is not exactly south of Alexandria, and that the Tropic of Cancer, 23.5° North Latitude, is actually 97 km (60 mi) south of Syene–and, of course, to variations in camel velocities! Otherwise his reasoning and calculations were correct. (See the inset map in Figure 1 of Focus Study 1.)

If you have arrangements with your local PBS affiliate and site licenses, then you can see video segments that will be useful. See Carl Sagan's *COSMOS* series, in Episode 1 of the television production, or Chapter 1 in the book of the same name. He visits the sites in Egypt and discusses what Eratosthenes accomplished in 247 B.C. Sagan presents a beautiful summary of the measurement that lasts about 5 minutes.

And, of course, there is the view of Earth from space–see the back cover of the textbook. Philosophers believed that when we had that image from a distance, actually photographed by a human space traveler, the impact on society would be profound. Of course, over 25 years have now passed and such powerful realizations still seem not to have penetrated our collective consciousness. Earth's picture at once makes us feel pride in an incredible technology and yet humble at the sight of our small, fertile sphere in the void of space. I still am amazed when I look at the photograph on the back cover and think that it was taken by a human with a camera and film that was brought back to Earth for development. The students can be told that by the end of the semester they will know the reasons for the pattern of clouds, land, and water in this photo by Apollo 12 astronauts.

LOCATION AND TIME ON EARTH

Great Circles and Small Circles

If you have a blank globe, suitable for chalk, you can mark these concepts on the surface. You can use a string or slide projector remote-control cord to plot the great circle route between two points after asking students to visually sight a possible route on a Mercator map in the classroom, e.g., San Francisco to London, England (Figure 1-21). If you also have a travel careers department at your school, borrow a map which shows the path of jet travel across the globe. Examining great and small circles will help students understand rhumb lines when discussing map projections and distortion.

Prime Meridian and Standard Time

Daniel J. Boorstin's *The Discoverers* (New York: Random House, 1983) has an excellent section on time. For a history of Greenwich and time determinations, see Derek Howse's *Greenwich Time and the Discovery of the Longitude* (London: Oxford University Press, 1980).

A good technique is to relate Earth's rotation to those who still have wrist watches with hands. I set the globe on my wrist and pretend to set it correctly as if it were a watch.

If you have access to a shortwave radio, I suggest playing the UTC time signal in class, or record it at an electronics store for classroom use. UTC is broadcast on radio worldwide (WWV, the Standard Frequency Station, broadcasts time-checks on 2.5, 5, 10, and 15 megahertz). The time broadcasts are continuous and include marine forecasts, tropical weather reports, and an hourly report of geophysical alerts such as solar activity and radio-wave propagation in the atmosphere. I have found that this, as well as the BBC, comes in clearly in the classroom with a minimal antenna. You may want to obtain the following publications: *NIST–Time and Frequency Services*, NIST Special Publication 432 (Revised 1990), Department of Commerce; *Canada's Time Service*, Institute for Measurement Standards, National Research Council, March 1991; *WWVH*, NIST Time and Frequency Services, Department of Commerce, August 1989.

A new section on *NIST-7* and the other primary standard clocks, three in Ottawa and two in Germany, is included in the third edition. For more information contact the National Institute for Standards and Technology in Boulder, CO, or the Institute for National Measurement Standards in Ottawa. See: James Jespersen and Jane Fitz-Randolph's *From Sundials to Atomic Clocks - Understanding Time and Frequency*, New York: Dover Publications, Inc. 1982. A.G. Mungall's *Canada's Atomic Standards of Time and Frequency* and *Ce-*

sium Atomic Clocks, Division of Physics, NRC, Ottawa, 1983.

International Date Line

This difficult teaching concept works well with a globe to show students the date line and the official day-change concept. This is an imaginary line extending between the poles dividing one calendar day from the next. I describe an imaginary power boat on the west side of the date line on a Monday at 2 P.M. local time. If we go west for 30 minutes and cross the line it then is 2:30 P.M. local time but it is Tuesday. The text mentions the consternation of Magellan's crew as to what day they arrived home. The complete Magellan expedition is described in "The Strait–and dire straits–of Magellan," by Simon Winchester in *Smithsonian*, Vol. 22, No. 1 (April 1991): pp. 84-95.

The use of Greenwich as the Prime Meridian makes possible the passage of the date line through underpopulated stretches of the Pacific Ocean on the opposite side of the globe, avoiding the political confusion that would have occurred if Washington, D.C. had been chosen, for the date line would then divide Asia.

On Friday August 20, 1993, residents of Kwajalein, Marshall Islands, went to bed and awoke the next day to a bright Sunday morning–there was no Saturday! The population decided to move their standard time calculation to the Western Hemisphere, east of the International Date Line, to better coordinate with mainland United States. The Marshall Islands include about 100 islets, of which Kwajalein is the largest, that fall west of the Date Line in the Eastern Hemisphere. Now they are on Western Hemisphere time.

MAPS, SCALES AND MAP PROJECTIONS

Thanks to institutions such as the National Geographic Society, wall maps up to mural size are available to public at low prices for home, office, and classroom.

I suggest getting the current catalog from the National Geographic Society, for they provide the most inexpensive, high-quality maps available (National Geographic Society, 17th and M Streets N.W., Washington, DC 20036; 1-800-638-4077). The political map, "The World" (#02690, Robinson projection, at 1:19,620,000) and the physical map, "World Ocean Floor" (#02683, Mercator projection) are both under $17 and are almost 1.5 m (5 ft) wide. And, for less than $64, "The World" (#02608, Robinson, at 1:11,990,000) political-mural map is available and is 3 m (10 ft) wide, suitable for classroom installation. These maps are constantly updated; for example, the latest version published January 1994 shows the division of Czechoslovakia into two countries, the new nations of the former Soviet Union, the breakup and new nations of the former Yugoslavia, a united Germany, united Yemen, inference of the Western Sahara status with Morocco, the return of the name Cambodia for Kampuchea, and Burma's new name, Myanmar.

PBS aired a 6-part series titled "The Shape of the World," produced by Granada Television and WNET–in both video and a companion text. The series covers the history of mapping and how the world was mapped. The coverage is from ancient times to the modern age of satellites, computers, and GIS. If this is available on your campus I suggest that you obtain a copy. There are many useful segments presented.

Many resource materials are available for your unit on maps and map projections. See the suggested readings section of Chapter 1 for a listing of sources. Two reasonably priced volumes from the Government Printing Office are: Snyder ,John P., *Map Projections–A Working Manual*, U.S. Geological Survey Professional Paper 1395, 1987; and Thompson, Morris M., *Maps for America*, 3rd ed., U.S. Geological Survey, 1987. These two books provide an extensive presentation of map projections and types of maps used in the U.S. Also, excellent chapters on topographic maps are included. The appendix of *Maps for America* contains a good section on latitude, longitude, and map projections.

The American Congress on Surveying and Mapping (210 Little Falls Street, Falls Church, Virginia 22046) has two publications that may be useful: "Which Map is Best? Projections for World Maps" and "Choosing A World Map Attributes, Distortions, Classes, Aspects." Also, from the National Council for Geographic Education, in their Topics in Geography series, see "Teaching Map Skills: An Inductive Approach" and "Map and Globe Skills, K-12 Teaching Guide" for basic introductory materials and ideas.

Also, Joel Making and Laura Bergen, eds., *The Map Catalog Every Kind of Map and Chart on Earth* (New York: Vintage Books, 1986), is a comprehensive guide to the vast inventory of maps available. And of course, National Geographic's *Atlas of the World*, Revised Sixth Edition, 1992, is essential for accurate place names and maps.

The Scale of Maps

Discussing a map of the classroom itself is one way to introduce scale–the idea being to do a map of the classroom at 1:1, then deciding what scale would be appropriate. With this in mind, expand the discussion to the community, the state or province, the country, and to the world.

If you have a 61 cm (24 in.) globe in class (or adjust the discussion for another size of globe), you can calculate the scale of the globe with the class, as described in the text.

Map Projections

Various teaching aids are available to demonstrate the three principal geometric shapes and how they are related to projections of the globe. You can make up overheads of examples from the USGS Professional Paper 1395 previously mentioned, so that there are specific examples to highlight the discussion. For those of you using multimedia in the classroom, there is an exceptional CD Rom titled, "Image of the World" published by the British Museum for about $20, or it is distributed in the U.S. by National Geographic for around $150. This CD Rom examines 10 map projections in detail, including maps from 1250 A.D. to current satellite images. These images are amazing to project using the liquid crystal display (LCD) which projects images from the computer onto your classroom screen. The program introduces students to each map and explains how it was created as well as the type of distortion it possesses. You can contact the British Library Publishing Office at 41 Great Russell Street, London, WC1B 3DG.

One other way to show students how projections are made is to find an old projection globe and have the students trace the image of the world onto a cylinder or cone. If you have a lab course, this may be a very good use of time.

Properties of Projections

Again with a globe, explain the unique properties that a globe possesses: equal area (equivalence), true shape (conformality), true direction (azimuthal), and true distance (equidistance). Students always seem in-trigued by the fact that absolute decisions must take place among these properties in the preparation of a flat map–you can't have it all!

You may want to refer to the way equal-area projections are used in the text when relative distributions are important. In other cases the Robinson projection is employed as a suitable compromise. All too often our textbooks use maps that are inappropriate in areal distortion to show distributions, such as temperature, pressure, ocean currents, etc.

The characteristics of a flattened globe can be demonstrated with an orange. Carefully peeling the orange to maintain as much of the whole peel as possible, flatten the spherical skin on a paper towel to show how the gores separate. The distortion in preparing a map projection takes place in deciding how to fill these gaps.

REMOTE SENSING AND GIS

REMOTE SENSING

This is a new section of *GEOSYSTEMS* for the third edition. We have expanded the previous discussion of Remote Sensing, explain how remote sensing occurs and highlighting its applicability to geographic studies (see News Report #2; GPS: A Personal Locator).

An important reference for your campus or department library is the ASP publication, Colwell, Robert N., ed. *Manual of Remote Sensing*. 2d ed. Vol. 1, "Theory, Instruments, and Techniques;" and, Vol. 2, "Interpretation and Application," VA: American Society of Photogrammetry, 1983.

No doubt you have experienced student interest in remote sensing imagery. The third edition of *GEOSYSTEMS* includes 79 remote-sensing images, 47 of them new. Given this, I am still amazed at how few of the students have been exposed to pictures and images from space, even including the dramatic planetary images of the past few years. I hope you have already put together a selection of slides from an orbital perspective to highlight these lectures; a little music as background can set the feeling. The new (October 1991) PBS series, "Spaceship Earth–A Global Geography," features dramatic satellite images of Earth's physical geography.

Geographic Information System (GIS)

Figure 1-27 demonstrates the physical structure of a GIS. Figure 1-26 illustrating a satellite image of Kliuchevskoi in Russia also shows students how one data plane may be created in the process of creating a GIS database. In Chapter 16, a discussion of Marco Island, Florida, presents a type of GIS analysis (Figure 16-18). News Report #1 examines the new job opportunities available through computer applications in geography. With Geo automobiles advertising the use of a personal locating system available in their cars by 1997, we may see a renewed interest in studying geography.

There are many new grants available for geography departments to develop courses which incorporate GIS technologies, such as Arc View 2 by ESRI (Environmental Systems Research Institute, Inc.) You can contact ESRI on the internet at http://www.ersi.com or call (800) 447-9778. Arc View 2 is an amazing resource which enables you or your students to get information of geographical phenomena, analyze that data, and make maps of such information.

On a field trip to the U.S. Geological Survey's western regional offices in Menlo Park, we were pleased to see the incredible level of GIS activity on a wide variety of topics. Aster Publishing now produces *GeoInfo Systems*, a magazine/journal published 10 months a year with Vol. 1, No. 1 released in December 1990 (859 Willamette Street, P.O. Box 10460, Eugene, Oregon 97440-2460). In the first editorial, Guy Maynard, Editor, says that, "The capabilities of these sets of tools we call GIS add up to the possibility that we can manage our environments with better sense in the future" (p. 10). Tremendous progress and a revolution has occurred since the time 20 years ago when Ian McHarg demonstrated relational environmental planning models that were manually constructed. See: Ian McHarg, *Design with Nature* (New York: Doubleday, 1971).

Mapping and Topographic Maps

The suggested approach is to show the class several examples of topographic maps representing different landscapes, including urban areas. Let them know where such maps might be obtained in your city. Show them the index map for your state or province. Show a quadrangle from lower latitudes and one from higher latitudes to demonstrate the relationship of the quads with the reference grid. Consult my new lab manual *Applied Physical Geography: Geosystems in the Laboratory*, Lab Exercises 9 and 10 for a detailed presentation on topographic maps.

In *GEOSYSTEMS*, several topographic maps are used: Figures 1-23, Figure 12-27, Figure 15-17, Figure 17-2, and Figure 17-15. The standard symbols for topographic maps are presented in a legend in Appendix B.

Various agents service the major USGS map repositories, e.g., Timely Discount Topos Inc., 9769 West 119th Drive, Suite 9, Broomfield, Colorado 80020; 1-800-821-7609, FAX 1-303-466-3780. They will get any USGS topographic map or image to you by regular or express mail. Timely Discount can also send you all state map indexes. The USGS does give some discount on bulk purchases. There are 11 Earth Science Information Centers (ESICs) that will handle all requests for information and USGS products. A partial listing of ESICs phone numbers:

Anchorage, AK, 907-786-7011
Fairbanks, AK, 907-456-0244
Menlo Park, CA, 415-329-4309
Denver, CO, 303-236-5829 or -7477
Washington, DC, 202-208-4047
Bay St. Louis, MS, 601-688-3544
Rolla, MO, 314-341-0851
Salt Lake City, UT, 801-524-5652
Reston, VA, 703-648-6045
Spokane, WA, 509-353-2524
EROS Data Center, Sioux Falls, SD, 605-594-6151
1-800-USA-MAPS

USGS sells topographic maps on computer discs. You can get topo maps of the United States for roughly $30. Also, the USGS is creating digital orthophotoquads, layering satellite images onto topo quads, and expects to complete orthophotoquads for the Bay Area by late 1996.

"Cartographic masterpiece" were the words used by Pierce Lewis in describing the digital terrain map of the United States prepared by the USGS and now available. (Pierce Lewis, "Introducing a Cartographic Masterpiece: A Review of the U.S. Geological Survey's Digital Terrain Map of the United States, by Gail Thelin and Richard Pike," *Annals*, AAG, Vol. 82, No. 2, June 1992, pp. 289-300.)

Dr. Lewis exclaims,

> **...a relief map of the coterminous United States which, in accuracy, elegance, and drama, is the most stunning thing since Erwin Raisz published his classic 'Map of the Landforms of the United States' in 1940.**

The details brought out in this digitized map are remarkable, not just the obvious mountain systems but the subtle topography of the plains emerge with new complexity and tangible texture. Imagine a digitized image combining 12 million spot-elevations, less than a kilometer apart, blended to produce a cloudless aerial view of the 48 states. The map–Map I-2206–is available from USGS for $10.00, plus shipping, at Denver, CO, Fairbanks, AK, Reston, VA, or Menlo Park, CA; also by phone at 1-800-USA-MAPS. A small portion of this digitized map taken from central Pennsylvania is shown in Figure 12-13 (a), p. 363.

The 1992 "Maps and Images Catalog" from Raven Maps (34 North Central Avenue, Medford, OR, 1-800-237-0798, FAX 503-773-6834), features their version of this fantastic new resource, derived from the public-domain negatives prepared by the USGS. Raven has added more balanced contrast, river and lake coloration, and thousands of unobtrusive name labels. Raven offers 2 versions: a black and white, 37" x 58", 1:3.5 million scale, for $60 laminated, $35 plain; and a hypsometric treatment in Raven's beautiful style (same size, same price).

Many lab manuals and texts are on the market to assist you with work on topographic maps, e.g., Miller, Victor C., and Mary E. Westerbrook, *Interpretation of Topographic Maps*, Columbus, OH: Merrill, 1989; or, Rabenhorst, Thomas D. and Paul D. McDermott, *Applied Topography Source Materials for Mapmaking*, Columbus, OH: Merrill, 1989.

Glossary Review for Chapter 1
(in alphabetical order)

abiotic
atmosphere
biosphere
biotic
cartography
closed system
Coordinated Universal Time (UTC)
daylight saving time
dynamic equilibrium
earth systems science
ecosphere
equal area
equilibrium
feedback loops
geodesy
geographic information systems (GIS)
geography
geoid
Goode's homolosine projection
great circle
Greenwich Mean Time (GMT)
human-Earth relationships
hydrosphere
International Date Line
latitude
lithosphere
location
longitude
map projection
meridian
midlatitude geographic zone
Miller cylindrical projection
model
movement
negative feedback
open system
parallel
physical geography
place
positive feedback
prime meridian
process
region
remote sensing
rhumb lines
Robinson projection
scale
small circles
spatial analysis
steady-state equilibrium
system
topographic maps
true shape

Annotated Chapter Review Questions

1. What is unique about the science of geography? Based on information in this chapter, define physical geography and review the geographic approach.

Geography is the science that studies the interdependence of geographic areas, places, and locations; natural systems; processes; and societal and cultural activities over Earth's surface. Physical geography involves the spatial analysis of Earth's physical environment. Various words denote the geographic context of spatial analysis: space, territory, zone, pattern, distribution, place, location, region, sphere, province, and distance. Spatial patterns of Earth's weather, climate, winds and ocean currents, topography,and terrestrial biomes are examples of geographic topics.

2. Assess your geographic literacy by examining available atlases and maps the next time you are in the reference room of your school library. What types of maps have you used – political? physical? topographic? Do you know what projections they employed? Do you know the names and locations of the four oceans, seven continents, and individual countries? Can you identify the new countries that have emerged since 1990?

An informal quiz might be appropriate here, as in, Ouagadougou is the capital of Burkina Faso. Where are Kuwait and Iraq and how many live there? What is Burma's proper name? What new nations have formed since 1990?

Where are warm and wet, hot and dry, cool and moist, cold and dry regions? Where are we right now in terms of latitude, longitude, elevation, etc.?

3. Suggest a representative example for each of the five geographic themes and use that theme in a sentence.

Suggestion: Use Figure 1-1, text and photographs, as cues for discussion of this question. The Association of American Geographers (AAG) and the National Council for Geographic Education (NCGE), in an attempt to categorize the discipline, set forth five key themes for modern geographic education: location, place, human-Earth relationships, movement, and region.

4. Have you made decisions today that involve geographic concepts discussed within the five themes presented? Explain briefly.

This may be a good assignment for the first day of class. Most students can easily complete a history, or chronological list, of a day's activities. Have students do a geography of their day so far. Listing locations they have been, how they got there and have them draw a map of their journey from home to school to work. This will also demonstrate how cartographers attempt to select information to include in their maps and how geographers attempt to select the most significant variables in a systems analysis.

5. Define systems theory as an organizational strategy. What are open systems and closed systems, negative feedback, and a system in a steady-state equilibrium condition? What type of system (open or closed) is a human body? A lake? A wheat plant?

Simply stated, a system is any ordered, interrelated set of objects, things, components, or parts, and their attributes, as distinct from their surrounding environment. A natural system is generally not self-contained: inputs of energy and matter flow into the system, whereas outputs flow from the system. Such a system is referred to as an open system: the human body, a lake, or a wheat plant. A system that is shut off from the surrounding environment so that it is self-contained is known as a closed system.

As a system operates, information is generated in the system output that can influence continuing system operations. These return pathways of information are called feedback loops. Feedback can cause changes that in turn guide further system operations. If the information amplifies or encourages responses in the system, it is called positive feedback. On the other hand, negative feedback tends to slow or discourage response in the system, forming the basis for self-regulation in natural systems and regulating the system within a range of tolerable performance. When the rates of inputs and outputs in the system are equal and the amounts of energy and matter in storage within the system are constant, or, more realistically, as they fluctuate around a stable average, the system is in a steady-state equilibrium.

6. Describe Earth as a system in terms of energy and of matter.

Most systems are dynamic because of the tremendous infusion of radiant energy from reactions deep within the Sun. This energy penetrates Earth's atmosphere and cascades through the terrestrial systems, transforming along the way into various forms of energy. Earth is an open system in terms of energy. In terms of physical matter–air, water, and material resources–Earth is nearly a closed system.

7. What are the three abiotic (nonliving) spheres that comprise Earth's environment? Relate these to the biotic sphere (living):the biosphere.

Earth's surface is the place where four immense open systems interface, or interact. Three nonliving, or abiotic, systems overlap to form the realm of the living, or biotic, system. The abiotic spheres include the atmosphere, hydrosphere, and lithosphere. The fourth, the biotic sphere, is called the biosphere and exists in an interactive position between and within the abiotic spheres.

8. Describe Earth's shape and size with a simple sketch.

Sketch after Figure 1-8 an oblate spheroid or geoid.

9. What are the latitude and longitude coordinates (in degrees, minutes, and seconds) of your present location? Where can you find this information?

Have students find a local or library source: bench marks, atlases, gazetteers, maps, many software programs, such as Global Explorer, and locations on the World Wide Web.

10. Define latitude and parallel and define longitude and meridian using a simple sketch with labels.

On a map or globe, lines denoting angles of latitude run east and west, parallel to Earth's equator. Latitude is an angular distance north or south of the equator measured from a point at the center of Earth. A line connecting all points along the same latitudinal angle is called a parallel.

On a map or globe, lines designating angles of longitude run north and south at right angles (90°) to the equator and all parallels. Longitude is an angular distance east or west of a surface location measured from a point at the center of Earth. A line connecting all points along the same longitude is called a meridian.

11. Identify the various latitudinal geographic zones that roughly subdivide Earth's surface. In which zone do you live?

Figure 1-12 portrays these latitudinal geographic zones, their locations, and their names: equatorial, tropical, subtropical, midlatitude, subarctic or subantarctic, and arctic or antarctic.

12. What does timekeeping have to do with longitude? Explain their relationship. How is Coordinated Universal Time (UTC) determined on Earth?

Earth revolves 360° every 24 hours, or 15° per hour, and a time zone of one hour is established for each 15° of longitude. Thus, a world standard was established, and time was set with the prime meridian at Greenwich, England. Each time zone theoretically covers 7.5° on either side of a controlling meridian and represents one hour. Greenwich Mean Time (GMT) is called Coordinated Universal Time (UTC); and although the prime meridian is still at Greenwich, UTC is based on average time calculations kept in Paris and broadcast worldwide. UTC is measured today by the very regular vibrations of cesium atoms in 6 primary standard clocks–the *NIST-7* being the newest placed in operation during 1994 by the United States.

13. What and where is the prime meridian? How was the location originally selected? Describe the meridian that is opposite the prime meridian on the Earth's surface.

In 1884, the International Meridian Conference was held in Washington, D.C. After lengthy debate, most participating nations chose the Royal Observatory at Greenwich as the place for the prime meridian of 0° longitude. An important corollary of the prime meridian is the location of the 180° meridian on the opposite side of the planet. Termed the International Date Line, this meridian marks the place where each day officially begins (12:01 A.M.) and sweeps westward across Earth. This westward movement of time is created by the planet's turning eastward on its axis.

14. Define a great circle, great circle routes, and a small circle. In terms of these concepts, describe the equator, other parallels, and meridians.

A great circle is any circle of Earth's circumference whose center coincides with the center of Earth. Every meridian is one-half of a great circle that crosses each parallel at right angles and passes through the poles. An infinite number of great circles can be drawn on Earth, but only one parallel is a great circle–the equatorial parallel. All the rest of the parallels diminish in length toward the poles, and, along with other circles that do not share Earth's center, constitute small circles.

15. Define cartography. Explain why it is an integrative discipline?

The part of geography that embodies map making is called cartography. The making of maps and charts is a specialized science as well as an art, blending aspects of geography, engineering, mathematics, graphics, computer sciences, and artistic specialties. It is similar in ways to architecture, in which aesthetics and utility are combined to produce an end product.

16. What is map scale? In what three ways is it expressed on a map?

The expression of the ratio of a map to the real world is called scale; it relates a unit on the map to a similar unit on the ground. A 1:1 scale would mean that a centimeter on the map represents a centimeter on the ground–certainly an impractical map scale, for the map would be as large as the area being mapped. A more appropriate scale for a local map is 1:24,000. Map scales may be presented in several ways: written, graphic, or as a representative fraction.

17. Identify the following ratios as to whether they are large scale, medium scale, or small scale: 1:3,168,000, 1:24,000, 1:250,000.

Small, large, and intermediate scales.

18. Describe the differences between the characteristics of a globe and those that result when a flat map is prepared.

Since a globe is the only true representation of distance, direction, area, shape, and proximity, the preparation of a flat version means that decisions must be made as to the type and amount of acceptable distortion. On a globe, parallels are always parallel to each other, evenly spaced along meridians, and decrease in length toward the poles. On a globe, meridians intersect at both poles and are evenly spaced along any individual parallel. The distance between meridians decreases toward poles, with the spacing between meridians at the 60th parallel being equal to one-half the equatorial spacing. In addition, parallels and meridians on a globe always cross each other at right angles. All these qualities cannot be reproduced on a flat surface. Flat maps always possess some degree of distortion.

19. What type of map projection is used in Figure 1-13? Figure 1-16?

Figure 1-13: a Robinson projection; Figure 1-16: a Mercator projection.

20. What is remote sensing? What are you viewing when you observe a weather image on TV or in the newspaper? Explain.

Remote sensing is the use of radar which can convert wavelengths into pixels of digital information and can be further processed, stored and analyzed by a computer. These pixels can be converted into images which may be enhanced to show a particular feature, for example a volcanic eruption (Figure 1-26). When viewing a weather satellite image on the TV or in the newspaper, you are seeing images, such as Doppler radar, that backscatters wavelengths to the Earth's surface in order to show atmospheric phenomena. Or the image may come from satellite GOES-8, which produces daily images of weather in the Western Hemisphere from space.

21. If you were in charge of planning for a large tract of land, how would GIS methodologies assist you? What if a portion of the tract in the GIS is a floodplain or prime agricultural land? How might this affect planning and zoning?

GIS would enable you to gather, manipulate and analyze vast amounts of geographic information from remote sensing or satellite technologies. The coordinates of a specific location may be digitized, and then geographers may add layers of remote sensing information to these coordinates. This allows geographers to overlay analytical information from more than one data plane and analyze complex relationships rapidly. This enables geographers to manipulate variables within their study very rapidly.

The usefulness of GIS in analyzing a floodplain would be that the user could examine the frequency of flooding, and the areal or monetary extent of damage in the last ten or more years. For agricultural land, the user could examine levels of salinity, soil fertility, crop production statistics for previous crops, and locate areas within their land that may need extra care or different methods of farming. GIS is applicable to urban and transport planning because it is able to consider and model areas of hazard, agricultural land values and routes of intensive transport use.

Overhead Transparencies

As an adopter you are provided with the following figures for overhead projector use.

- Figure 1-1: Five themes of geographic science
- Figure 1-2: The content of geography
- Figure 1-4: A leaf is a natural open system
- Figure 1-8: Earth's Four Spheres
- Mounted together: Figure 1-11- Parallels of latitude (top); and, Figure 1-14-Meridians of longitude (bottom)
- Figure 1-19: From globe to a flat map
- Figure 1-20: Classes of map projections
- Figure 1-25: Remote sensing technologies
- Figure 1-27: A geographic information system model

PART ONE:
The Energy-Atmosphere System

Overview--Part One

GEOSYSTEMS begins with the Sun and solar system to launch the first of four parts. Our planet and our lives are powered by radiant energy from the star closest to Earth–the Sun. Each of us depends on many systems that are set into motion by energy from the Sun. These systems are the subjects of Part One.

Part One exemplifies the systems organization of the text: It begins with the origin of the solar system and the Sun. Solar energy passes across space to Earth's atmosphere varying seasonally in its effects on the atmosphere (Chapter 2). Insolation then passes through the atmosphere to Earth's surface (Chapter 3). From Earth's surface, and surface energy balances (Chapter 4) generate patterns of world temperature (Chapter 5) and general and local atmospheric circulations (Chapter 6). Each part contains related chapters with content arranged according to the flow of individual systems or in a manner consistent with time and the flow of events.

2

Solar Energy to Earth

Overview--Chapter 2

The ultimate spatial inquiry is to discern the location of Earth in the universe. To properly set the stage for a course in the physical geography of Earth, slides, videography, and posters can be used to establish the location and place of our planetary home. Our immediate home is North America, a major continent on planet Earth, the third planet from a typical yellow star in a solar system. That star, our Sun, is only one of billions in the Milky Way Galaxy, which is one of millions of galaxies in the universe. This chapter examines the nature of the flow of energy and material from the Sun to the outer reaches of Earth's atmosphere. A brief section on remote sensing and GIS closes the chapter.

The scientific method is featured in a focus study in this chapter. For an interesting discussion of science and the scientific method, including principal criticisms, see Chapter 2 (pp. 10-34) of Stanley A. Schumm's *To Interpret the Earth–Ten Ways to be Wrong*, New York (London): Cambridge University Press, 1991.

The *Student Study Guide* presents 20 "Learning Objectives" to guide the student in reading the chapter. The *Applied Physical Geography* lab manual has one exercise with four steps that involve aspects of this chapter.

New to the Third Edition

(Note: This section highlights major changes, new features, and additions in the third edition. This does not describe all the rewrite and recast of the text.)

1. A list of key learning concepts begins the chapter.

2. New discussion of the use of the Hubble Space Telescope in aiding astronomers to observe the condensation of material in other locations of the galaxy. A new method of examining planetesimals.

3. Expanded discussion of sunspots and auroras from information collected by the Voyager and Pioneer satellites in the 1970's.

4. Discussion of current research on sunspots and the predicted effect of sunspots upon climatic patterns in 1997.

5. A revised focus section on the Scientific Method with greater emphasis placed upon global changes caused by humans. A new hypothesis examines the damage which human-made chemicals cause in the stratosphere. Included in Figure 1 of the Focus Study, is a new model of the Scientific Method flow chart.

6. News Report #1: The Nature of the Order is Chaos. This report suggests that chaos may be caused by ordered principles and examines the chaos theory in detail.

7. Figure 2-12 is a new representation of the angle of incidence and its impact upon the temperature distributions on Earth. This graph may enable the student to more easily comprehend how the Sun's delineation determines the intensity of heat received at a specific location.

8. Figure 2-13 Daily insolation received at the top of the troposphere, includes graphs which illustrate the declination of the sun in four specific locations throughout the year. The student will be able to correlate the changing angle of the Sun to each location's annual receipt of energy.

9. Figure 2-20 illustrates seasonal changes in the angle of incidence.

10. Figure 2-21 depicts photos of a pear tree in fall, winter, spring and summer, to demonstrate the changing energy needs of vegetation throughout the year.

11. Figure 2-22 demonstrates the seasonal changes captured by satellite.

12. A new summary and review section ends the chapter.

Key Learning Concepts

1. *Distinguish* among galaxies, stars, and planets, and locate Earth.

2. *Overview* Earth's origin, formation, and development, and that of the atmosphere. Specifically, *construct* Earth's annual orbit around the Sun.

3. *Describe* the Sun's operation and *explain* the characteristics of the solar wind and the electromagnetic spectrum of radiant energy.

4. *Portray* the intercepted solar energy and its uneven distribution at the top of the atmosphere.

5. *Define* solar altitude, solar declination, and daylength; *describe* the annual variability of each-Earth's seasonality.

Expanded Outline Discussion

The following headings (boldfaced) match some of the first, second, and third order headings in Chapter 2. The narrative under each heading contains information, sources, and anecdotal facts relating to portions of the chapter. Not all text headings are discussed.

SOLAR SYSTEM, SUN, AND EARTH

Solar System Formation and Structure

The role of origins in geographic education–the genesis of processes and systems–is one that is debated by geographers as to possible relevance in geographic content. I included this section in Chapter 2 for several reasons; namely, the point made earlier that for many students this is their only science class, and the origin of the Universe, galaxies, stars, and planets are at the core of a scientific curriculum. Secondly, I think that enough is established about Earth's physical origins for us to be able to describe the dynamic *spatial* evolution of abiotic and biotic systems as the crust solidified, water collected on the surface, and prebiotic synthesis began, followed by chemosynthesis, photosynthesis, and the development of Earth's early atmospheres.

Interesting background and graphics are in "The Early Life of Stars" by Steven W. Stahler, *Scientific American* 265, no. 1 (July 1991), pp. 48-55. Also, refer to the twenty volume set titled *Voyage Through the Universe*, by the Editors of Time-Life Books, 1988-1991. Specific volumes cover *The Sun*, *Stars*, *The Third Planet*, among others. These are a great source of illustrations and graphics.

My experience is that few students have been exposed to the dimensions of the universe, or the relationship among galaxies, solar systems, stars, planets, and where we are in all this vastness. I express my shock at the results of continued informal polling of students, but try this: Ask your classes if they have ever seen the Milky Way Galaxy. A response of only 15% is about average. If your physical geography course is their only science class, then this lecture is the only place they will receive information that serves to put our world in such spatial perspective. The openness of science and the scientific method really comes across with this discussion.

An excellent poster for this discussion comes as a supplement with the June 1983 *National Geographic* magazine (Vol. 163, No. 6, p. 704A), "Journey into the Universe Through Time and Space." (Back issues are available from NGS.) The same illustration appears modified in the editions of the *Atlas of the World* by National Geographic, now in its Revised 6th edition (p. 119, "The Universe, Nature's Grandest Design").

For the students to imagine a starry night in the mountains, desert, or at the coast is useful–that myriad of stars visible to the unaided eye represents only a few thousand of nearly 400 billion stars that reside in our galaxy alone. The Milky Way in turn is located in our local group of 17 objects; there are evidently tens of thousands of local groups with ours being very small. All the local groups in this vicinity together form a local supercluster of local groups; there are evidently thousands of local superclusters–and so on–and so on.

Dimensions

The distance from the Sun in terms of kilometers and the speed of light is basic to placing Earth in the solar system and setting the stage for the flow of solar energy and material across space to Earth. In terms of light speed, the Moon is 1.28 seconds from Earth, the Sun 8 minutes and 20 seconds from Earth, the Solar System is about 11 hours in diameter, and the Milky Way is 100,000 light years across (edge to edge).

Earth's Orbit

Definitions of the plane of the ecliptic, perihelion, and aphelion are important. Also, mention the variability of Earth's orbit over long time-spans as shown in Chapter 17 (Figure 17-23) amounting to more than a 17.7 million km (11 million mi) difference during a 100,000 year cycle. The Aristotelian model of perfection and the prevailing religious views of several hundred years ago were quite incorrect. It is good for students to see the variability of the established numbers they learn. The figure also shows long-term variability in Earth's tilt and precessional wobble.

A comparison of Earth with the other planets is interesting. Use the table included in Figure 2-3 to compare distance, size, tilt, revolution, and rotation. In the second edition I include a count of natural satellites for each planet. The average distance from the Sun to Earth is referred to as an astronomical unit, or A.U., and is used as a basic measurement unit in the Solar System. The average distance from the Sun in terms of astronomical units are as follows: Mercury = 0.39, Venus = 0.72, Earth = 1.0, Mars = 1.52, Jupiter = 5.20, Saturn = 9.54, Uranus = 19.18, Neptune = 30.70, and Pluto 39.46. Note: Pluto's motion is so skewed from the plane of the ecliptic and eccentric in orbit that Neptune actually is further from the Sun than Pluto from 1979 until 1999. In 1999, Pluto will cross the orbit of Neptune and again be the most distant planet.

Earth's Development; Earth's Past Atmospheres

I reduced the content here to a brief section and a summary in Table 2-1. Notably absent from the second distinct atmosphere was free oxygen. The oxygen that did exist was derived from photodissociation, a process whereby ultraviolet light from the Sun splits the water molecule (H_2O) apart, although photodissociation produced only a fraction of today's oxygen concentration. Ultraviolet light was streaming unfiltered through the primitive atmosphere (no ozonosphere), and added to the radiation level created by radioactive decay from Earth's crust.

What little oxygen there was rapidly oxidized surface materials and combined with rocks and minerals on Earth's surface, particularly with iron (ferrous ions). There is little rust (iron oxides) in ancient sediments (older than 3.2 billion years ago), thought to be proof of very low oxygen levels. Interestingly, higher oxygen levels in the early atmosphere would have been poisonous to any evolving biotic processes and could have prevented the initiation of life. The environment was without significant oxygen, or was anaerobic, during the beginning of the evolutionary atmosphere. An interesting process-response system was in operation in the evolutionary atmosphere relative to oxygen. This system response was hypothesized by Harold Urey. As photodissociation freed more oxygen, some of it migrated up in the atmosphere, where it absorbed ultraviolet radiation, a process that oxygen performs in the atmosphere today. With less ultraviolet coming through, the rate of photodissociation slowed, thus producing less free oxygen to intercept ultraviolet. This increased ultraviolet raised photodissociation rates in the lower atmosphere, which freed more oxygen, thus repeating the cycle. This is an excellent example of negative feedback information regulating a system.

As the percentage of oxygen increased, some of it migrated upward in the still-forming atmosphere and reacted with ultraviolet light; there, another variant of the oxygen molecule called ozone (O_3) formed and actively absorbed ultraviolet. The effect of less ultraviolet reaching the surface permitted lifeforms to rise to shallower depths in the water, safe from dangerous ultraviolet radiation. More photosynthesis–more oxygen–more oxygen producing more ozone to absorb more ultraviolet–less ultraviolet radiation reaching Earth's surface–existence of life possible closer to the water's surface–all created conditions that allowed photosynthesizing organisms to reach the surface to maximize needed light for greater cell operation.

The *Science* journal from AAAS for February 12, 1993, (259, no. 5097) has a special section devoted to "Evolution of Atmospheres," pp. 906-41. Origin and evolution of Earth's atmosphere and an in-depth article on "The Global Carbon Dioxide Budget" are presented.

SOLAR ENERGY: FROM SUN TO EARTH

Solar Fusion

The fusion process is simply explained in the text–the level of detail will depend on the level of your class and the depth of your approach. An interesting book about the Sun with wonderful illustrations is *The Sun* by the Editors of Time-Life Books, Voyage Through the Universe series, Alexandria, VA: 1990.

A few basics might be of assistance to support possible questions. An atom is the smallest particle of matter that still has distinct properties and can enter into chemical combination. The central core of an atom is called the nucleus and is composed of tightly packed particles called protons (positive charge) and neutrons (neutral charge). Many other smaller particles reside in the nucleus as well. Electrons are small, almost weightless, negative electrical charges that exist about the nucleus at varying distances (energy levels). The nucleus in the center of the atom is surrounded by vast amounts of empty space. A neutral atom has the same number of electrons and protons, thereby creating a balanced neutral charge. An ion is an ionized atom that has an electric charge with a gained or lost electron. Hydrogen in its most common form (isotope) has one electron and one proton. Oxygen possesses eight protons with eight orbiting electrons, and so on up the periodic table of elements. Two or more atoms bound together form a molecule; molecules together form compounds.

Hydrogen has one proton and one electron in its principal form; add a neutron to the nucleus and it becomes a deuterium hydrogen atom. Helium has two protons and two electrons. Under the conditions of temperature and pressure in the Sun's core, pairs of hydrogen nuclei, the lightest of all the natural elements, are forced to fuse together in various isotopic combinations. This process of forcibly joining positively charged nuclei is called fusion. An unstable form of helium, Helium-3, is formed and a neutron is liberated as energy. Helium-3 fuses further with another deuterium to form Helium-4, a stable form of helium, and in this case a proton is liberated as energy. The Sun has converted about 8 percent of its mass to helium overall. The text gives a simple equation of the fusion reaction with hydrogen nuclei forming helium, the second lightest element in nature, and liberating enormous quantities of energy in the form of free protons, neutrons, and electrons that are propelled out from the core of the Sun. The liberated energy takes approximately 10 million years to migrate through the Sun's gaseous densities to its surface!

Every second a tremendous conversion of hydrogen to helium and liberated energy takes place. During each second of operation the Sun consumes 657 million tons of hydrogen, converting it into 652.5 million tons of helium. The difference of 4.5 million tons is the quantity that is converted directly to energy–literally disappearing solar mass just as Einstein postulated in his $E = mc^2$ equation, where the nuclear energy [E] in matter can be calculated for a given mass [m] by multiplying mass times the speed of light [c] squared. Energy and matter are interchangeable.

Solar Wind

The solar wind of charged particles (principally electrons and protons) travels in all directions from the Sun's surface at less than the speed of light, or at about 50 million km (31 million mi) a day taking approximately 3 days to reach Earth. I include this topic in the text because of the spatial implications of solar wind to Earth's systems: auroral activity, radio propagation, and possible weather-related effects.

Sunspots and Solar Activity

As massive as the Sun is, we must remember that it is still a gaseous object. The Sun rotates eastward, or left to right, as do all the planets except Uranus and Venus. The differential speeds of rotation are probably a product of convection currents below the Sun's surface. Sunspots generally are not seen further from the solar equator than 40° latitude on the Sun, and as they develop over their cycle they are seen progressively closer to the Sun's equator.

The years 1989 through 1991 were a particularly active period on the Sun, the solar maximum was reached in March and April 1990. See for an overview "When all Hell Breaks Loose on the Sun, Astronomers Scramble to Understand," by Stephen P. Maran, *Smithsonian* 20, No. 12 (March 1990): pp. 33-41. One of the record-setting and remarkable events of 1991 is reported in "Fortnight of Flares Dazzles Astronomers," by R. Cowen, *Science News*, Vol. 139 (22 June 1991): p. 388. A dramatic photograph of aurora australis taken by shuttle Discovery astronauts appears in "Earth Almanac," *National Geographic* Vol. 180, No. 4 (October 1991): p. 146.

The National Geophysical Data Center (NGDC) publishes a "Solar-Terrestrial Data" on the bulletin board. For more information call 303-497-6346. The bulletin board number is 303-497-7319. The Solar-Terrestrial Physics Division (E/GC2), 325 Broadway, Boulder, CO, offers slide sets of auroras with captions; 303-497-6131.

Sunspot Cycles

The text makes the point that the solar maximum in 1990 was a record-setting cycle with over 200 sunspots. Even into 1991 the Sun remained active, with auroras visible as far south as New York City in June 1991. It will be interesting to watch weather patterns over the next several years with the variables in operation as of the fall of 1991, all with basically unknown consequences: an active Sun and record solar maximum, continued increases in greenhouse gases and global warming, with 1990 the warmest year in instrumental history, continuing loss of stratospheric ozone, an emerging El Niño in the Pacific Ocean, and volcanic eruptions in the Philippines and Japan. The effects of the Mount Pinatubo eruption was outlined in the introduction to this manual. Imagine the spatial complexity of all these variables as they interplay in the general circulation models of the atmosphere-ocean system.

Earth's Magnetosphere

In Chapter 11, the section describing the outer core of Earth's interior tells about magnetic reversals and the phasing out, disappearance, phasing in, and reversal of Earth's magnetic field and magnetosphere. This sets a record in the rocks of magnetic particle alignment that has been important to unlocking the plate tectonics theory.

The *Dynamics Explorer I* satellite images that are available include a full selection from above the North Pole to near the equator. The image I choose for Figure 2-7a is from above 60° north latitude.

Solar Wind Effects
Auroras

The solar wind excites certain molecules and atoms in the ionosphere with the absorbed energy being emitted as light. When traveling at higher latitudes, students can check with a local college, planetarium, or other information source as to the possibility that auroral activity is expected.

CBS broadcast the auroras that were active over Norway during the 1994 Winter Olympics. Many students will be interested in this following such popular attention. This author saw the aurora borealis in 1972 in Winnipeg, Manitoba, Canada. An explosive flare rose off the Sun during early August generating enormous amounts of energy and particles. A few evenings later, after local radio stations faded to disrupted static, we moved outside and witnessed a most beautiful sight. The sky was filled with a curtain of blue-green light in irregular folds, gently waving and moving across the sky. Flashes of yellow and green shimmered, came and went, and migrated in the night sky for hours! We again saw faint auroral activity (reddish glows and flashes in the sky) in Sacramento in March 1989, following three days of a powerful solar outbreak.

Weather Effects

An interesting anthology derived from a symposium/workshop held at Ohio State University is *Solar-Terrestrial Influences on Weather and Climate*, edited by Billy D. McCormac and Thomas A. Seliga, Dordrecht, Holland: D. Reidel Publishing, 1979. (In particular, in this volume see J. Murray Mitchell, NOAA, Charles W. Stockton and David M. Meko, University of Arizona, "Evidence of a 22-Year Rhythm of Drought...," pp. 125-143.)

Dr. Harry van Loon of the National Center for Atmospheric Research described the pursuit of connections between sunspot cycles and weather in Earth's atmosphere: "It's been kind of a fad for many scientists, since it's such a precarious topic to deal with, those who tried to delve into it were thought crazy." Present theory concerns a connection between the sunspot cycle and the shifting of stratospheric winds above the equator, which tend to shift every 12 to 15 months. These winds are known as the quasi-biennial-oscillation (QBO). When they are westerly near a peak in the sunspot cycle, certain weather effects occur. The initial connection was described by Karin Labitzke of the Free University of Berlin.

Whatever the cause and linkages between solar activity and terrestrial weather, or if there is any connection, be sure to read the paragraph on p. 47 that begins, "A remarkable failure in current planning worldwide is the lack of attention given these evident cyclical patterns of drought and wetness, regardless of their cause...."

Electromagnetic Spectrum of Radiant Energy

Figure 2-9, is included in the overhead transparency set and can be used to illustrate some of the familiar aspects of the spectrum.

NASA has produced a great chart of the radiant spectrum, including electromagnetic energy, that is suitable for mounting and is available from one of the U.S. Government Printing Office Bookstores; please see Appendix B for an address.

Experiencing the Sun's electromagnetic energy outside Earth's atmosphere would be fatal. Our society experiences some harmful shortwave electromagnetic energy in various ways here at Earth's surface. Fortunately, the atmosphere screens out most of these wavelengths as discussed in the text in Chapter 3. The shortest wavelengths, less than 0.03 angstrom [Å] of gamma radiation, are penetrating and cause serious cell damage. (An "Å" equals 0.0001 micrometer [Σm] in length, which in turn is a ten-thousandths of a millimeter.) Society has passed strict measures in attempts to control gamma radiation in the environment, although controversy surrounds what indeed constitutes "safe" levels of exposure. At stake are millions of dollars of expense for affected industries if restrictions and regulations are tightened. So far, a known lower limit of safety, or threshold, has <u>not</u> been determined.

At a little longer wavelength (0.03Å to 100Å), X-rays are used for diagnostic treatment by the medical professions and in some specialized industrial applications. Each X-ray should be analyzed as to its physical cost (some damage) vs. its benefit (medical diagnosis). Imagine being unprotected outside of the atmosphere.

Still a bit longer in wavelength (100Å to 4000Å), ultraviolet energy creates burns when skin is exposed to the Sun, or sunlamps, for prolonged periods of time. An increasing amount of ultraviolet energy today is surviving passage through the atmosphere, owing to a reduced level of stratospheric ozone. Increased ultraviolet exposures are of growing concern and will be discussed in detail in Chapter 3.

ENERGY AT THE TOP OF THE ATMOSPHERE

The thermopause should be thought of as a conceptual surface, used for convenience in the determination of the solar constant. This level in the atmosphere is above most all detectable gases in the atmosphere, and the intercepted solar energy is close to 100% intensity.

Intercepted Energy Solar Constant

The text discusses the Solar Maximum Mission (1980-1989), dubbed *Solar Max*. The irony is that the satellite was brought down by the upper atmosphere's thermal expansion brought on by the extremely active solar cycle, which is what the satellite was sent up to monitor! The Earth Radiation Budget (ERB) sensors aboard *Nimbus-7* determined that solar irradiance, or luminous brightness of solar radiation, is correlated to sunspot cycles. The following graph illustrates this relationship. Note the clearly evident 11-year sunspot cycle.

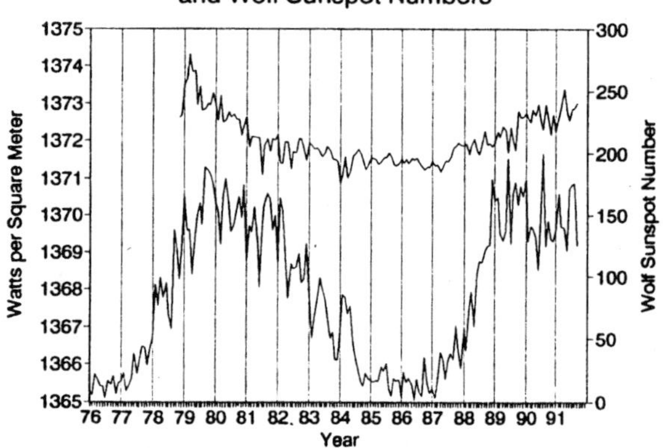

(From Figure 2 in Philip E. Ardanuy, H. Lee Kyle, and Douglas Hoyt, "Global Relationships among the Earth's Radiation Budget, Cloudiness, Volcanic Aerosols, and Surface Temperature," *Journal of Climate*, 5 no. 10, October 1992: 1120-39.)

Uneven Distribution of Insolation

The determination of the imbalance of energy input related to Earth's sphericity sets the stage for the discussion of dynamic motions in the atmosphere that follow in Chapters 3 through 9. As in any system, an analysis of inputs is necessary to understand resultant actions and the eventual outputs. Figure 2-11 demonstrates the angle of incidence of the Sun's rays as they vary with latitude.

Figure 2-12 is particularly useful in discussing the energy patterns at the thermopause; you will find it as one of the overhead transparencies. Students can trace the energy receipts along latitudes 0°, 40°, 60°, and at the poles. This figure can be referred to when analyzing seasons later in the chapter. In the lab manual, *Applied Physical Geography*, students work with a graphing exercise and this figure.

ERB sensors provided the data that appear in the new map in Figure 2-13. This map can act as a preview to the pages that follow culminating in the summary illustration in Figure 4-14. See p. 82 in Office of Research and Applications, National Environmental Satellite, Data, and Information Service (NESDIS) *Research Programs*, Washington, DC: Department of Commerce, October 1989, for a description of the ERB program. *Nimbus-7* has three instrument packages: ERB (Earth Radiation Budget), TOMS (Total Ozone Mapping Spectrometer), and SAM (Stratospheric Aerosol Monitor).

Additional ERB maps appear in Figure 4-7 showing January and July surface albedo and Figure 4-13 presenting outgoing longwave radiation. See the overview article by V. Ramanathan, *et al.*, "Climate and the Earth's Radiation Budget," *Physics Today*, May 1989: 22-32. And, H. Lee Kyle, *et al.*, *Atlas of the Earth's Radiation Budget as Measured by Nimbus-7: May 1979 to May 1980*, NASA Reference Publication 1263, May 1991, Greenbelt, MD: Scientific and Technical Information Program, NASA.

For the combined Earth-atmosphere system as measured at the top of the atmosphere (thermopause, 480 km altitude) the graph to the right illustrates average patterns of insolation, longwave (LW) and shortwave (SW) radiation, and net radiation (Net R) as measured by instruments aboard *Nimbus-7*. Net radiation roughly matches top-of-the-atmosphere insolation, although at a reduced rate.

There are many good Web sites for locating temperature maps and satellite images of insolation. Use the search words, "temperature maps", and you will be connected to more than ten different sites, each with maps that can be downloaded and used as class projections. Another great source for maps of all kinds is the University of Buffalo, Geographic Information and Analysis Lab. U of B is a source center for connections to NASA, Xerox and many other map servers. Contact webmaster@geog.buffalo.edu.

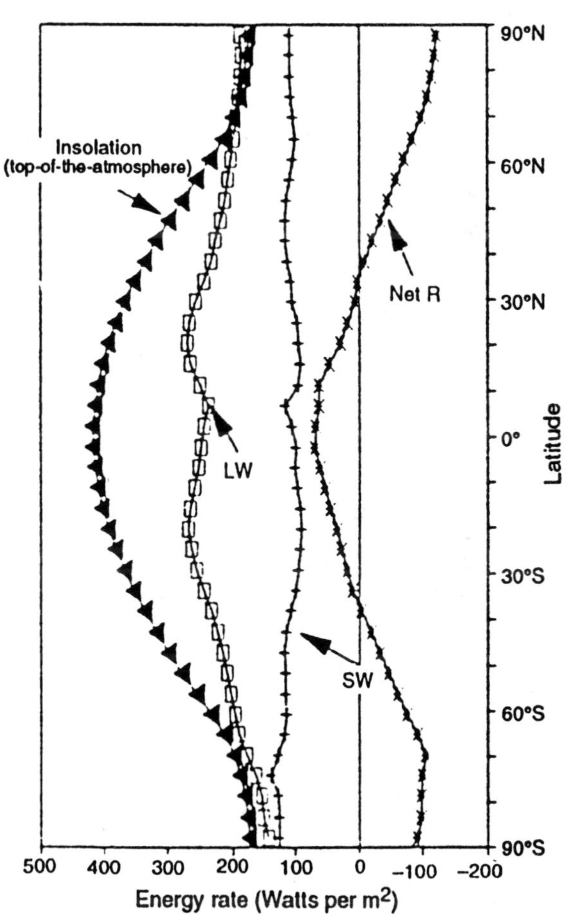

THE SEASONS

This is a new placement for this first-order heading dealing with seasons.

A lot can be done with the calendar to highlight the concept of seasons. The need for an accurate calendar to schedule planting and other activities was recognized early as a necessity. As early as 2500 B.C. the Babylonians were able to measure the length of a year, complete with the concept of weeks and months, and to establish a calendar based on lunar cycles. The Egyptians had established a calendar by 423 B.C. that was based on daily and annual solar cycles. The Mayan calendar in the New World, based on a complex of interacting cycles, was extremely accurate and assigned specific names to every day, week, and month.

In 46 B.C., Julius Caesar established a modified version of the Egyptian effort. And, of course the names of some months were eventually thrown off by two months (Dec<u>ec</u>ember is the 12th month, <u>Sep</u>tember is the 9th month, etc.) because Julius and Augustus both added months in their own names–July and August. The Julian calendar, by A.D. 1582, through accumulated errors, differed by more than 10 days from what the Sun was doing in the sky. Consequently, on 4 October 1582, Pope Gregory XIII instituted a new calendar that would correct many of the flaws in the Julian calendar. He adjusted for past errors by declaring that for the year 1582, "October 4th shall be followed by October 15th." The Gregorian calendar was not adopted in Britain until 1752, Russia in 1919, and China in 1949. In this high-technology era of atomic clocks, corrections of seconds are made occasionally, with leap days normally taken every four years (29 February), so that clocks and calendars are kept in sync with the heavens.

Seasonality

Depending on their own living circumstances, I ask students to place a small mark on a wall at their residence where the sunlight hits. Doing this on one or two Saturdays during a month, and maybe longer through the year, will reveal a pattern of seasonality. Or, they might notice that during the year the Sun is in their eyes at the supper table whereas at other times of the year it strikes someone else. Our backyard faces west so we have painted equinox and solstice markers on the back fence where we sight the Sun at sunset.

Our local National Weather Service office prepares a weather calendar that is published in the *Sacramento Bee* once a month. I keep a copy and a transparency of the current calendar with my notes and have one taped to the lectern in the classroom so that I can give them the daylength and reinforce the concept of seasonality and the changes going on each day. The farther north you are, the greater the daily change you can report. As an example, a sample Weather Calendar for March 1994 for 40° north latitude is attached at the end of this chapter (p. 41). Consult with your local newspaper or NWS to see if something like this is available, or perhaps they already are printing one in your local newspaper.

Astronomical Phenomena for the Year is prepared annually as a joint publication of the Nautical Almanac Office, U.S. Naval Observatory and Her Majesty's Nautical Almanac Office, Royal Greenwich Observatory. It is available directly from the U.S. Government Printing Office or Her Majesty's Stationary Office. The publication includes all seasonal information about the Sun, Moon, planets, calendar considerations, Gregorian and Julian numbers, equation of time, time zones, and eclipses.

Reasons for Seasons

In discussing the contributing physical factors (revolution, rotation, tilt of Earth's axis, and axial parallelism in Table 2-2), I like to take a moment and discuss what Earth would be like if each of these were faster or slower, greater or lesser, etc. If Earth were tilted completely over on its side so that the axis was parallel to the plane of the ecliptic, then there would be maximum seasons–every place would experience a 24-hour day and night at some time during the year. If on the other hand, Earth were not tilted from a perpendicular to the plane, we would have a perpetual equinox.

In Chapter 17, the variability of tilt, axial alignment, and orbital configuration are illustrated. Students can look ahead at this material, which augments the Milankovitch model of astronomical causes behind the ice ages.

An analemma is presented in Figure 2-16 and is explained in the text. Students sometimes want to know why the analemma is shaped the way it is. Reasons for the curious figure-8 shape are that Earth's orbit about the Sun is elliptical producing a variation in orbital speed during the year (faster during perihelion and slower during aphelion). This is explained by Kepler's first and second laws of planetary motion. A radius vector (line from the Sun to Earth) passes over equal areas as Earth revolves in its elliptical orbit during an equal amount of time. For this to be true the longer radius vector at aphelion must move slower than the shorter radius vector during perihelion. Usually a globe will feature an analemma printed in the Pacific Ocean.

Annual March of the Seasons

I find that most students have little idea how all the contributing factors combine to create seasons. As elementary as this sounds, I take a small bare-bulbed lamp and a globe with two pencils for an axis and I work my way around the lamp, beginning in December and pausing at each key anniversary date, as I expose another line of Table 2-5 on the overhead projector. Along with slides of Stonehenge and the midnight Sun (for the June position, see Figure 2-19), I show how Earth rotates west to east so that the Sun appears to move east to west, and I demonstrate sunrise at the North Pole (March), sunrise at the South Pole (September), and the changing circle of illumination throughout the year. The classroom demonstration, slides, and overhead, make a good summary of all this material. The annual march of the seasons in Figure 2-18 is included in your color overhead set.

Seasonal Observations

The 47° annual variability in the Sun's altitude can be easily measured and demonstrated with a protractor. The *Applied Physical Geography* lab manual has specific exercises that has the student measure and graph these seasonally changing conditions.

Imagine living in Barrow, Alaska, where the night lasts for 65 days from sunset on 19 November until sunrise at 1:09 P.M. local time 23 January. Scientists are studying what the effects of this are on an individual's biological rhythms, moods, and behavioral patterns. Add to this that the outside temperature is around –57°C (–70°F), so that people are indoors more in winter. The effects on people in Fairbanks, where night is about 21 hours in length, are being studied. Effects on those living in higher latitudes are being studied as well. Some actually develop a seasonal disorder related to these long periods without the Sun. Treatment involves "phototherapy," which involves full-spectrum light exposure. This light simulates the Sun's spectrum at the surface at about the time of the onset of twilight. Whatever the results of such research, it appears that to stay active and busy is important, that women are affected more than men, that Caucasians are affected more than native Eskimos, and that adults are affected more than children–some 35 million people are possibly affected.

Refer to the following: Bruce Bower, "Here Comes the Sun," *Science News*, 142, July 25, 1992: 62-3. This reviews scientific efforts to understand winter depression known as "seasonally affective disorder," or SAD.

Another source is Dr. Norman E. Rosenthal, National Institute of Mental Health, Bethesda, MD.

Dawn and Twilight Concepts

The U.S. Navy Oceanographic Office, in Publication No. 5175, presents a chart of the duration of dawn and twilight in hours for all latitudes. You may want to obtain the chart to illustrate this seasonally changing concept of valuable work time for humans. (U.S. Navy Oceanographer, U.S. Naval Observatory, Massachusetts Avenue and 34th Street, N.W., Washington, DC 20392-1800.)

Glossary Review for Chapter 2

(in alphabetical order)

altitude
aphelion
auroras
autumnal (September) equinox
axial parallelism
axis
chemosynthesis
circle of illumination
daylength
declination
electromagnetic spectrum
evolutionary atmosphere
fusion
gravity
insolation
living atmosphere
magnetosphere
Milky Way Galaxy
modern atmosphere
nebula
perihelion
plane of the ecliptic
planetesimal hypothesis
primordial atmosphere
revolution
rotation
scientific method
solar constant
solar wind
speed of light
subsolar point
summer (June) solstice
sunrise
sunset
sunspots
thermopause
Tropic of Cancer
Tropic of Capricorn
vernal (March) equinox
wavelength
winter (December) solstice

Annotated Chapter Review Questions

1. Describe the Sun's status among stars in the Milky Way Galaxy. Describe the Sun's location, size, and relationship to its planets.

Our Sun is both unique to us and commonplace in our galaxy. It is only average in temperature, size, and color when compared with other stars, yet it is the ultimate energy source for almost all life processes in our biosphere. Planets do not produce their own energy. Our Sun is located on a remote, trailing edge of the Milky Way Galaxy, a flattened, disk-shaped mass estimated to contain up to 400 billion stars. See the table included in Figure 2-3.

2. If you have seen the Milky Way at night, briefly describe it. Use specifics from the text in your description.

From our Earth-bound perspective in the Milky Way, the galaxy appears to stretch across the night sky like a narrow band of hazy light. On a clear night the naked eye can see only a few thousand of the nearly 400 billion stars.

3. Briefly describe Earth's origin as part of the Solar System.

According to prevailing theory, our solar system condensed from a large, slowly rotating, collapsing cloud of dust and gas called a nebula. As the nebular cloud organized and flattened into a disk shape, the early proto-Sun grew in mass at the center, drawing more matter to it. Small accretion (growing) eddies– the protoplanets – swirled at varying distances from the center of the solar nebula. The early protoplanets, or planetesimals, were located at approximately the same distances from the Sun that the planets are today. The beginnings of the Sun and the solar system are estimated to have occurred more than 4.6 billion years ago.

4. Define the scientific method and give an example of its application.

See Focus Study 2-1. The scientific method is simply the application of common sense in an organized and objective manner

5. How far is Earth from the Sun in terms of light speed? In terms of kilometers and miles? Relate this distance to the shape of Earth's orbit during the year.

Earth's orbit around the Sun is presently elliptical–a closed, oval-shaped path (Figure 2-2). Earth's average distance from the Sun is approximately 150 million km (93 million mi). Earth is at perihelion, its closest position to the Sun, on January 3 at 147,255,000 km (91,500,000 mi) and at aphelion, its farthest distance from the Sun, on July 4 at 152,083,000 km (94,500,000 mi). This seasonal difference in distance from the Sun results in a variation of 3.4 percent in the solar output that is intercepted by Earth.

6. How many distinct atmospheres have there been on Earth? How does Earth's present atmosphere compare to past atmospheres?

A principal component of Earth's history is the evolution of the modern atmosphere. The original constituents of Earth's atmosphere were derived from the original solar nebula. The first two atmospheres–primordial and evolutionary atmospheres–are thought to have persisted for relatively short periods, whereas the living atmosphere and the modern atmosphere have existed over much greater time spans. The summary is in Table 2-1 and compares the present and past atmospheres.

7. Within which of Earth's atmospheres, past and present, did photosynthesis begin?

In the living atmosphere, 3.3 to 0.6 years ago, the first photosynthesis occurred in cyanobacteria (at 3.3 billion years ago). Heavy global rains filling ocean basins provided the initial protection to organisms so that the slow evolution of gaseous constituents developed toward the modern atmosphere.

8. How does the Sun produce such tremendous quantities of energy? Write out a simple fusion reaction, using the quantities (in tons) of hydrogen involved.

The solar mass produces tremendous pressure and high temperatures deep in its dense interior region. Under these conditions, pairs of hydrogen nuclei, the lightest of all the natural elements, are forced to fuse together. This process of forcibly joining positively charged nuclei is called fusion. In the fusion reaction hydrogen nuclei form helium, the second lightest element in nature, and liberate enormous quantities of energy in the form of free protons, neutrons, and electrons. During each second of operation, the Sun consumes 657 million tons of hydrogen, converting it into 652.5 million tons of helium. The difference of 4.5 million tons is the quantity that is converted directly to energy–resulting in literally disappearing solar mass.

9. What is the sunspot cycle? At what stage in the cycle were we in 1997?

A regular cycle exists for sunspot occurrences, averaging 11 years from maximum peak to maximum peak; however, the cycle may vary from 7 to 17 years. In recent cycles, a solar minimum occurred in 1976, whereas a solar maximum took place in 1979, when over 100 sunspots were visible. Another minimum was reached in 1986, and an extremely active solar maximum occurred in 1990 with over 200 sunspots, 11 years from the previous maximum, in keeping with the average. In 1997, we will again be at a solar minimum.

10. Describe Earth's magnetosphere and its effects on the solar wind and the electromagnetic spectrum.

Earth's outer defense against the solar wind is the magnetosphere, which is a magnetic force field surrounding Earth, generated by dynamo-like motions within our planet. As the solar wind approaches Earth, the streams of charged particles are deflected by the magnetosphere and course along the magnetic field lines. As shown in Figure 2-5, the extreme northern and southern polar regions of the upper atmosphere are the points of entry for the solar wind stream.

11. Summarize the presently known effects of the solar wind relative to Earth's environment.

The interaction of the solar wind and the upper layers of Earth's atmosphere produces some remarkable phenomena relevant to physical geography: the auroras, disruption of certain radio broadcasts and even some satellite transmissions, and possible effects on weather patterns.

12. Describe the segments of the electromagnetic spectrum, from shortest to longest wavelength. What wavelengths are mainly produced by the Sun? Which are principally radiated by Earth to space?

See Figure 2-10. All the radiant energy produced by the Sun is in the form of electromagnetic energy and, when placed in an ordered range, forms part of the electromagnetic spectrum. The Sun emits radiant energy composed of 8 percent ultraviolet, X-ray, and gamma ray wavelengths; 47 percent visible light wavelengths; and 45 percent infrared wavelengths. Wavelengths emitted from the Earth back to the Sun are of lower intensity and composed mostly of infrared wavelengths.

13. What is the solar constant? Why is it important to know?

The average value of insolation received at the thermopause (on a plane surface perpendicular to the Sun's rays) when Earth is at its average distance from the Sun. That value of the solar constant is 1372 watts per square meter (W/m^2). A *watt* is equal to one joule (a unit of energy) per second and is the standard unit of power in the SI-metric system. (See the inside back cover of *GEOSYSTEMS* for more information on measurement conversions.) In nonmetric calorie heat units, the solar constant is expressed as approximately 2 calories (1.968) per cm^2 per minute, or 2 langleys per minute (a langley being 1 cal per cm^2). A *calorie* is the amount of energy required to raise the temperature of one gram of water (at 15° C) one degree Celsius and is equal to 4.184 joules.

Knowing the amount of insolation intercepted by Earth is important to climatologists and other scientists as a basis for atmospheric and surface energy measurements and calculations.

14. Select 40° or 50° north latitude on Figure 2-13, and plot the amount of energy in W/m^2 per day characteristic of each month throughout the year. Compare this to the North Pole; to the equator.

See Figure 2-13. Note the watts per m^2 received at each month for specific latitudes.

15. If Earth were flat and oriented perpendicular to incoming solar radiation (insolation), what would be the latitudinal distribution of solar energy at the top of the atmosphere?

The atmosphere is like a giant heat engine driven by differences in insolation from place to place. If Earth were flat there would be an even distribution of energy by latitude with no differences from place to place and therefore little motion produced.

16. In what possible way does the Stonehenge monument relate to seasons?

Many ancient societies demonstrated a greater awareness of seasonal change than do modern peoples, and they formally commemorated natural energy rhythms with festivals and monuments. At Stonehenge, stones weighing 25 metric tons (28 tons) were hauled 480 km (300 mi) and placed in patterns that evidently mark seasonal changes, specifically sunrise on or about June 21 at the summer solstice. Other seasonal events and eclipses of the Sun and the Moon are apparently predicted by this 3500 year-old calendar monument.

17. The concept of seasonality refers to what specific terms? How do these two aspects of seasonality change with latitude? 0°? 40°? 90°?

Seasonality refers to both the seasonal variation of the Sun's rays above the horizon and changing daylengths during the year. Seasonal variations are a response to the change in the Sun's altitude, or the angular difference between the horizon and the Sun. Seasonality also means a changing duration of exposure, or daylength, which varies during the year depending on latitude. People living at the equator always receive equal hours of day and night, whereas people living along 40° N or S latitude experience about six hours' of difference in daylight hours between winter and summer; those at 50° N or S latitude experience almost eight hours of annual daylength variation. At the polar, extremes the range extends from a six-month period of no insolation to a six-month period of continuous 24-hour days.

18. Differentiate between the Sun's altitude and its declination at Earth's surface.

The Sun's altitude is the angular difference between the horizon and the Sun. The Sun is directly overhead at zenith only at the subsolar point. The Sun's declination–that is, the angular distance from the equator to the place where direct overhead insolation is received–annually migrates through 47 degrees of latitude between the two tropics at 23.5° N and 23.5° S latitudes.

19. For the latitude at which you live, how does daylength vary during the year? How does the Sun's altitude vary? Does your local newspaper publish a weather calendar containing such information?

See reference materials in this resource manual chapter to guide the students. See Table 2-4.

20. List the five physical factors that operate together to produce seasons.

Table 2-2 details these factors: Earth's revolution and rotation, its tilt and fixed-axis orientation, and its sphericity.

21. Describe revolution and rotation, and differentiate between them.

The structure of Earth's orbit and revolution about the Sun is described in Figure 2-2. Earth's revolution determines the length of the year and the seasons. Earth's rotation, or turning, is a complex motion that averages 24 hours in duration. Rotation determines daylength, produces the apparent deflection of winds and ocean currents, and produces the twice daily action of the tides in relation to the gravitational pull of the Sun and the Moon. Earth's axis is an imaginary line extending through the planet from the North to South geographic poles.

22. Define Earth's present tilt relative to its orbit about the Sun.

Think of Earth's elliptical orbit about the Sun as a level plane, with half of the Sun and Earth above the plane and half below. This level surface is termed the plane of the ecliptic. Earth's axis is tilted 23.5° from a perpendicular to this plane.

23. Describe seasonal conditions at each of the four key seasonal anniversary dates during the year. What are the solstices and equinoxes and what is the Sun's declination at these times?

See Tables 2-4, 2-5 and Figure 2-18.

Overhead Transparencies

As an adopter you are provided with the following figures for overhead projector use.

- Figure 2-2: Earth's orbit and the solar system
- Figure 2-9: The electromagnetic spectrum of radiant energy
- Figure 2-12: Insolation receipts and Earth's curved surface
- Figure 2-13: Total daily insolation received at the top of the atmosphere
- Figure 2-17: The plane of Earth's orbit - the ecliptic
- Figure 2-18: Annual march of the seasons
- Figure 2-20: Seasonal observations-sunrise, noon, sunset

An example of a "Weather Calendar" prepared with data from the National Weather Service. Daylength, Moonrise and Moonset times, temperatures, and precipitation data for Sacramento, CA, 40° N latitude are presented. Check to see if your local paper can present a feature like this each month specific to your city. (Note the effect of atmospheric refraction on daylength times; check the times at the equinox.)

DECEMBER 1994 WEATHER CALENDAR

DAY	SUN			MOON		TEMPERATURES							PRECIPITATION		
	Rise AM	Set PM	Day length Hrs.:Mins.	Rise	Set	Records Max.	Year	Min.	Year	Normal Max.	Min.	Avg.	Record Amt.	Year	Last year
1	7:05	4:45	9:40	5:27a	3:55p	71	1959	32	1929	57	42	50	1.70	1952	
2	7:06	4:45	9:39	6:38a	4:51p	69	1959	30	1906	57	42	50	2.05	1880	
3	7:07	4:45	9:38	7:44a	5:54p	71	1958	32	1918	57	42	49	2.00	1890	
4	7:07	4:45	9:38	8:34a	7:01p	71	1958	29	1909	56	42	49	1.41	1881	
5	7:08	4:45	9:37	9:34a	8:10p	72	1979	32	1972	56	42	49	.78	1889	
6	7:09	4:45	9:36	10:18a	9:17p	70	1989	29	1891	56	41	49	.96	1950	
7	7:10	4:45	9:35	10:56a	10:21p	68	1979	28	1891	56	41	48	.98	1889	.05
8	7:11	4:45	9:34	11:30a	11:23p	71	1988	27	1972	55	41	48	1.23	1909	.55
9	7:12	4:45	9:33	12:01p	–	69	1893	23	1932	55	41	48	1.87	1954	.25
10	7:13	4:45	9:32	12:31p	12:22a	68	1958	22	1932	55	41	48	1.92	1937	
11	7:13	4:45	9:32	1:01p	1:20a	71	1958	17	1932	54	41	47	1.39	1906	.47
12	7:14	4:45	9:31	1:32p	2:17a	71	1958	21	1932	54	40	47	1.09	1922	
13	7:15	4:45	9:30	2:05a	3:13a	71	1958	23	1932	54	40	47	1.73	1915	.14
14	7:16	4:46	9:30	2:41p	4:09a	69	1958	23	1940	54	40	47	1.56	1929	.52
15	7:16	4:46	9:29	3:21p	5:03a	72	1958	26	1932	54	40	47	1.18	1957	
16	7:17	4:46	9:29	4:05p	5:56a	70	1958	26	1892	54	40	47	.95	1957	
17	7:18	4:47	9:29	4:53p	6:46a	69	1958	28	1928	53	40	46	1.33	1884	
18	7:18	4:47	9:28	5:45p	7:33a	68	1958	28	1924	53	40	46	1.40	1955	
19	7:19	4:47	9:28	6:40p	8:16a	66	1929	25	1924	53	40	46	2.41	1955	
20	7:20	4:48	9:28	7:38p	8:56a	66	1981	27	1928	53	40	46	1.32	1884	
21	7:20	4:48	9:28	8:38p	9:32a	63	1969	22	1990	53	40	46	2.81	1885	
22	7:21	4:49	9:28	9:38p	10:06a	65	1914	18	1990	53	39	46	1.94	1955	
23	7:21	4:49	9:28	10:40p	10:38a	66	1964	21	1990	53	39	46	1.38	1884	
24	7:21	4:50	9:29	11:43p	11:10a	66	1964	23	1990	53	39	46	2.21	1983	
25	7:22	4:50	9:29	–	11:43a	64	1967	26	1891	53	39	46	2.42	1884	
26	7:22	4:51	9:29	12:48a	12:19p	65	1967	25	1879	52	39	46	1.58	1955	.06
27	7:23	4:52	9:29	1:56a	12:58p	68	1953	27	1878	52	39	46	1.96	1931	
28	7:23	4:52	9:29	3:05a	1:42p	72	1967	26	1930	52	39	46	1.25	1992	
29	7:23	4:53	9:30	4:14a	2:33p	66	1989	24	1878	52	39	46	1.47	1933	
30	7:23	4:54	9:31	5:21a	3:32p	60	1970	28	1990	52	39	46	1.32	1913	
31	7:24	4:55	9:31	6:24a	4:36p	61	1979	24	1915	52	39	46	1.07	1913	
									Normal Temp:	54	40	47	**Normal:** 2.76"		

3

Earth's Modern Atmosphere

Overview

Earth's atmosphere is a unique reservoir of gases, the product of billions of years of development. This chapter examines its structure, function, and composition, starting from outer space and descending through the various layers and regions of the atmosphere to Earth's surface. A consideration of our modern atmosphere must also include the spatial aspects of human-induced problems that affect it, such as air pollution, the stratospheric ozone predicament, and the blight of acid deposition.

The *Student Study Guide* presents 20 "Learning Objectives" to guide the student in reading the chapter. The *Applied Physical Geography* lab manual has one exercise with seven steps that involve aspects of this chapter.

New to the Third Edition

(Note: This section highlights major changes, new features, and additions in the third edition. This does not describe all the rewrite and recast of the text.)

1. A list of key learning concepts begins the chapter.

2. Discussion of air pressure, density and how pressure is determined, has been moved to Chapter 6.

3. Figure 3-3, illustrates the path of solar radiation through the atmosphere, to demonstrate the amount of UV, infrared and visible light that reach the surface.

4. The stable components of the atmosphere are portrayed in a pie-chart format to reflect relative proportions of nitrogen, oxygen and CO_2, Figure 3-4.

5. Figure 3-6, temperature profile of the troposphere, has been updated to illustrate the elevational difference of the tropopause at the equator compared to the midlatitudes, and what impact this has upon global temperatures.

6. An inset map of Mexico has been added to Figure 3-10 to ease location of Mexico City.

7. News Report #1: ""Earth's Atmosphere: Unique Among Planets" compares the Earth to Mars, Venus and Mercury using satellite and remote sensing images to assess atmospheric components on each planet.

8. Updated maps in Figures 3-12 and 3-14, illustrate the distribution of emissions and wet deposition for NO and SO_2 in 1991.

9. News Report #2: "New UV Index Announced to Save Your Skin". This article explains the UV index, which recommends the amount of exposure various skin types can endure without damage. This information has great significance as the ozone continues to thin.

10. News Report #3: "1995 Nobel Chemistry Prize for Ozone Depletion". This article summarizes research which was awarded the first Nobel prize for atmospheric studies. Three men discovered human-produced chemicals proven to destroy the stratospheric ozone. These findings have great significance for future policies concerning the creation and release of nitrogen oxides, nitrogen dioxide and CFC's.

11. News Report #4: "Air Pollution Abatement-Cost vs. Benefits" compares the costs of reducing auto emissions and ozone-depleting chemicals to the benefits gained with reduced pollution levels upon crops, health, cities and the economy.

12. New discussion on sulfate aerosols and their connection to the greenhouse effect and acid rain.

13. Expanded discussion regarding the source of air pollution; updated statistics concerning carbon monoxide contributions in the U.S., the production of SO_2 from Hawaiian and Icelandic volcanic eruptions, as well as the impact of such pollution upon the environment, crops and health, including the death of tennis star Vitas Gerulaitis.

14. Focus Study 3-1, contains an updated image of *Nimbus* 7 taken October 1995. Figure 3-1-1 illustrates the changes in ozone depletion since 1979, and Figure 3-1-2 demonstrates the relationship between levels of ClO and ozone depletion between September 1991 and 1992.

15. Elaborate discussion about strategies which may reduce air pollution, including policies introduced in Mexico City.

16. A new summary and review section ends the chapter.

Learning Objectives

1. *Construct* a general model of the atmosphere based on composition, temperature, and function and diagram this model in a simple sketch.

2. *List* the stable components of the modern atmosphere and their relative percentage contributions by volume, and *describe* each.

3. *Describe* conditions within the function and status of the ozonosphere (ozone layer).

4. *Distinguish* between the natural and anthropogenic variable gases and materials in the lower atmos phere.

5. *Describe* the sources and effects of CO, NO_2, and SO_2, and *construct* a simple equation that illustrates photochemical reactions that produce O_3 and PAN, and HNO_3 and H_2SO_4.

Expanded Outline Discussion

The following headings (boldfaced) match some of the first, second, and third order headings in Chapter 3. The narrative under each heading contains information, sources, and anecdotal facts relating to portions of the chapter. Not all text headings are discussed.

ATMOSPHERIC STRUCTURE AND FUNCTION

The organization of the content is in the context of the atmospheric profile from the thermopause to troposphere through to Earth's surface. Chapter 4 then picks up the flow of insolation through the lower atmosphere as it cascades to the surface and establishes surface energy budgets. The process-system flow of insolation is supported by this section on the atmosphere's structure and function. Hopefully this proves useful as a teaching sequence.

The dramatic and poignant quote from Lewis Thomas's *Lives of the Cell* comparing the presence and function of the atmosphere to a cell membrane is a wonderful metaphor. Dr. Thomas passed away in 1993 at the age of 81, active until the end and still alive in everyone who has read his works. Thomas said, "Statistically the probability of any one of us being here is so small that you'd think the mere fact of existing would keep us all in a contented dazzlement of surprise." Maybe through the perspective of physical geography we can raise students to a greater sense of wonder about our life-supporting Earth. George Will in a December 1993 *Washington Post* column said, "A quiet but insistent voice....He watched in amazement undiminished over the years."

The students should be directed to the large integrated illustration of the atmosphere shown in Figure 3-2. The three organizing criteria are portrayed along the left-hand margin of the figure. This important figure is included in your overhead transparency package and is also in the *Student Study Guide* in outline form for them to label and color. The best approach may be to redraw Figure 3-2 on the board or follow along on the overhead as you work through the following section, thus recreating the diagram in your own frame of reference using colored chalk or the overhead transparency using colored pens.

Atmospheric Composition, Temperature, and Function

To facilitate this flow of topics the atmosphere is conveniently classified using three criteria: composition, temperature, and function.

- Based on chemical *composition*, the atmosphere is divided into two broad regions: the *heterosphere* and the *homosphere*.
- Based on *temperature*, the atmosphere is divided into four distinct zones: the *thermosphere*, *mesosphere*, *stratosphere*, and *troposphere*.
- Finally, two specific zones are identified on the basis of *function* relative to their role of removing most of the harmful wavelengths of solar radiation–these are the *ionosphere* and the *ozonosphere*, or *ozone layer*.

Heterosphere–Thermosphere–Ionosphere

Note the discussion of *kinetic energy* and *sensible heat* relative to high temperatures in the thermosphere. Compare this to the concept of heat in the lower atmosphere.

When a "shooting star" streaks across the sky, the glow is caused by a teaspoon-sized piece of nickel-iron material encountering the atmosphere in the upper mesosphere. The friction between the small meteor and the very thin atmosphere creates tremendous heating that causes the entering object to glow.

Ask the students to tune in a faraway station tonight on the AM dial. They may find themselves listening to some distant station, even though they are far removed from the transmitter–KNX, WBZ, WLS, KOMO, KSL. Why does the station come in strong at times and fade out at other times? Imagine the great distances covered by these signals from their source to the radio and the role of ionospheric propagation of these signals. FM radio and television wavelengths pass right through these ionospheric regions unaffected. They may have noticed that FM and TV transmissions operate across line-of-sight to receiver sets, whereas AM radio communications depend on this functional region of the atmosphere to propagate radio signals to distant areas.

Homosphere–Mesophere

Early exploration of the atmosphere was accomplished by piloted and unpiloted balloons, with some missions going as high as 34 km (21 mi), well into the stratosphere. Note the mention of the new *Upper Atmosphere Research Satellite* (*UARS*) that became operational in late 1991 as a modern orbiting observation platform. New discoveries about vast wind systems in the mesophere are intriguing scientists.

The picture in Figure 3-5, showing air being mined from the sky, always proves to be of interest because we usually do not connect the tank of oxygen at the hospital with its source at a factory that extracts air from the sky for cryogenic processing.

Between 1987 and 1991, carbon dioxide increased at an average annual rate of 0.5% (1.71 ppm), with a maximum increase in 1987 of a full 1% (measured at the Climate Monitoring and Diagnostic Laboratory of NOAA, Boulder, Colorado). Note Table 10-1, that gives CO_2 concentrations back to 1825.

Stratosphere and Ozonosphere–The Ozone Predicament

See an interesting article and illustrations in Owen B. Toon and Richard P. Turco's, "Polar Stratosphere Clouds and Ozone Depletion," *Scientific American* 264, No. 6 (June 1991): pp. 68-74.

If scientific prediction is related to aspects of business activities, the issue will obviously be politically volatile, and the "ozone war" is no exception. The U.S. Senate voted in 1986, 80 in favor to 2 against, that "there is no safe or acceptable level of ozone depletion." Despite the growing body of empirical evidence of stratospheric ozone depletion the U.S. Administration through the first eight years of the 1980s cut related spending for research by more than half (from $10 million to under $5 million a year). The U.S. Environmental Protection Agency was successfully sued to force them to implement certain mandatory laws and related research. Of course, the world has now moved forward with several historic treaties on this issue and deadlines for stopping all production of the damaging CFCs.

Today we find an emerging problem of serious implication to human skin and eyes, crop production and net photosynthetic rates, oceanic phytoplankton and coral survival, and possible climate change. In Canada, an "Ozone Watch" was reported (*Climatic Perspectives*, 15, July 1993: 10-1). During February 1993, one of Canada's longest-running ozone observatories in Edmonton reported record seasonal-low values of stratopheric ozone and resultant higher levels of ultraviolet radiation arriving at the surface. The report states, "The persistently low ozone values that have been observed over the last four months can be attributed in part to the buildup of industrial chemicals in the atmosphere." The following graph was presented on p. 11 of this issue of *Climate Perspectives*.

Environment Canada is now reporting a "UV Index" on a regular basis–a relative scale from 0 to 10. A "1993 Ozone Review" by Bob Saunders was presented in *Climatic Perspectives* presented in v. 15, December 1993, p. 6-10. Stratospheric ozone was 7.7% thinner than normal in 1993 with the greatest thinning between January and April when it dropped to 14% below normal. Extreme lows of 22 to 25% were hit in March over Toronto, Edmonton, and Resolute!

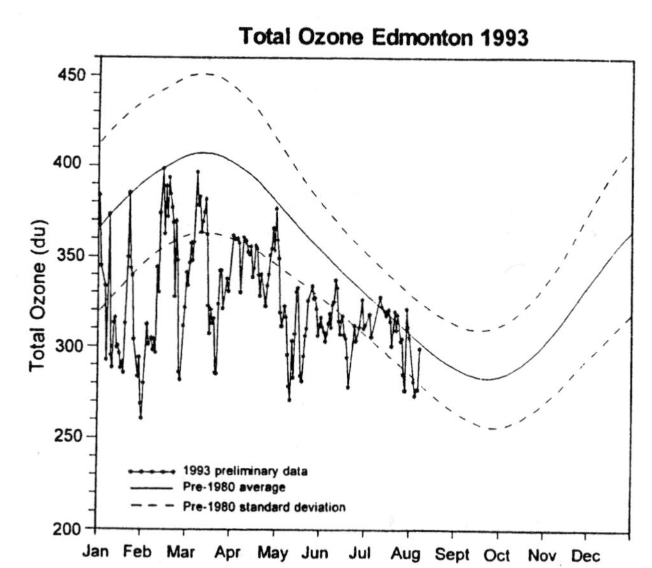

Record seasonal-low values of ozone in southern Canada were observed at Edmonton during February 1993, near record seasonal-low values persisting through March and April.

Environment Canada has launched a new scientific observatory to study the ozone layer over Canada. This high-Arctic facility now is operational at a remote weather station near Eureka on Ellesmere Island, N.W.T., about 1000 km from the North Pole. The Arctic ozone depletion affects concentrations at lower latitudes and thus should draw the United States into further cooperation.

Despite all this scientific activity, measurement, and confirmation, there remain a few media personalities who have for unexplained reasons captured popular interest by declaring that none of these events in the ozone layer are occurring–that the entire thing is a hoax put on by government and private scientists, institutions, and liberal academics. The provocation became so intense that the American Association for the Advancement of Science (AAAS) ran a strong condemnation of the media circus. **Please see: Gary Taubes, "The Ozone Backlash," *Science*, 260 June 11, 1993: 1580-83.** The article details the misrepresentation and misunderstanding by the pop-critics as they formulated their ill-founded and loudly stated views that everything is all right in the stratosphere. I have found this to be an effective handout for classes and it tends to counter the vocal nonscientific critics.

Reported in *Science News*, 144, December 4, 1993: 382, are measurements made by James B. Kerr and C. Thomas McElroy of Environment Canada that show increased ultraviolet light reaching the surface in North America. The summary stated: "The new findings play into the debate launched by some conservative commentators who downplay the importance of ozone depletion....The measurements from Toronto show that UV light levels have, in fact, climbed there and most likely have increased over other regions where ozone concentrations are dropping. This question is no longer disputable."

Over the years, many of my students have used the topic of ozone depletion and the growing predicament in the stratosphere for research papers, speech presentations, discussion groups, and for other class assignments. I assume your experience is similar to mine. This is the reason I expanded the focus study in *GEOSYSTEMS*. Hopefully, it will provide a valuable resource on the subject to which the latest findings can be added. The spatial implications of the loss of ozone are profound thus integrating it through many aspects of physical geography.

VARIABLE ATMOSPHERIC COMPONENTS

Air pollution standards. The *Pollutant Standards Index* (PSI) in the U.S. is tied to health considerations as related to five basic criteria pollutants: total suspended particulates (TSP), sulfur dioxide, carbon monoxide, ozone, and nitrogen dioxide. Pollutant level standards are listed in milligrams per cubic meter (m^3). The health effect label is what you most often hear reported. The EPA presently lists twenty-eight municipal regions in the U.S. that are chronically listed as being unhealthful to very unhealthful on the PSI.

The present National Ambient Air Quality Standards (NAAQS) for the five variable components discussed above (criteria pollutants) are set to protect the most sensitive humans and represent standards targeted by current law. Still to be determined are the overall costs required to meet these standards and the relationship of these costs to exact benefits obtained by meeting the standard. Industries affected by each standard are fighting for a strict cost-benefit type of assessment for each standard.

Large descriptive tables for both of these standards (PSI and NAAQS) are explained and presented in *Environmental Trends*, by the Council on Environmental Quality (co-sponsored by USGS, EPA, Interior and State departments), Washington, U.S. Government Printing Office: July 1981, pp. 271-293. Despite the date, this 350 page report is generally available because it was one of the last of its type.

For the effects of industrial emissions and sulfate aerosols on the atmosphere and climate change see: Robert J. Charlson and Tom M. L. Wigley, "Sulfate Aerosol and Climate Change," *Scientific American*, February 1994: 48-57.

A Case Study:

An opportunity to examine health effects in the region near a Utah steel plant occurred in the late 1980s. Utah Valley acts to trap air and is augmented by winter high pressure anti-cyclonic systems and temperature inversions that worsen the air pollution situation. The plant's operation produces about 70% of the region's industrial pollution offering scientists a unique natural experiment (a portion of this plant is shown in Figure 3-13). A labor dispute closed the coking ovens and open-hearth furnaces for 13 months allowing scientists to compare the respiratory health of residents before and during closure, and after it reopened. Visible changes in air quality were remarkable and therefore commented on by many residents.

Researchers correlated the smallest particulates, 10 microns or less (designated as PM_{10}), with increased rates of hospitalization for bronchitis, asthma, pneumonia, and pleurisy, especially in children. Numbers of children entering the hospital during closure were less than half the previous number and were a third of those that entered after the plant reopened. Similar studies have now been completed in seven other cities with children affected by related illnesses at a rate twice as high between the dirtiest air studied and the city with the cleanest air. One of the references for these studies is in the suggested readings for this chapter, see Pope, Schwartz, and Ransom.

Photochemical Smog Reactions

The combustion of fuels for transportation is at the core of this topic on anthropogenic pollution. So, I have included some background here on the history of photochemistry and the automobile.

As the air grew worse in Los Angeles through the 1940s, various measures were taken to curtail outdoor incineration and to reduce evaporation of liquid fuels during transport from refineries. Despite these initial clean-up attempts, the air problem worsened. The L.A. Air Pollution Control District (LAAPCD) was established in 1947 to monitor these activities and report on the air. Dr. A.J. Haagen-Smit of the California Institute of Technology worked with Gordon Larson, the first director of the LAAPCD. Following many experiments, success came from an unlikely source: neighboring farms were experiencing specific leaf damage to spinach crops. Haagen-Smit and Larson took the spinach plants and tried to duplicate the damage. When sunlight, specifically ultraviolet rays, automobile exhaust, and spinach plants were placed in a controlled experiment, the experimenters discovered complex chemical reactions that killed the plants. Results of the investigation were published in 1953, and were quickly followed by the firing of Gordon Larson. Such was the political-corporate environment in those important early days of the highway-real-estate-money-lending lobby.

Efforts to cut back on car exhaust make cars more efficient, or curb automobile use have run into stiff opposition. Standards that have been established are chronically delayed and even when implemented represent "output" fixes, rather than the "input" fixes needed at the root of the problem. For four years in the mid-1980s, the Reagan Administration unsuccessfully called for an end of the Corporate Annual Fuel Economy (CAFE) law. Efficiencies were targeted for 27.5 mpg in 1987 but were rolled back to 26 mpg, the minimum allowable, without further congressional approval. The CAFE law is very popular and had led to an increase of fuel economy from a fleet (all autos produced) average of only 14 miles per gallon (mpg) in 1975 to over 27 mpg in 1987, a savings of $500 a year for the driver of an average car. Ironically, the call to reduce automobile efficiency even further and increase fuel consumption came from President Bush at the height of the 1991 Persian Gulf War, which supposedly was being conducted to guarantee dependable fuel sources for the country. Interestingly, only a 10% cutback in mileage driven in America would more than equal the total imported oil quantity from Iraq and Kuwait before the conflict began in 1990!

The January 1994 Northridge earthquake and aftershocks demonstrated the vulnerability of the interstate freeway system to such hazards. Such a chaotic time as this might present the opportunity to build alternative, less-polluting, more energy efficient transportation systems.

The Global Environment Monitoring System (GEMS) monitors air and water quality under the authority of the UNEP and WHO. See their report "Air Pollution in the World's Megacities," *Environment*, 36, no. 2, March 1994: 4-13+. The photochemical smog problem is a feature of the air urban residents breathe worldwide.

Solutions

Automobiles consume fuel at a rate influenced by vehicle weight, engine size, and driver behavior. Average weight of cars dropped by over one-half a ton between 1975 and 1987, and engine sizes decreased. Both were changes demanded by a public faced with higher fuel prices. More rapid transit transport systems would reduce emissions greatly. Even land-use patterns that maintain urban limit lines to encourage in-filling in urban areas instead of sprawl could reduce air pollution.

One reason the subject of air pollution is controversial relates to the debate as to what constitutes a harmful level of contamination. Many of the pollutants are now rising to significant health-threatening levels. Owing to sophisticated monitoring technology, the determination of what is detrimental and how much pollution is present is becoming more precise. Society appears to be causing pollution both in the air and water at a rate which exceeds our ability to adapt. However, only time and research will prove or disprove the relationship between pollution and our increasing rates of cancer, birth defects, genetic drift, and disease. Overall, we can ask, who is willing to pay the price? The problem of increasing air pollution is an evolving situation, arising with the advent of the industrial revolution in the 1700s, and becoming ever more noticeable and costly to society in the modern era.

A last thought:

Informed society finds itself with a classic question of hazard perception, whether it is judged at the level of the individual assessing personal risk or by society as a whole. Question: Do we err on the side of caution, or do we err on the side of risk? How do we assess hazardous risk in our society? How do we establish accountability for those who put us at risk? Whatever the answers here, the atmosphere is being subjected to an enormous and unprecedented experiment both with air pollution in the lower atmosphere and ozone depletion in the stratosphere. There will always be some degree of uncertainty in the vast atmospheric system. And even if we vote for caution the persistence of these compounds will still be a problem for future generations, although, we can reduce the potential disruption with informed action now.

Glossary Review for Chapter 3
(in alphabetical order)

air pressure
aneroid barometer
anthropogenic atmosphere
carbon dioxide
carbon monoxide (CO)
chlorofluorocarbons (CFCs)

environmental lapse rate
exosphere
heterosphere
homosphere
industrial smog
ionosphere
kinetic energy
lapse rate
mercury barometer
mesosphere
nitrogen dioxide (NO_2)
noctilucent clouds
normal lapse rate
ozone layer
ozonosphere
peroxyacetyl nitrates (PAN)
photochemical smog
sensible heat
stratosphere
sulfate aerosols
sulfur dioxide
temperature inversion
thermosphere
troposphere

Annotated Chapter Review Questions

1. What is air? Where did the components in Earth's present atmosphere originate?

Earth's atmosphere is a unique reservoir of gases, the product of billions of years of development. The modern atmosphere is probably Earth's fourth general atmosphere. The principal substance of this atmosphere is air, the medium of life as well as a major industrial and chemical raw material. Air is a simple additive mixture of gases that is naturally odorless, colorless, tasteless, and formless, blended so thoroughly that it behaves as if it were a single gas.

The primordial, evolutionary, and living atmospheres are discussed in Chapter 2. The outgassing hypothesis is discussed in Chapter 7. This modern atmosphere is, in reality, a gaseous mixture of ancient origins–the sum of all the exhalations and inhalations of life on Earth throughout time.

2. In view of the analogy by Lewis Thomas, characterize the various functions the atmosphere performs that protect the surface environment.

A membrane around a cell regulates the interactions of the delicate inner workings of that cell and the potentially disruptive outer environment. Each membrane is very selective as to what it will and will not allow to pass. The modern atmosphere acts as Earth's protective membrane. The atmosphere absorbs and interacts with harmful wavelengths of electromagnetic radiation, the streams of charged particles in the solar wind, and natural and human-caused space debris–all protecting the delicate aspects of the biosphere in the lower troposphere. As critical as the atmosphere is to us, it represents only a thin-skinned envelope amounting to less than one-millionth of Earth's total mass.

3. What three distinct criteria are employed in dividing the atmosphere for study?

The atmosphere is conveniently classified using three criteria: composition, temperature, and function. Based on chemical *composition* the atmosphere is divided into two broad regions: the heterosphere and the homosphere. Based on *temperature* the atmosphere is divided into four distinct zones: the thermosphere, mesosphere, stratosphere, and troposphere. Finally, two specific zones are identified on the basis of *function* relative to their role of removing most of the harmful wavelengths of solar radiation–these are the ionosphere and the ozonosphere, or ozone layer. See the left-hand column in Figure 3-2.

4. Describe the overall temperature profile of the atmosphere and list the four layers defined by temperature.

The temperature profile in Figure 3-2 shows that temperatures rise sharply in the thermosphere, up to 1200°C (2200°F) and higher at the top of the atmosphere. Individual molecules of nitrogen and oxygen and atoms of oxygen are excited to high levels of vibration from the intense radiation present in this portion of the atmosphere. This *kinetic energy*, the energy of motion, is the vibrational energy stated as temperature, although the density of the molecules is so low that little heat is produced. Heating in the lower atmosphere closer to Earth differs because the active molecules in the denser atmosphere transmit their kinetic energy as *sensible heat*. The mesosphere is the coldest portion of the atmosphere, averaging –90°C (–130°F), although that temperature may vary by 25C° to 30C° (45F° to 54F°).

Temperatures increase throughout the stratosphere from –57°C (–70°F) at 20 km (tropopause), warming to freezing at 50 km (stratopause). The tropopause is defined by an average temperature of –57°C (–70°F), but its exact location varies with the seasons of the year, latitude, and sea level temperatures and pressures.

Tropospheric temperatures decrease with increasing altitude at an average of 6.4C° per km (3.5F° per 1000 ft), a rate known as the normal lapse rate. The actual lapse rate at any particular time and place under local weather conditions is called the environmental lapse rate, and may vary greatly from the normal lapse rate. The lab manual and student study guide both have students work with these concepts.

5. Describe the two divisions of the atmosphere based on composition.

Heterosphere and the homosphere

6. What are the two primary functional layers of the atmosphere and what do each do?

Troposphere and stratosphere

7. Name the four most prevalent stable gases in the homosphere. Where did each originate? Is the prevalence of any of these changing at this time?

Symbol	%ByVol.	Parts permillion
N_2	78.084	780,840
O_2	20.946	209,460
Ar	0.934	9,340
CO_2	0.036	360

The atmosphere is a vast reservoir of relatively inert nitrogen, principally originating from volcanic sources. In the soil, nitrogen is bound up by nitrogen-fixing bacteria and is returned to the atmosphere by denitrifying bacteria that remove it from organic materials. Oxygen, a by-product of photosynthesis, is essential for life processes. Along with other elements, oxygen forms compounds that compose about half of Earth's crust. Argon is a residue from the radioactive decay of a form of potassium (^{40}K). A slow process of accumulation accounts for all that is present in the modern atmosphere. Argon is completely inert and as a noble gas is unusable in life processes. Although it has increased over the past 200 years, carbon dioxide is included in this list of stable gases. It is a natural by-product of life processes, and the implications of its current increase are critical to society and the future. The role of carbon dioxide in the gradual warming of Earth is discussed in Chapters 5 and 10.

8. Why is stratospheric ozone (O_3) so important? Describe the effects created by increases in ultraviolet light reaching the surface.

Ozone absorbs wavelengths of ultraviolet light and subsequently reradiates this energy at longer wavelengths, as infrared energy. Through this process, most harmful ultraviolet radiation is converted, effectively "filtering" it and safeguarding life at Earth's surface. See Focus Study 3-1 for greater discussion of the effects caused by ozone depletion.

9. Summarize the ozone predicament and the present trends, and any treaties that intend to protect it.

See Focus Study 3-1.

10. Evaluate Crutzen, Rowland and Molina's use of the scientific method in investigating stratospheric ozone depletion.

The possible depletion of the ozone layer by human activity was first suggested during the summer of 1974 by University of California, Irvine professors F. Sherwood Rowland and Mario J. Molina. Rowland and Molina hypothesized that the stable, large CFC molecules remained intact in the atmosphere, eventually migrating upward and working their way into the stratosphere. CFCs do not dissolve in water and do not break down biologically. The increased ultraviolet light encountered by the CFC molecule in the stratosphere dissociates, or splits, the CFC molecules, releasing chlorine (Cl) atoms and forming chlorine oxide (ClO) molecules. At present, after much scientific evidence and verification of actual depletion of stratospheric ozone by chlorine atoms (1987-1990), even industry has had to admit that the problem is real.

The scientific method is discussed in Focus Study 2-1. Recall in the chapter when Dr. Rowland stated his frustration: "What's the use of having developed a science well enough to make predictions, if in the end all we are willing to do is stand around and wait for them to come true....Unfortunately, this means that if there is a disaster in the making in the stratosphere, we are probably not going to avoid it." (Roger B. Barry, "The Annals of Chemistry," *The New Yorker*, 9 June 1986, p. 83.)

11. Why are anthropogenic gases more significant to human health than those produced from natural sources?

Natural sources produce a greater quantity of nitrogen oxides, carbon monoxide, and carbon dioxide than do certain human-made sources. However, any attempt to diminish the impact of human-made air pollution through a comparison with natural sources is irrelevant, for we have co-evolved and adapted to the presence of certain natural ingredients in the air. We have not evolved in relation to the concentrations of anthropogenic (human-caused) contaminants found in our metropolitan areas.

12. In what ways does a temperature inversion worsen an air pollution episode? Why?

A temperature inversion occurs when the normal decrease of temperature with increasing altitude reverses at any point from ground level up to several thousand feet. Such an inversion most often results from certain weather conditions, for example when the air near the ground is radiatively cooled on clear nights or when cold air drains into valleys. The warm air inversion in Figure 3-8 prevents the vertical mixing of pollutants with other atmospheric gases. Thus, instead of being carried away, the pollutants are trapped below the inversion layer. During the winter months in eastern and midwestern regions of the United States, high pressure areas created by subsiding cold air masses produce inversion conditions that trap air pollution. In the West, summer subtropical high pressure systems also cause inversions and produce air stagnation.

13. What is the difference between industrial smog and photochemical smog?

The air pollution associated with coal-burning industries is known as industrial smog. The term smog was coined by a London physician at the turn of this century to describe the combination of fog and smoke containing sulfur, an impurity found in fossil fuels. The combination of sulfur and moisture droplets forms a sulfuric acid mist that is extremely dangerous in high concentrations.

Photochemical smog is another type of pollution that was not generally experienced in the past but developed with the advent of the automobile. Today, it is the major component of anthropogenic air pollution. Photochemical smog results from the interaction of sunlight and the products of automobile exhaust. Although the term *smog* is a misnomer it is generally used to describe this phenomenon. Smog is responsible for the hazy appearance of the sky and the reduced intensity of sunlight in many of our cities.

14. Describe the relationship between automobiles and the production of ozone and PAN in city air. What are the principal negative impacts of these gases?

The nitrogen dioxide derived from automobiles and power plants to a lesser extent is highly reactive with ultraviolet light, which liberates atomic oxygen (O) and a nitric oxide (NO) molecule (Figure 3-11). The free oxygen atom combines with an oxygen molecule (O_2) to form the oxidant ozone (O_3); the same gas that is beneficial in the stratosphere is an air pollution hazard at Earth's surface. In addition to forming O_3, the nitric oxide (NO) molecule reacts with hydrocarbons (HC) to produce a whole family of chemicals generally called peroxyacetyl nitrates (PAN). PAN produces no known health effects in humans but is particularly damaging to plants, which provided the clue for discovery of these photochemical reactions.

15. How are sulfur impurities in fossil fuels related to the formation of acid in the atmosphere and acid deposition on the land?

Certain anthropogenic gases are converted in the atmosphere into acids that are removed by wet and dry deposition processes. Nitrogen and sulfur oxides (NO_x, and SO_x) released in the combustion of fossil fuels, can produce nitric acid (HNO_3) and sulfuric acid (H_2SO_4) in the atmosphere. Figures 3-12 and 3-14 depict the patterns of NO_x and SO_x emissions in the United States and Canada. Precipitation as acidic as pH 2.0 has fallen in the eastern United States, Scandinavia, and Europe. By comparison, vinegar and lemon juice register slightly less than 3.0. Aquatic life perishes when lakes drop below pH 4.8.

16. In your opinion, what are the solutions to increasing problems in the anthropogenic atmosphere?

Solutions to our growing air pollution problem pose no mystery; technologies to eliminate or abate emissions are known and in some cases are in operation at this time. In addition, since energy production is at the heart of pollution production, it stands to reason that conservation (reduction in energy use) and efficiency (using energy with less waste) are key strategies, especially with the level of energy waste and inefficiency in the United States. However, even without such fundamental changes, stack emissions can be cleaned with precipitators, scrubbers, filters, and separators; car exhaust can be cleaned with more efficient carburation and emission controls, and the use of the automobile can be reduced through development and use of alternative transportation systems.

Certainly, society cannot simply halt two centuries of industrialization; the resulting economic chaos would be devastating. However, neither can it permit pollution production to continue unabated; catastrophic environmental feedback will inevitably result. Thus, we are confronted with a paradox characterized by the dichotomy portrayed in Figure 1-2, a metaphor for these philosophical and economic problems.

Overhead Transparencies

As an adopter you are provided with the following figures for overhead projector use.

- Figure 3-2: Profile of the modern atmosphere
- Figure 3-11: Photochemical reactions

4

Atmosphere and Surface Energy Balances

Overview

Earth's biosphere pulses daily, weekly, and yearly with flows of energy. Think for a moment of the annual pace of your own life, your wardrobe, gardens, and lifestyle activities–all reflect shifting seasonal energy patterns. Thus begins the culmination of the passage of energy from the Sun, across space, to the top of the atmosphere, down through the layers of the atmosphere to Earth's surface. This entire process should be thought of as a vast flow-system with energy cascading through fluid Earth systems. The beginning lecture here can mention this flow-system and the journey of the past three chapters.

The *Student Study Guide* presents 20 "Learning Objectives" to guide the student in reading the chapter. The *Applied Physical Geography* lab manual has several steps that involve aspects of this chapter.

New to the Third Edition

(Note: This section highlights major changes, new features, and additions in the third edition. This does not describe all the rewrite and recast of the text.)

1. A list of key learning concepts begins the chapter.

2. Figure 4-1, is a simplified flow diagram of shortwave and longwave radiation in the Earth-atmosphere system. This is a good construct for the more sophisticated version in Figure 4-12.

3. Figure 4-3, illustrates the distorted appearance of the Sun near sunset over the ocean, caused by the refraction of the Sun's image in the atmosphere.

4. Focus Study 1, Figure 1, displays new photographs of the utility of solar cookers in developing countries and their ability to reduce fuelwood demand that causes deforestation.

5. News Report #1, "Earthshine-A Diagnostic Tool for Analysis of Earth's Energy Budget", describes the use of measuring the Earth's reflection of light to space to analyze energy budgets and climatic change on Earth. The amount of Earthshine on the Moon can be used as a measure of the Earth's albedo; therefore, changes in the Earth's albedo will be reflected in the amount of Earthshine that is observed, and may allow us to predict changes in global temperature.

6. A more elaborate discussion of the effect of clouds and air pollution in the atmosphere, including; description of albedo-forcing clouds and greenhouse-forcing clouds, and the effect of sulfate aerosols in the atmosphere (connection to Chapter 3). Figure 4-9 is photograph of how air pollution affects the color of the atmosphere, and Figure 4-8, an illustration of how clouds and air pollution affect shortwave and longwave radiation.

7. Expanded discussion of how energy moves through the atmosphere, describing conduction, convection and advection in greater detail. This section adds an illustration of the process of energy flow using a stove as a conductor, causing convection through the latent heat transfer of water.

8. Figure 4-12, Earth-atmosphere energy balance, has been updated to give greater clarity to the flow of shortwave and longwave radiation through the atmosphere.

9. Figure 4-19, the global pattern of latent heat of evaporation (LE), and Figure 4-20, the global pattern of Sensible Heat (H), add greater clarity in the description of microclimatology and surface energy budgets.

10. Table 4-1, comparison between rural and urban environments, has been improved to list the quantifiable differences in temperature, cloudiness, wind, relative humidity, radiation and contaminants.

11. Figure 4-13 adds a new photograph of Chicago, demonstrating the wind channeling properties of urban morphology.

12. A new summary and review section ends the chapter.

Learning Objectives

1. *Identify* the pathways of solar energy through the troposphere to Earth's surface: transmission, albedo (reflectivity), scattering, diffuse radiation, conduction, convection, and advection.

2. *Analyze* the effect of clouds and air pollution on solar radiation received at ground level and *describe* what happens to insolation when clouds are in the atmosphere.

3. *Review* the energy pathways in the Earth-atmosphere system, the greenhouse effect and the patterns of global net radiation.

4. *Plot* the daily radiation curves for Earth's surface and *label* the key aspects of incoming radiation, air temperature, and the daily temperature lag.

5. *Portray* typical urban heat island conditions and *contrast* the microclimatology of urban areas with surrounding rural environments.

Expanded Outline Discussion

The following headings (boldfaced) match some of the first, second, and third order headings in Chapter 4. The narrative under each heading contains information, sources, and anecdotal facts relating to portions of the chapter. Not all text headings are discussed.

ENERGY BALANCE IN THE TROPOSPHERE

The key here is Figures 4-1 and 4-12, which illustrate the Earth-atmosphere energy balance, which is referred to at the beginning of this section. This figure is reproduced in outline form without labels in the *Student Study Guide*. Students are instructed to fill in information, add labels, and even color in the illustrations as they move along through this discussion. If you use a chalkboard, this is an opportunity to draw the energy budget along with the students as you lecture. The illustration is also included in the overhead transparency packet with this chapter.

LAB

Energy in the Atmosphere: Some Basics– Albedo and Reflection

The suggestion is for the students to perform an albedo check of wardrobes, house and apartment wall colors, and automobile exterior and interior colors–important because albedo relates to absorption and heat production.

Figure 4-7 was provided by Wayne L. Darnell, Radiation Services Branch, Atmospheric Sciences Division, NASA Langley Research Center. Some related articles and studies are listed in the suggested readings section at the end of the chapter. Dr, H. Lee Kyle, NASA/Goddard Space Flight Center is also researching, among others, ERB scanner and wide field-of-view maps.

Scattering (Diffuse Reflection)

So, why is the sky blue? And, how high up does the blue sky go? And, why is the SR-71 high-altitude plane (over 30,000 km, 100,000 ft) painted black? The idea that a pilot would see stars during the day at high altitude, above the enhanced scattering at lower altitudes, may be difficult to comprehend at first. This discussion helps draw material in from Chapter 3.

Earth Reradiation and the Greenhouse Effect

This discussion is important to portions of the rest of the text relative to the greenhouse effect and global climate change. The operation of Earth's greenhouse sets the stage for the discussion of future temperature trends and global warming, and future climate patterns and consequences of climatic warming in Chapter 10, sea level changes in Chapter 16, the Antarctic ice sheet in Chapter 17, ecosystem stability and climate change in Chapter 19, and the summary overview comments in Chapter 21. I chose to present global change in this matter instead of in an isolated focus study because this present trend is so spatially pervasive through many of Earth's systems. I hope this integrated approach will assist the student in seeing the complex interconnections that link all Earth systems and the human population.

Global Net Radiation

Figure 4-14 summarizes the latitudinal distribution of net radiation. This completes a full cycle of discussion that began in Chapter 2 with energy measurements at the top of the atmosphere. This is the net radiation portrait of the entire Earth-atmosphere system. A useful analogy for the students could be some version of the following:

Imagine that you are dealing with money instead of energy, so that at the equator there is the Tropical Branch Bank, and at the pole the Polar Branch Bank. You are in charge of accounts at both of these bank branches, and for reasons unknown to you, more deposits are made at the Tropical Branch than you are withdrawing. At the Polar Branch, you make more withdrawals than you do deposits. Fix in your mind that both accounts act as open-flow systems, with yourself in charge of the inputs and outputs. The resultant dollar amounts in your two accounts pose an interesting problem: one is full of surplus cash while the other account is in a deficit, with checks bouncing. Your solution? To balance your financial situation, you must transfer excess deposits from the surplus account to the deficit account. Of course this is what happens in the Earth-atmosphere system, with the transfer of energy and mass (water and water vapor).

ENERGY BALANCE AT EARTH'S SURFACE

You may want to look through a publication from the AAG: Miller, David H., "A Survey Course: The Energy and Mass Budget at the Surface of the Earth," Association of American Geographers, Community College Geography Publication No. 7, Washington D.C.: 1968, 142 pgs. The volume includes extensive references. Also, see Dr. Miller's publication noted in the suggested readings for this chapter in the text as an excellent overview of this entire field of study.

Daily Radiation Curves

The student's own experience with temperatures will relate to these temperature and insolation curves–the lag of the warmest time of day several hours after noon and the timing of the lowest temperatures at sunrise.

Simplified Surface Energy Balance

Lecturing through this conceptual equation will draw concepts from earlier chapters and again act as a valuable summary. Depending on the level of your students you may want to use some actual quantities in comparing various surfaces and environments. Data are available in langleys per day for each of the factors in the surface energy budget equation.

Net Radiation

As students travel about during the day (perhaps not as far afield as these sample stations), they can note the different surfaces and imagine each of the energy balance components. Ask them to consider the pathways for the expenditure of net radiation available at each surface: turbulent transfer, latent heat of evaporation, photosynthesis, conduction into the soil, or conduction and convection processes in bodies of water. And, consider various alterations to those surfaces: clearing, paving, reforestation, and tilling. The students might speculate about what changes take place in the net radiation balance because of these alterations.

As a contrasting example, work through the equation as if we were applying it to the lunar surface. The NET R at the surface of the Moon is totally expended for *G*. The surface increases in temperature to almost the boiling point of water and decreasing in temperature an equal amount below freezing in shadow or at night. An astronaut in a spacesuit will receive total *G* on his sun-facing side and total loss of *G* on his shady side as he stands in the direct sunlight.

Solar Energy Collection and Concentration (Focus Study 4-1)

Relative to solar energy applications, a knowledge of this simple surface energy budget is useful: letting in the "SW" and trapping the "LW" wavelengths in the collector. A flat-plate collector may include tubing that is painted black and placed in a shallow box. Such a collector is usually covered by glass to trap heat in a greenhouse fashion to further increase heating. It is amazing to me that simple solar collectors are not being more actively distributed by the United States and the United Nations. In some villages, they are walking miles to gather enough twigs to bake a loaf of bread. For these people, and for western urbanites, simple solar box cookers could handle the bulk of cooking and baking requirements during summer months. In the tropics, such cookers could be used year-round, see Focus Study 4-1, Figure 1.

I recommend that you write to the nonprofit organization listed below for their literature and designs. We purchased a simple box-cooker kit and find that it does extremely well for something made out of cardboard, one pane of glass, foil, and a reflector lid. The insulation in the walls is provided by crumpled newspaper and all the other materials are from recycled sources. You need about 15 minutes of direct sunlight per hour to cook food. We have cooked everything from pasta to bread to a whole turkey during the seven month period (April to October) at 40° N latitude. The only problem we have had was during the solar eclipse on July 11, 1991. With 59% of the Sun blocked by the Moon in Sacramento, CA, temperature in the solar cooker dropped by 25°C (45°F).

For a copy of "Your Own Solar Box–How to Make and Use," "Teachers Guide Fun with the Sun," and "Leaders Guide, Spreading Solar Cooking," (all three for $12.) contact the following nonprofit group: Solar Cookers International (SCI), 1724 11th Street, Sacramento, California 95814; Ph-916-444-6616; FAX-916-444-5379. Their Second World Conference On Solar Cooker Use and Technology was at Heredia, Costa Rica, July 12-15, 1994. A major effort is underway to place solar box cookers throughout the Third World. SCI also has video tapes, other teaching tools, and a newsletter dealing with solar cooking. Donations are tax deductible.

Relative to solar energy and the focus study with this chapter, there are some key generalizations that help organize energy resources for analysis. A first step is to analyze the actual end-use energy demand in the United States, or any country. The pattern of consumer demand most accurately dictates what methods and modes should be selected to supply and meet that energy need. Presently in the United States, 58% of end-use energy need is for heat roughly split one-half above and one-half below the boiling temperature of water. Another 38% is for mechanical motion, including 4% for electric motors. And finally, 4% is classified as necessary electrical. If the taxpayer is worried about taxes and big government, and is basically conservative, then having such a localized consumer-driven energy plan would seem wisest and most conservative of capital, political power, and economic control.

Secondly, let's briefly examine two categories of energy resources and their characteristics. The essays in the text on wind and solar resources should be viewed in the context of these two sets of energy characteristics. These two paradigms can form the basis of a class discussion if there is time.

Energy Resource Characteristics

Centralized energy	Decentralized energy
Nonrenewable sources	Renewable sources
Indirect to the consumer	Direct to the consumer
Capital intensive	Labor intensive
Limited domestic supply	Unlimited domestic supply
Large foreign imports	No foreign imports
Monopolies/cartels	Widely available
"Big" government control	Home and neighborhood control
Concentrated tax advantages	Diffused tax advantages

In the focus study in Chapter 6, I present the USGS estimates for domestic oil and gas supplies in the United States–something else to keep in mind in your assessment of energy resources and alternatives to fossil fuels.

The Urban Environment

For more depth, a model designed as a teaching aid for microclimatology and ecology courses is presented in Leonard O. Myrup's "A Numerical Model of the Urban Heat Island," *Journal of Applied Meteorology* 8 (December 1969): pp. 908-918.

In addition to the work of Helmut Landsberg (a principal source listed in the suggested readings for this chapter), the work of Stanley A. Changnon of the Illinois State Water Survey should not be overlooked, including his studies of the La Porte weather anomaly in 1969 and 1970 and of St. Louis in the 1970s in a project called METROMEX. All of these efforts were focused on defining the impact of urbanized landscapes on the weather of the city and the area downwind from metropolitan regions. For the St. Louis region, published findings conclude that cloudiness increased 10%, summer rainfall increased 30%, and frequency of severe thunderstorms increased 10 to 100%. These effects were noted for the city and a fan-shaped region downwind from the city.

An earlier work of Dr. Landsberg's was presented at the historic conference of 1955. See "The Climate of Towns," in *Man's Role in Changing the Face of the Earth*, edited by William L. Thomas, Jr.

Glossary Review for Chapter 4 (in alphabetical order)

absorption
advection
albedo
cloud-albedo forcing
cloud-greenhouse forcing
conduction
convection
diffuse radiation
dust dome
greenhouse effect
microclimatology
net radiation (NET R)
reflection
refraction
scattering
sunset
transmission
urban heat islands

Annotated Chapter Review Questions

1. Diagram a simple energy balance for the troposphere. Label each shortwave and longwave component and the directional aspects of related flows.

From Figures 4-1 and 4-12, in the text and featured in the *Student Study Guide*.

2. Define refraction. How does this relate to daylength? To a rainbow? To the beautiful colors of a sunset?

When insolation enters the atmosphere, it passes from one medium to another (from virtually empty space to atmospheric gas) and is subject to a bending action called *refraction*. In the same way, a crystal or prism refracts light passing through it, bending different wavelengths to different degrees, separating the light into its component colors to display the spectrum.

A rainbow is created when visible light passes through myriad raindrops and is refracted and reflected toward the observer at a precise angle. Another example of refraction is a *mirage*, an image that appears near the horizon where light waves are refracted by layers of air of differing temperatures (densities) on a hot day.

An interesting function of refraction is that it adds approximately eight minutes of daylight for us. The Sun's image is refracted in its passage from space through the atmosphere, and so, at sunrise, we see the Sun about four minutes before it actually peeks over the horizon. Similarly, at sunset, the Sun actually sets but its image is refracted from over the horizon for about four minutes afterward. To this day, modern science cannot predict the exact time of sunrise or sunset within these four minutes, because the degree of refraction continually varies with temperature, moisture, and pollutants.

3. List several types of surfaces and their albedos. Explain the differences. What determines the reflectivity of a surface?

A portion of arriving energy bounces directly back to space without being converted into heat or performing any work. This returned energy is called reflection, and it applies to both visible and ultraviolet light. The reflective quality of a surface is its albedo, or the relationship of reflected to incoming insolation expressed as a percentage. In terms of visible wavelengths, darker colors have lower albedos, and lighter colors have higher albedos. On water surfaces, the angle of the solar beam also affects albedo values; lower angles produce a greater reflection than do higher angles. In addition, smooth surfaces increase albedo, whereas rougher surfaces reduce it. See Table 4-1 for specific values.

4. Using Figure 4-7, explain the seasonal differences in albedo

values for each hemisphere. Be specific, using the table of albedos (Table 4-3) when appropriate.

Figure 4-7 portrays total albedos for July 1985 (a) and January 1986 (b) as measured by the Earth Radiation Budget Experiments (ERBE) aboard several satellites. These patterns are typical of most years. As compared to July, note the higher January albedos poleward of 40°N caused by snow and ice covering the ground. Tropical forests are characteristically lower in albedo (15%) whereas generally cloudless deserts have higher albedos (35%). The southward-shifting cloud cover over equatorial Africa is quite apparent on the January map. Note the regions poleward of the Antarctic Circle in July and the Arctic Circle in January–this can be tied into the seasons discussion.

5. Why is the lower atmosphere blue? What would you expect the sky color to be at 50 km (30 mi) altitude? Why?

The principle known as Rayleigh scattering–named for English physicist Lord Rayleigh, who stated the principle in 1881–relates wavelength to the size of molecules or particles that cause the scattering. The general rule is the shorter the wavelength, the greater the scattering, and the longer the wavelength, the less the scattering. Shorter wavelengths of light are scattered by small gas molecules in the air. Thus, the shorter wavelengths of visible light, the blues and violets, are scattered the most and dominate the lower atmosphere. And, because there are more blue than violet wavelengths in sunlight, a blue sky prevails. As the atmosphere thins with altitude, there are fewer molecules to scatter these shorter wavelengths, and the sky darkens. At 50 km, even though it may be daylight, the sky appears as at night and the stars become visible along with the Sun.

6. Define the concepts transmission, absorption, diffuse radiation, conduction, and convection.

See the Glossary section in text, and Figure 4-10 for illustration.

7. What role do clouds play in the Earth-atmosphere radiation balance? Is cloud type important? Compare high, thin cirrus clouds and lower, thick stratus clouds.

A major uncertainty in the tropospheric energy budget, and therefore in refining climatic models, is the role of clouds. Clouds reflect *insolation*, thus cooling Earth's surface. Yet, clouds act as *insulation*, thus trapping longwave radiation and raising minimum temperatures. Important is not merely the percentage of cloud cover but the cloud type, height, and thickness (water content and density). High-altitude, ice-crystal clouds have albedos of about 50%, whereas thick, lower cloud cover reflects about 90%. Understanding the nature of global cloud cover is crucial in refining computer models of predicted global warming. A section on clouds is in Chapter 7. Note the references under suggested readings at the end of the chapter.

8. In what way do the presence of sulfate aerosols affect solar radiation receipt at ground level? Affect cloud formation?

Sulfate aerosols act as particles in the atmosphere which may scatter incoming shortwave radiation, and reduce the receipt of solar radiation at ground level. This connects nicely to discussion of albedo-forcing clouds. If sulfate aerosols scatter shortwave radiation, they increase the albedo of the atmosphere. On the other hand, if sulfate aerosols may become condensation nuclei and increase cloud formation. This could produce thick, low level clouds and intensify the greenhouse effect, as well as increase albedo rates. In either case, sulfate aerosols reduce the amount of solar energy receipt at the ground level.

9. What are the similarities and differences between an actual greenhouse and the gaseous at-

mospheric greenhouse? Why is Earth's greenhouse changing?

In the greenhouse analogy, the glass is transparent to shortwave insolation, allowing light to pass through to the soil, plants, and wood planks inside. The absorbed energy is then radiated as infrared energy back toward the glass, but the glass effectively traps the longer infrared wavelengths and warm the air inside the greenhouse. Thus, the glass allows the light in but does not allow the heated air out. In the atmosphere, the greenhouse analogy is not fully applicable because infrared radiation is not trapped as it is in a greenhouse. Rather, its passage to space is delayed as the heat is radiated and reradiated back and forth between Earth's surface and certain gases and particulates in the atmosphere. The present warming is associated with an increase in radiatively active greenhouse gases.

10. In terms of energy expenditures for latent heat of evaporation, describe the annual pattern as mapped in Figure 4-19.

Latent heat of evaporation is the energy that is stored in water vapor, becoming latent heat as water changes from a liquid to a gas. Energy expenditures for latent heat are greatest at the equator and over the oceans, due to the high amount of solar radiation that is received at ground level. This energy is absorbed by land surfaces near the equator due to the high amount of insolation that occurs in the tropics. High amounts of latent heat transfer occur over oceans in the subtropic latitudes where warm water comes into contact with hot, dry air. Latent heat of evaporation is lowest near the poles, due to the small amount of solar radiation that is absorbed at the surface in the poles, and also the temperature of water in the poles is much colder, which would require great amounts of energy to warm the water and stimulate the water's change of state from liquid to gas.

11. Generalize the pattern of global net radiation. How might this pattern drive the atmospheric weather machine? (Figure 4-14 and 4-18.)

In the equatorial zone, surpluses of energy dominate, for in those areas more energy is received than is lost. Sun angles there are high, with consistent daylength. However, deficits exist in the polar regions, where more energy is lost than gained. At the poles the Sun is extremely low in the sky, surfaces are light and reflective, and for six months during the year no insolation is received. This imbalance of net radiation from the equator to the poles drives the vast global circulation of energy and mass.

12. In terms of surface energy balance, explain the term net radiation (NET R).

Net radiation is the net all-wave radiation available at Earth's surface; it is the final outcome of the entire radiation balance process discussed in this chapter. Net radiation (NET R) is the balance of all radiation, shortwave (SW) and longwave (LW), at Earth's surface.

13. What are the expenditure pathways for surface net radiation? What kind of work is accomplished?

Output paths at the surface for the principal expenditures of net radiation from a nonvegetated surface include H (turbulent sensible heat transfer), LE (latent heat of evaporation), and G (ground heating and cooling). See bulleted items in the text chapter.

14. What is the key role played by latent heat in surface energy budgets?

Latent heat refers to heat energy that becomes stored in water vapor as water evaporates. Large quantities of latent heat are absorbed into water vapor during its change of state from liquid to gas. Conversely, this heat is released in its change of state back to a liquid (see Chapter 7). Because the evaporation of water is the principal method of transferring and dissipating heat surpluses vertically into the atmosphere, latent heat is the *dominant expenditure* of Earth's entire NET R. Latent heat links Earth's energy and water (hydrologic) systems and for most landscapes is the key component in surface energy budgets, see Figure 4-19.

15. Compare the daily surface energy balances of El Mirage and Pitt Meadows. Explain the differences.

El Mirage, California, at 35°N, is a hot desert location characterized by bare, dry soil with very little vegetation (Figure 4-21 a). The summer day selected was clear, with a light wind in the late afternoon. El Mirage has little or no expenditure of energy for LE. With an absence of water and plants, most of the available radiant energy is dissipated as turbulent sensible heat, warming air and soil to high temperatures. The G component is higher in the morning, when winds are light and turbulent transfers are reduced.

The NET R at this desert location is quite similar to that of midlatitude, vegetated, and moist Pitt Meadows, British Columbia (Figure 4-21 b). The energy balance data for Pitt Meadows, at 49°N, are plotted for a cloudless summer day. The Pitt Meadows landscape is able to retain much more of its energy because of a lower albedo (less reflection), the presence of more water and plants, and lower surface temperatures. The higher LE values are attributable to the moist environment of rye grass and irrigated mixed-orchard ground cover for the sample area, contributing to the more moderate sensible heat levels throughout the day.

16. Why is there a temperature lag between the highest Sun altitude and the warmest time of day? Relate your answer to the insolation and temperature patterns during the day.

Incoming energy arrives throughout the illuminated part of the day, beginning at sunrise, peaking at local noon, and ending at sunset. As long as the incoming energy exceeds the outgoing energy, temperature continues to increase during the day, not peaking until the incoming energy begins to diminish in the afternoon as the Sun loses altitude. The warmest time of day occurs not at the moment of maximum insolation but at that moment when a maximum of insolation is absorbed. Thus, this temperature lag places the warmest time of day three to four hours after solar noon as absorbed heat is supplied to the atmosphere from the ground. Then, as the insolation input decreases toward sunset, the amount of heat lost exceeds the input, and temperatures begin to drop until the surface has radiated away the maximum amount of energy, just at dawn.

The annual pattern of insolation and temperature exhibits a similar lag. For the Northern Hemisphere, January is usually the coldest month, occurring after the winter solstice, the shortest day in December. Similarly, the warmest months of July and August occur after the summer solstice, the longest day in June.

17. What is the basis for the urban heat-island concept? Describe the climatic effects attributable to urbanization as compared with nonurban environments.

The surface energy characteristics of urban areas possess unique properties similar to energy balance traits of desert locations. See the 6 items of analysis in the chapter and the contents of Figure 4-11, and Table 4-1.

18. Which of the items in Table 4-1 have you yourself experienced? Explain.

Personal analysis and response.

19. Assess the potential for solar energy applications in our society. Negatives? Positives?

Solar energy systems can generate heat energy of an appropriate scale for approximately half the present population in the United States (space heating and water heating). In marginal climates, solar-assisted water and space heating is feasible as a backup, even in New England and the Northern Plains, solar-efficient collection systems prove effective. Kramer Junction, California, about 140 miles northeast of Los Angeles, has the world's largest operating solar electric-generating facility. The facility converts 23% of the sunlight it receives into electricity during peak hours. Roof-top photovoltaic electrical generation is now cheaper than power line construction to rural sites. Obvious drawbacks of both solar-heating and solar electric systems are periods of cloudiness and night, which inhibit operation.

The success of solar energy appears tied to the political arena, for without tax incentives and formal encouragement, in amounts at least equal to those given the fossil-fuel industry, it is difficult to operate such a plant.

Overhead Transparencies

As an adopter you are provided with the following figures for overhead projector use.

- Figure 4-1: Global average annual solar radiation
- Figure 4-12: Detail of Earth-atmosphere energy balance
- Figure 4-14: Energy budget by latitude
- Figure 4-15: Daily radiation curves

5

World Temperatures

Overview

Air temperature has a remarkable influence upon our lives, both at the microlevel and at the macrolevel. A variety of temperature regimes worldwide affect entire lifestyles, cultures, decision-making, and resources spent. Global temperature patterns presently are changing in a warming trend that is affecting us all and is the subject of much scientific, geographic, and political interest. Our bodies sense temperature and subjectively judge comfort, reacting to changing temperatures with predictable responses.

This chapter presents principles and concepts that are synthesized on the January and July temperature maps, and on the annual range of temperature map. The chapter then relates these temperature patterns and concepts directly to the student with a discussion of apparent temperatures–the wind chill and heat index charts. Focus Study 5-1 introduces some essential temperature concepts to begin our study of world temperatures. Later, Chapter 10 takes a look at how Earth's temperature system appears to be in a state of dynamic change as concerns about global warming and potential episodes of global cooling are discussed

The *Student Study Guide* presents 21 "Learning Objectives" to guide the student in reading the chapter. The *Applied Physical Geography* lab manual has one exercise with six steps that involve aspects of this chapter.

New to the Third Edition

(Note: This section highlights major changes, new features, and additions in the third edition. This does not describe all the rewrite and recast of the text.)

1. A list of key learning concepts begins the chapter.

2. Description of how normal body temperature was first derived.

3. Figure 5-1 illustrates the extremes in temperature which North America can experience, using winter in Canada and summer in along the Gulf Coast as examples.

4. Figure 5-3, a picture of high elevation farming, is included to demonstrate how moderate temperatures (48°F) can be found in tropical regions that have high elevations.

5. News Report #1: "The Heat Index Can Kill", summarizes the impact of high temperatures upon several American cities during 1995. Over 1000 people died during the summer heat spells of 1995, including 700 in Chicago alone. Also see Figure 5-17.

6. News Report #2: "Record Temperatures Suggest Greenhouse Warming" discusses the impact of fossil fuels and other radiatively active gases upon Earth systems. Effects include the melting of the Earth's glaciers, a resulting rise in sea level, intensified thunderstorms and weather extremes (such as the high temperatures discussed in News Report

#1 above), changes in vegetation, and reduced crop yields.

7. More elaborate discussion on oceanality. The third edition connects the concept of oceanality to related issues of latent heat of evaporation (see map 4-19), cloud formation and greenhouse effects.

8. A detailed discussion of kinetic energy, with reference to human examples of the latent heat of evaporation and cooling.

9. Recent statistics on temperature extremes, and description of how climatic data is obtained.

10. Comparison of the effect of ocenality and continentality along the same degree of latitude (50 N), including discussion on the effect of altitude upon temperature.

11. A new summary and review section ends the chapter.

LAB

Key Learning Concepts

1. *Define* the concepts of temperature, kinetic energy, and sensible heat, and *distinguish* among Kelvin, Celsius and Fahrenheit scales and how they are measured.

2. *List* and *review* the principal controls and influences that produce global temperature patterns.

3. *Review* the factors that produce marine and continentality differences as they influence temperatures and *utilize* several pairs of stations to illustrate these differences.

4. *Interpret* the pattern of Earth's temperatures from their portrayal on January and July temperature maps, and on a map of annual temperature ranges.

5. *Define* apparent temperature or sensible temperature, and relate specific physiological effects of low temperature and high temperature on the human body; *utilize* the wind-chill and heat-index charts to determine some apparent temperatures.

Expanded Outline Discussion

The following headings (boldfaced) match some of the first, second, and third order headings in Chapter 5. The narrative under each heading contains information, sources, and anecdotal facts relating to portions of the chapter. Not all text headings are discussed.

PRINCIPAL TEMPERATURE CONTROLS

Focus Study 5-1 discusses temperature and the way it is measured. An article by Philip D. Jones and Tom M. L. Wigley describes in detail temperature measurements and preciseness on land as well as at sea and the methods used to avoid error in these measurements that might be induced by urbanization, ship exhaust, discharges at sea, and groundcover alterations (see "Global Warming Trends," *Scientific American*, August 1990, pp. 84-91).

Latitude, Altitude + aspect

These two controls draw on material presented in Chapters 1 and 2 (latitude) and Chapter 3 (altitude effects). The energy portrayal in Figure 2-13, "Total daily insolation received at the top of the atmosphere," can be used at this point to suggest the variable energy input by latitude. The global solar radiation map shown in Figure 4-4 can also be referenced.

Figure 5-2 is in the overhead transparency packet and features a comparison of temperature patterns in La Paz and Concepción, Bolivia. The fact that the people of La Paz are farming at elevations that approach 4103 m (13,461 ft) integrates the concepts of both low latitude and high elevation. I find it intriguing that almost one million people work at such an altitude. Human adaptation and lung capacity are incredible. The *Applied Physical Geography* lab manual has students graph and analyze these two stations.

Cloud Cover

Clouds are presented in Chapter 8 (Figures 8-17 and in Table 8-1) and can be mentioned at this point. Using an Earth photo (poster, slide, or the back-cover of *GEOSYSTEMS*), the role of cloud cover, and its albedo and energy absorption can be discussed. The difficulty in modeling such a variable component is obvious. The International Satellite Cloud Climatology Project is under the partial sponsorship of the United Nations Environment Programme (see the listing in Appendix B).

See: Wayne L. Darnell and W. Frank Staylor, *et al.*, "Seasonal Variation of Surface Radiation Budget Derived from International Satellite Cloud Climatology Project C1 Data," *Journal of Geophysical Research*, 97, no. D14, October 20, 1992: 15,741-760. And, George Tselioudis, et al., "Potential Effects of Cloud Optical Thickness on Climate Warming," *Nature*, 366 December 16, 1993: 670-72. Numerous articles covering the effects of clouds and cloud climatology appear in the *Journal of Climate* published monthly by the American Meteorology Society, 45 Beacon Street, Boston, MA 02108.

Land-Water Heating Differences

Evaporation, transmissibility, specific heat, movement, and ocean currents are suggested in Figure 5-4, which is included in the overhead transparency packet. If you have a soil thermometer available, a quick measurement can be made to illustrate the transmissibility/opacity principle. One architectural strategy is to partially build or insulate construction with soil–thereby creating subterranean insulation.

Because scientific terminology is so foreign to students, I give them an activity to become more familiar with the concept of specific heat. First, I have students list strategies they use for staying cool at the beach. Secondly, I have them define the terms albedo, specific heat, and heat capacity. Finally, I encourage students to utilize these terms to discuss their strategies for keeping cool. It's a fun activity, and it demonstrates to students that they do have a working knowledge of these concepts.

Summary of Marine vs. Continental Conditions

You probably have your own favorite sets of stations to portray continental and marine characteristics. I present San Francisco and Wichita, Vancouver and Winnipeg, and Trondheim and Verkhoyansk and have included three pairs of cities in the overhead transparency packet. In simplest terms, these demonstrate continentality of the interior stations and the greater moderation of temperature characteristics of the coastal stations. With a more sophisticated analysis, cities in the western Basin and Range Province such as Elko, Nevada, probably exhibit the least maritime influence in the United States measured by indicator formulas, i.e., the greatest degree of continentality.

As stated in the text in several places, northcentral Siberia exhibits extreme temperature ranges and continentality. Temperatures for Verkhoyansk are plotted in Figure 5-12. A climograph including temperature and precipitation for Verkhoyansk appears in Chapter 10, Figure 10-22. And a complete set of data for Oymyakon (Dwd) and Tobol'sk (Dwc), Siberia, Russia, are in Appendix A. The students could graph the data for these two stations. The study guide and lab manual include such activities.

This discussion sets up the annual temperature range map in Figure 5-14, and the maps for January and July average temperatures in Figures 5-11 and 5-13.

EARTH'S TEMPERATURE PATTERNS

(January and July Temperatures, and Annual Range of Temperatures Maps)

These maps are prepared on Robinson projections to diminish the distortion common to temperature maps presented on Mercator projections. The maps feature the thermal equator and coloration suggestive of temperature patterns. Note that the cities mentioned in the text as examples are noted on the maps for convenience.

Many sources of climatic data are available. A few are suggested below:

The National Weather Service maintains the *Climatology of the United States* for each of the 50 states and *World Weather Records*, updated in 1979. Local weather service offices prepare reports for metropolitan areas. One example is Tony Martini's "Climate of Sacramento, California," NOAA Technical Memorandum NWS WR-65 Sacramento: Weather Service Office, April 1990, 70 pp. Check to see if your local or state climatologist has prepared such a report. For Canada, contact the Atmospheric Environment Service, *Climatic Normals* publications. See Appendix B for specific addresses.

Landsberg, Helmut E., ed. *World Survey of Climatology*, 15 volumes published 1969-1984. New York: Elsevier, North Holland.

Pearce, E.A. and C.G. Smith. *The World Weather Guide*. London: Hutchinson and Company, 1984.

Riordan, Pauline and Paul G. Bourget. *World Weather Extremes*. Fort Belvoir, VA: U.S. Army Corp of Engineers, 1985, available from USGPO.

Rudloff, Willy. *World Climates*. Stuttgart, Germany: Wissenschaftliche Verlagsgesellschaft, 1981.

Ruffner, James A. and Frank E. Blair, eds. *The Weather Almanac*. New York: Avon Books, 1977.

Wernstedt, Frederick L. *World Climatic Data*. Lemont, PA: Climatic Data Press, 1972.

Willmott, Cort J., John R. Mather, and Clinton M. Rowe. *Average Monthly and Annual Surface Air Temperature and Precipitation Data for the World*. Part 1, "The Eastern Hemisphere," and Part 2, "The Western Hemisphere." Elmer, NJ: C. W. Thornthwaite Associates and the University of Delaware, 1981.

For a complete survey of CD-ROM resources see: Clifford F. Mass, "The Application of Compact Discs (CD-ROM) in the Atmospheric Sciences and Related Fields: An Update," *Bulletin of the American Meteorological Society*, 74 no. 10, October 1993: 1901-08. The article reviews CD-ROM technology and has two appendices: "Currently Available CD-ROM Titles in the Atmospheric Sciences and Related Disciplines," and "Contact Information for CD-ROM Vendors." The article references several data sets for climatological information.

Air Temperature and the Human Body

This section is presented in an effort to relate temperature concepts to the student personally. I have prepared a wind chill chart that appears with C and F conversions for your convenience. The heat index chart is also presented with conversions. The year 1993 again experienced record heat index readings in the United States and Europe. For additional material see Kalkstein Laurence S. and Robert E. Davis. "Weather and Human Mortality: An Evaluation of Demographic and Interregional Responses in the U.S.," *Annals* of the Association of American Geographers, Vol. 79, No. 1, March 1989: 44-64.

Glossary Review for Chapter 5 (in alphabetical order)

apparent temperature
continentality
Gulf Stream
isotherm
land-water heating differences
marine
specific heat
temperature
thermal equator
transparency
wind chill factor

Annotated Chapter Review Questions

1. Distinguish between sensible heat and sensible temperature.

The amount of heat energy present in any substance can be measured and expressed as its temperature. Temperature actually is a reference to the speed of movement of the atoms and molecules that make up a substance. Our bodies sense temperature and subjectively judge comfort, reacting to changing temperatures with predictable responses. Our perception of temperature is described by the terms apparent temperature or sensible temperature. This perception of temperature varies among individuals and cultures.

2. What does air temperature indicate about energy in the atmosphere?

Air temperature, a measure of sensible heat energy present in the atmosphere, indicates the average kinetic energy of individual molecules within the atmosphere.

3. Compare the three scales that express temperature. What is the basic assumption for each?

The three scales that express temperature are Kelvin, Celsius and Farenheit. Each scale was derived for a specific use, which differentiates the location of absolute zero in each scale. The Kelvin scale was created to measure kinetic energy. Kelvin starts at absolute zero and readings are proportional to the actual change in kinetic energy of the substance. Celsius was created to reflect the decimal scale, with 100 degrees between water's melting and boiling temperatures. And lastly, Farenheit was created in response to measuring body temperature, 98.6°F. See chart below.

Scale	H_2O melting	H_2O boiling	Difference
Kelvin	273K	373K	100°
Celsius	0 C	100 C	100°
Farenheit	32 F	212 F	180°

4. What is your source of daily temperature information? Describe the highest temperature you have experienced and the lowest temperature. From what we have discussed in this chapter, can you identify the factors that may have contributed to these temperatures?

Personal analysis and response.

5. Explain the effect of altitude upon air temperature. Why is air at higher altitudes lower in temperature? Why does it feel cooler standing in shadows at higher altitudes?

Air temperatures in the troposphere decrease with increasing elevation above Earth's surface (recall that the normal lapse rate of temperature change with altitude is 6.4°C/1000 m or 3.5°F/1000 ft). Thus, worldwide, mountainous areas experience lower temperatures than do regions nearer sea level, even at similar latitudes. Temperatures may decrease noticeably in the shadows and shortly after sunset.

Surfaces both heat rapidly and lose their heat rapidly at higher altitudes.

6. What noticeable effect does air density have on the absorption and radiation of heat? What role does altitude play in that process?

The density of the atmosphere also diminishes with increasing altitude, as discussed in Chapter 3. As the atmosphere thins, its ability to absorb and radiate heat is reduced. The consequences are that average air temperatures at higher elevations are lower, nighttime cooling increases, and the temperature range between day and night and between areas of sunlight and shadow also increases.

7. How is it possible to grow moderate-climate crops such as wheat, barley, and potatoes at an elevation of 4103 m (13,461 ft) near La Paz, Bolivia, so near the equator?

The combination of elevation and equatorial location guarantees La Paz nearly constant daylength and moderate temperatures, averaging about 9°C (48°F) for every month. Such moderate temperature and moisture conditions lead to the formation of more fertile soils than those found in the warmer, wetter climate of Concepción.

8. Describe the effect of cloud cover with regard to Earth's temperature patterns. From the last chapter, review the albedo-forcing and greenhouse-forcing influence of different cloud types.

Clouds are moderating influences on temperature, producing lower daily maximums and higher nighttime minimums. Acting as insulation, clouds hold heat energy below them at night, preventing more rapid radiative losses, whereas during the day, clouds reflect insolation as a result of their high albedo values. The moisture in clouds both absorbs and liberates large amounts of heat energy, yet another factor in moderating temperatures at the surface.

9. List the physical aspects of land and water that produce their different responses to heating. What is the specific effect of transparency in a medium?

Figure 5-4 summarizes the operation of all these land-water temperature controls: evaporation, transmissibility, specific heat, movement, and ocean currents. The physical nature of the substances themselves–solid rock and soil vs. water–is the reason for these land-water heating differences. More of the energy arriving at the ocean's surface is expended for *evaporation* than is expended over a comparable area of land. The transmission of light obviously differs between soil and water; solid ground is opaque, water is transparent. Consequently, light striking a soil surface does not penetrate but is absorbed, heating the ground surface. That heat is accumulated during times of exposure and is rapidly lost at night or in shadows. This is the property of *transmissibility*. In contrast, when light reaches a body of water it penetrates the surface because of water's transparency, transmitting light to an average depth of 60 m (200 ft) in the ocean. Water can hold more heat than can soil or rock, and therefore water is said to have a higher *specific heat*. A given volume of water represents a more substantial heat reservoir than an equal volume of land. Land is a rigid, solid material, whereas water is a fluid and capable of *movement*. Differing temperatures and currents result in a mixing of cooler and warmer waters, and that mixing spreads the available heat over an even greater volume. Ocean currents of differing temperatures are an influence on temperature patterns.

10. What is specific heat? Compare specific heat of water and soil.

Specific heat is the term used to express the ability of a substance to absorb energy, specifically; the amount of energy required to raise 1 gram of substance by 1°C. The specific heat of land and water are very different which is the main determinant in the temperature variation between land and water locations. The specific heat of land is low, which means that it takes a relatively less energy to raise the temperature of land by 1°C. Water has a high specific heat. Due the transparency of water, water can absorb a high amount of energy before it will effect its temperature. See Figures 5-4, 5-5, 5-6, and 5-7. For geographic examples of differential heating see Figures 5-9, 5-10 and 5-12. The temperature variation caused by specific heat differences between land and water explain the annual variation in temperature, as seen on Figure 5-14.

11. Describe the pattern of sea-surface temperatures (SSTs) as determined by satellite remote sensing. Where is the warmest ocean region on Earth?

In an air mass, water vapor content is affected by ocean temperatures, for warm water tends to energize overlying air through high evaporation rates and transfers of latent heat.

The **Gulf Stream** (described in Chapter 6) moves northward off the east coast of North America, carrying warm water far into the North Atlantic (Figure 5-7). As a result, the southern third of Iceland experiences much milder temperatures than would be expected for a latitude of 65° N, just below the Arctic Circle (66.5°).

In the western Pacific Ocean, the *Kuroshio* or Japan Current, similar to the Gulf Stream, functions much the same in its warming effect on Japan, the Aleutians, and the northwestern margin of North America. In contrast, along midlatitude west coasts, cool ocean currents influence air temperatures. When conditions in these regions are warm and moist, fog frequently forms in the chilled air over the cooler currents.

TIROS-N and *NOAA* satellites provided scientists at the University of Delaware with a 10-year record of SSTs. Mean annual SSTs increased from 1982 to 1991 with some fluctuations after 1987. Figure 5-8 displays these satellite data. The red and orange area in the southwestern Pacific Ocean with temperatures above 28°C (82.4°F), occupying a region larger than the United States, is called the Western Pacific Warm Pool.

12. What effect does sea-surface temperature have upon air temperature? Describe the negative feedback mechanism created by higher sea-surface temperatures and evaporation rates.

Sea surface temperatures (SSTs) have a great impact on the amount of evaporation which occurs on a global basis. With increased SSTs the latent heat of evaporation will increase, increasing the temperature of the air, and enabling greater amounts of longwave radiation to be absorbed by the atmosphere. As the air temperature rises, the air will have a greater ability to hold water (a lower relative humidity), and the water in the atmosphere will increase the amount of longwave radiation which is scattered and reflected resulting in greenhouse effects and warming the lower atmosphere. Increased absorption of water in the atmosphere will lead to greater cloud formation and this will create a negative feedback mechanism due to the increased albedo of clouds. So, less shortwave radiation will be absorbed by the atmosphere and the surface (in this case, the ocean). This reduced amount of insolation will lower air and water temperatures which will reduce the rate of evaporation and the ability of the air to hold such moisture (higher relative humidity). This is a great example of how the Earth maintains a temperature equilibrium.

13. Differentiate between marine and continental temperatures. Give geographical examples of each from the text: Canada, United States and Norway/Russia.

The term marine, or maritime, is used to describe locations that exhibit the moderating influences of the ocean, usually along coastlines or on islands. Continentality refers to the condition of areas that are less affected by the sea and therefore have a greater range be-

Boulder vs SF

tween maximum and minimum temperatures diurnally and yearly. The Canadian cities of Vancouver, British Columbia, and Winnipeg, Manitoba exemplify these marine and continental conditions (Figure 5-9). Vancouver has a more moderate pattern of average maximum and minimum temperatures than does Winnipeg. A similar comparison of San Francisco and Wichita, Kansas, is presented in Figure 5-10.

14. What is the thermal equator? Describe the location of the thermal equator in January and in July. Explain why it shifts position annually.

The thermal equator, a line connecting all points of highest mean temperature (red line on the maps in Figure 5-11 and 5-13), in January trends southward into the interior of South America and Africa, indicating higher temperatures over landmasses. The thermal equator shifts northward in July with the high summer Sun and reaches the Persian Gulf-Pakistan-Iran area (relate to specific heat).

15. Observe trends in the pattern of isolines over North America and compare the January average temperature map with the July map. Why do these patterns shift locations?

Isolines over North America vary with the movement of the ITCZ. As the ITCZ shifts toward the Southern Hemisphere, North America receives less solar radiation, due to reduced daylength and a lower angle of incidence this time of year. This correlates to reduced temperatures experienced during January in North America. As the ITCZ shifts toward the Northern Hemisphere, North America receives more solar radiation, due to increased daylength and a greater angle of incidence. Temperatures in North America will be more extreme. Due to the specific heat properties of land, North America will experience greater temperature extremes, the continent will lose energy rapidly in January due to land's low heat capacity, and the continent will heat rapidly in July due to the low amount of energy that is required to heat land.

16. Explain the extreme temperature range experienced in north-central Siberia between January and July.

As you might expect, the largest temperature ranges occur in subpolar locations in North America and Asia, where average ranges of 64°C (115°F) are recorded. The Verkhoyansk region of Siberia is probably the greatest example of continentality on Earth. The coldest area on the map is in northeastern Siberia in the Soviet Union. The cold experienced there relates to consistent clear, dry, calm air, small insolation input, and an inland location far from any moderating maritime effects. Verkhoyansk and Omakon, Siberia, Russia, each have experienced a minimum temperature of –68°C (–90°F) and a daily average of –50.5°C (–58.9°F) in January. Verkhoyansk experiences at least seven months of temperatures below freezing, including at least four months below –34°C (–30°F)! July temperatures in Verkhoyansk average more than 13°C (56°F), which represents a 63C° (113F°) seasonal variation between winter and summer averages.

17. Where are the hottest places on Earth? Are they near the equator or elsewhere? Explain. Where is the coldest place on Earth?

The hottest places on Earth occur in Northern Hemisphere deserts during July. These deserts are areas of clear and dry skies and strong surface heating, with virtually no surface water and few plants. Locations such as portions of the Sonora Desert area of North America and the Sahara of Africa are prime examples. Africa has recorded shade temperatures in excess of 58°C (136°F), such as a record set on 13 September 1922 at Al-Aziziyah, Libya (32° 32' N; 112 m or 367 ft elevation). The highest maximum and annual average temperatures in North America occurred in Death Valley, California, where the Greenland Ranch

Station (37° N; –54.3 m or –178 ft below sea level) reached 57°C (134°F) in 1913.

Outside of the polar regions, see the description under Question 16. July is a time of 24-hour-long nights in Antarctica. The lowest natural temperature reported on Earth occurred on 21 July 1983 at the Russian research base at Vostok, Antarctica (78°27 S, elevation 3420 m or 11,220 ft): a frigid –89.2°C (–128.56°F). For comparison, such a temperature is 11°C (19.8°F) colder than dry ice (solid carbon dioxide)!

18. From the maps in Figures 5-11, 5-13, and 5-14, determine the average temperature values and annual range of temperatures for your present location.

Personal analysis and response.

19. Identify the different responses of the human body to low-temperature and high-temperature stress.

See Table 5-1 for specific effects. The lab manual sets a discussion on this topic in Lab Exercise 4.

20. Describe the interaction between air temperature and wind speed and their affect on skin chilling. Select several temperatures and wind speeds from Figure 5-15 and determine the wind chill index.

The wind chill index is important to those who experience winters with freezing temperatures. The wind chill factor indicates the enhanced rate at which body heat is lost to the air. As wind speeds increase, heat loss from the skin increases. For example, if the air temperature is –1°C (30°F) and the wind is blowing at 32 kmph (20 mph), skin temperatures will be –16°C (4°F). The colder wind chill values present a serious danger to exposed flesh.

21. Define the heat index. What characteristic patterns are experienced throughout the year where you live relative to the graph in Figure 5-16?

A measured index of the human body's reaction to air temperature and water vapor is called the heat index (HI). Assume that the amount of water vapor in the air affects the evaporation rate of perspiration on the skin, because the more water vapor in the air (the higher the humidity), the less water from perspiration the air can absorb through evaporation. The heat index indicates how the air feels to an average person–in other words, its apparent temperature.

Overhead Transparencies

As an adopter you are provided with the following figures for overhead projector use.

- Figure 5-2: A comparison of temperature patterns in La Paz and Concepción Bolivia
- Figure 5-4: The differential heating of land and water
- Mounted together: Figure 5-9 Comparison of Vancouver and Winnipeg (top); and, Figure 5-10-comparison of San Francisco and Wichita (bottom)
- Mounted together: Figure 5-11: Mean sea level global temperatures for January (top); and Figure 5-13: Mean sea level global temperatures for July (bottom)
- Figure 5-14: Annual range of global temperatures

6

Atmospheric and Oceanic Circulations

Overview

Earth's atmospheric circulation is an important transfer mechanism for both energy and mass. In the process, the energy imbalance between equatorial surpluses and polar deficits is partly resolved, Earth's weather patterns are generated, and ocean currents are produced. Human-caused pollution also is spread worldwide by this circulation, far from its points of origin. In this chapter we examine the dynamic circulation of Earth's atmosphere that carried Tambora's debris and Chernobyl's fallout worldwide and carries the everyday ingredients oxygen, carbon dioxide, and water vapor around the globe. We also consider Earth's wind-driven oceanic currents.

The keys to this chapter are in several integrated figures: the portrayal of winds by the *SEASAT* image (Figure 6-4), the three forces interacting to produce surface wind patterns and winds aloft (Figure 6-10), and the buildup through (Figure 6-15) to produce the geostrophic wind (Figures 6-18 through 6-20)

The *Student Study Guide* presents 20 "Learning Objectives" to guide the student in reading the chapter.

New to the Third Edition

(Note: This section highlights major changes, new features, and additions in the third edition. This does not describe all the rewrite and recast of the text.)

1. A list of key learning concepts begins the chapter.

2. Discussion of air pressure has been moved from Chapter 3, including Figures 6-5, 6-7, and 6-8, for greater correlation with atmospheric properties and movement.

3. Figure 6-6, a newly illustrated profile of air pressure at varying altitudes.

4. Elaborate description of the constant isobaric surface, explaining ridges and troughs more thoroughly. Figure 6-18 has been added to illustrate contours in the geostrophic wind as it migrates across the North American continent.

5. Revised explanation of the forces which determine atmospheric movement. More detailed discussion of Coriolis force, with images added to Figures 6-10 and 6-11. Greater explanation of the pressure gradient force, including a detailed map of isobars superimposed on the United States to demonstrate how isobar spacing reflects wind speed.

6. News Report #1: "Coriolis, a Forceful Effect on Drains?" critically examines the ability of the Coriolis force to effect the direction of water's drainage in sinks and toilets on a global basis.

7. News Report #2: "Jet Streams Affect Flight Times" details the history of how headwinds from the jet stream hinders air travel, and consequently, how air travel aided our knowledge of the jet stream.

8. Figure 6-20 illustrates the appearance of the jet stream and also gives geographical reference to the location of jet streams across North America. This is a great connection to Figures 6-18 and 6-12.

9. Greater discussion of surface winds, and their use in global trade.

10. News Report #3: "A Message in a Bottle and Rubber Ducks". This report summarizes how the circulation of ocean gyres stimulated global travel, and continue to be used today as illustrated by the diffusion of message bottles and rubber duckies along the tradewinds and westerlies in the Pacific Ocean.

11. A new summary and review section ends the chapter.

Key Learning Concepts

1. *Define* wind and *describe* how wind is measured, wind direction is determined, and how winds are named.

2. *Define* the concept of air pressure and *portray* the pattern of global pressure systems on isobaric maps.

3. *Explain* the three driving forces within the atmosphere-pressure gradient, Coriolis, and friction forces-and describe the primary high and low pressure areas.

4. *Describe* upper air circulation and its support role for surface systems and *define* the jet streams.

5. *Explain* several types of local winds: land-sea breezes, mountain-valley breezes, katabatic winds, and the regional monsoons.

6. Discern the basic pattern of Earth's major surface and deep ocean currents.

Expanded Outline Discussion

The following headings (boldfaced) match some of the first, second, and third order headings in Chapter 6. The narrative under each heading contains information, sources, and anecdotal facts relating to portions of the chapter. Not all text headings are discussed.

WIND ESSENTIALS

Wind: Description and Measurement

The imbalance in net energy from equator to pole is established in Figure 2-14 and 4-14. The poleward transport of heat and mass (water and water vapor) are important functions of atmospheric circulation that maintain Earth's energy balance. Figures 6-2 and 6-3 show an anemometer and wind vane, and a wind compass with compass designations for wind directions. A modified version of the Beaufort wind scale is presented in Table 6-1 with observed effects at sea and on land. This should provide the student with a "feel" for winds of various speeds.

The student can also refer to Figure 8-24 and the standard weather map presentation that includes wind speed and direction information and the wind speed legend with Figure 6-12.

Global Winds

In the introduction to this chapter I give several dramatic examples of the impact of global circulation: the Tambora eruption of 1815 and the Chernobyl nuclear disaster in 1986. The eruption of Mount Pinatubo that began in June 1991 is providing further evidence of this dynamic circulation. This eruption exceeded any other this century, including Mount

Katmai, Alaska, which extruded 12 km^3 (2.88 mi^3) of pyroclastics. Also remember beyond these dramatic examples, global circulation mixes oxygen from principal production areas, presses against sails and kites, and guides and drives Earth's weather machine. See an initial mention by Richard A. Kerr, "Huge Eruption May Cool the Globe," in *Science* Vol. 252, No. 5014 (28 June 1991): p. 1780, and the new references under suggested readings at the end of the chapter. You no doubt noticed the afterglow (past sunset) during the fall of 1991 resulting from high-altitude ash and mist–more intense at lower latitudes than higher latitudes. Review the opening discussion of the spatial implications of this eruption in the introduction to this manual.

Examples abound for use in this introductory lecture, e.g., a U.S. satellite, with a plutonium-238 power plant on board, failed to achieve orbit following an April 1964 launch. The satellite burned on reentry, spreading minute amounts of radiation to monitoring stations throughout the northern hemisphere. Or, in 1976, an above-ground nuclear test at the Lop Nor Chinese Test site sent radioactive debris into the atmosphere. Some 10 days later, fallout occurred over the United States in measurable quantities.

As the text states, "Our atmosphere makes all the world a spatially linked society–one person's or nation's exhalation is another's breath." In the ongoing aftermath of the 1991 Persian Gulf war, the media have aired various versions of concern about what the winds will do with the soot and sulfur from the oil-well fires and war-related smoke and ash. A widespread global effect did not occur, whereas regional consequences were serious–local daytime temperatures have been as much as 15°C (27°F) cooler in the immediate region of the war. Soot from the war was montitored in samples collected at the Mauna Loa Observatory.

The Tambora eruption is unequaled in historic times for its output of ash and sulfur compounds. The volcano did trigger a global signature–this is undeniable. A direct connection with the cool summer of 1816 is still elusive and unproven despite several articles and books describing the effects. A couple of sources to consult:

Stothers, Richard B., "The Great Tambora Eruption in 1815 and Its Aftermath," *Science* Vol. 224, No. 4654 (15 June 1984): 1191-1198.

Stommel, Henry and Elizabeth Stommel, "The Year Without A Summer," *Scientific American* (June 1979): 176-186; also, by the same authors in *Volcano Weather–The Story of the Year Without a Summer*, Newport: Seven Seas Press, 1983.

With weather related crop failures, the price of wheat soared in the years after the eruption to a level not reached again until 1972 and the Soviet grain shortage!

The aftermath of the Chernobyl accident continues to unfold. Some 200,000 people in Byelorussia and Ukraine have been evacuated and another 110,000 await relocation. Yet, 5 million people live in the overall affected area. Contamination is continuing to spread through wind, air, water, soil, and food chains. Unstable isotopes of plutonium, americium, cesium, and strontium pose a distinct threat to the population. (Source: Dr. Yevgeny F. Konoplya, Director of the Radiobiology Institute of the Byelorussian Academy of Sciences.)

Air Pressure

The mine drainage problem that Torricelli was working on was in Czechoslovakia in A.D. 1643. Evidently, Galileo's suggestions and guidance were important. Several sources stress the importance of Torricelli's synthesis of previously determined concepts in the solution of his mine drainage problem and the development of the mercury barometer.

I hope you have a barometer available for demonstration and reference throughout the semester. We have a small glass-doored instrument cabinet in our geography classroom. During the semester, it is instructive to refer to air pressure on days of significant change. And of course, we all walk around with an aneroid barometer in our head that leaks (inner ear drum-eustachian tubes).

The *Student Study Guide* asks the student to observe and record atmospheric pressure for a sequence of days.

DRIVING FORCES WITHIN THE ATMOSPHERE:

Pressure Gradient Force, Coriolis Force, and Frictional Force

Figure 6-10 a, b, and c takes each force one at a time and builds a model of wind flow circulation associated with high- and low-pressure areas. Once pressure gradient is defined (Figure 6-4) then its initial place in Figure 6-10 can be referred to and the pattern of vertical and horizontal flows established. Figure 6-12 shows an upper-air weather chart that can be used as an example of geostrophic winds. Professor Buys-Ballot (1817-1890) established a governing law, which bears his name, Buys-Ballot's Law: If you stand with your back to the wind in the Northern Hemisphere, the low pressure will always be to your left. Check this law against the upper air isobaric map shown in Figure 6-12.

Do we call Coriolis a *force* or an *effect* ? The proper way to refer to the Coriolis force has plagued physical geography texts for years, yet in other academic areas such ambivalence and confusion do not seem to appear. I chose to add the reference to Newton's definition of a force to give you a way of explaining the usage in *GEOSYSTEMS*. Think of a deflective force that causes these apparent effects. For several basic examples of the typical treatment of Coriolis in other works consult the following among many. I briefly quote these sources to give you an idea of usage.

"Coriolis force," *Encyclopedia Britannica*, Micropædia Vol. 3, Chicago (1989 ed.): 632. "...in classical mechanics, an inertial force..."

Barry Roger B. and Richard J. Chorley, *Atmosphere, Weather, and Climate*, 5th ed., London: Methuen, 1987, 116-118. "The Coriolis force arises from the fact that the movement of masses over the earth's surface is usually referred to as a moving coordinate system...."

Petterssen, Sverre, *Introduction to Meteorology*, 3rd ed., New York: McGraw-Hill Book Company, 1969, 153-155. "The deviating force...."

Neiburger, Morris and James G. Edinger, William D. Bonner, *Understanding Our Atmospheric Environment*, San Francisco: W.H. Freeman, 1973, 99-104. "The effect of the rotation of the earth on any moving object is to make it appear as though a force is acting on the object. This apparent force is called the Coriolis force...." (A quantified explanation and description of angular velocity principles is presented).

ATMOSPHERIC PATTERNS OF MOTION

Understanding the model of atmospheric circulation can be difficult for the student because it is spatial and dimensional. Therefore, I constructed this section in distinct stages. The set-up begins in Figure 6-15 with 2 views of the Hadley cell and the equatorial circulation including a cross-section view from the equator to the North Pole. The locations of the jet streams are also located here for later discussion. The students should be able to recreate a simple version of Figure 6-15 in their notes, building the diagram as the illustration and discussion build. Note that the sections that follow begin at the equator, flow to the subtropics, then on to the subpolar low and polar high.

Primary High-Pressure and Low-Pressure Areas

Table 6-2 details the four hemispheric pressure areas that will be included in the integrated illustration discussed above. These can be located on the seasonal pressure maps in Figure 6-13.

At the end of this chapter please consult two new pressure maps prepared by the National Climatic Data Center, *Monthly Climatic Data for the World*, 46, no. 1 (January and July 1993). Prepared in cooperation with the World Meteorological Organization, Washington, D.C.: National Oceanic and Atmospheric Administration.

UpperAtmospheric Circulation–Jet Streams

"A.M. Weather," originating early in the morning from Maryland Public Television and presented daily on most Public Broadcasting System (PBS) stations offers a detailed depiction of the upper-air circulation at several altitudes, specifically developing the jet streams' location and speed. This can be valuable in illustrating this section of the text.

Upper-air circulation is stronger in the winter hemisphere when greater thermal differences exist; therefore, westerly migrating surface weather systems travel faster in winter.

Note the evolution of longwave development shown in Figure 6-19. The formation of anticyclonic and cyclonic curvature in (b) and the breakout of a cold air mass in (c).

Local Winds

Lab.

This section presents land-sea breezes, mountain-valley breezes, katabatic winds, and Santa Ana winds. This sets the placement for Focus Study 6-1 "Wind Power: An Energy Resource."

Wind Power: An Energy Resource

defn in lecture

Exploitation of this viable, inexhaustible, nonpolluting resource is only prevented through political (extrinsic) action. The various realities pointed out in the focus study will force increased deployment. The public utility in Sacramento, California, is embarking on wind-farm development in the Coast Ranges of California at this time.

Please refer to the section of this resources manual in Chapter 4 covering solar energy applications. The same political environment that plagues solar energy is also hampering the deployment of wind-energy applications. Especially refer to the table "Energy Resource Characteristics" (centralized and decentralized) presented in Chapters 3 and 4 of this manual. You can see that wind fits the decentralized list and therefore does not possess any of the traits desired in a production-oriented economy. Wind most clearly fits a consumer-oriented economy more suited to end-use energy needs. The truth of the matter is that at ideal sites, wind-generated electricity is actually cheaper than are some traditional alternatives, especially when longterm and marginal costs are calculated.

In California, 1985 costs per installed kilowatt were estimated at $2450 at the Diablo Canyon nuclear power plant, whereas wind costs were at about $900 per installed kilowatt in 1989. You might want to check the following article for more detail about energy costs: Harold M. Hubbard, "The Real Cost of Energy," *Scientific American* Vol. 264, No. 4 (April 1991): 36-42. In this era of capital shortage, high deficits, negative trade balances, and fossil fuel-related wars, it is still amazing to me that the 1991 National Energy Strategy of the Bush Administration asks that we invest in the most expensive sources. The "strategy" asks that we reduce conservation and mileage efficiency that are still the cheapest strategies at only a penny or two per kwh.

Note the description of an experiment completed more than 10 years ago: The Japanese oil tanker Shin Aitoku Maru installed two computer-guided sails of 300 m^2 (3228 ft^2) each. Fuel consumption was reduced by 50% in early tests (focus study 6-1).

Monsoonal Winds

Monsoonal flows occur over Australia, southeast Asia, and Africa, although each is different in intensity and behavior. For instance, the hot, pre-monsoon season of India is not repeated in China. The role of the upper-air circulation pattern is critical for the Indian wet-summer monsoon to reach its downpour condition. Upper-air easterly winds, opposite in direction to lower atmospheric southwesterly flows into the ITCZ, must be in place for full development of the summer monsoon pattern (Figure 6-23). Note, that there is still uncertainty as to the exact interaction of the conditions that produce the monsoon.

In the figure note that as the winds cross the equator the Coriolis deflection shifts to the opposite direction. The pressure patterns in the upper atmosphere are not depicted on these two maps but are important in creating this circulation.

The quote given in the text about the effect of the monsoon on Indian culture can be emphasized with a brief listen to some Indian classical music. The patter of raindrops, strikes of lightning, and thunder of the storm seem to be represented by the instrumental sounds.

Related orographic rainfall patterns for India are discussed in Chapter 8, using Cherrapunji as an example. Climatic data for Calcutta, India are presented in Appendix A.

OCEANIC CURRENTS

Deep Currents

The interrelationship that exists between Earth's atmosphere and oceans makes the inclusion of oceanic currents appropriate in this chapter.

Similar to a rising column of air above the equator, where a loss of friction allows the Coriolis force to gradually strengthen against the pressure gradient force, a pattern forms with depth in the sea. The friction of the wind and surface Coriolis movements produce an ocean current at a 45° angle from the wind. As this direction is carried to deeper and deeper layers in the sea, the frictional forces near the surface are reduced, allowing the Coriolis force to divert the flow further to the right. The net transport of water below the surface will exceed 90° to the right of the wind direction. This spiral-like arrangement with depth is named the *Ekman spiral* after its author V.W. Ekman (1905).

Glossary Review for Chapter 6 (in alphabetical order)

Aleutian low
anemometer
aneroid barometer
Antarctic high
anticyclone
Azores high
Beaufort wind scale
Bermuda high
constant isobaric surface
Coriolis force
cyclone
doldrums
downwelling current
equatorial countercurrent
equatorial low-pressure trough
friction force
geostrophic winds
gyres
Hadley cell
horse latitudes
Icelandic low
intertropical convergence zone (ITCZ)
isobars
jet stream
katabatic winds
mercury barometer
monsoon
Pacific high
polar easterlies
polar front

polar high-pressure cells
polar jet stream
pressure gradient force
Rossby waves
subpolar low-pressure cells
subtropical high-pressure cells
subtropical jet stream
trade winds
upwelling current
westerlies
western intensification
wind
wind vane

Annotated Chapter Review Questions

1. What is a possible explanation for the beautiful sunrises and sunsets during the summer of 1816 in New England? Relate your answer to global circulation.

Early in April 1815, on an island named Sumbawa in present-day Indonesia, the volcano Tambora erupted violently, releasing an estimated 150 km^3 (36 mi^3) of material, an 80-times greater volume than that produced by the 1980 Mount Saint Helens eruption in Washington State. Some materials from Tambora–the aerosols and acid mists–were carried worldwide by global atmospheric circulation, creating a stratospheric dust veil. The result was both a higher atmospheric albedo and absorption of energy by the particulate materials injected into the stratosphere. Remarkable optical and meteorological phenomena resulting from the spreading dust were noted for months and years after the eruption. Beautifully colored sunsets and periods of twilight were enjoyed in London, Paris, New York, and elsewhere. In the summer of 1816, one year later, farmers in New England and Europe were shocked to experience frosts every month, and in some places, every week. Mean temperatures in the Northern Hemisphere apparently had dropped by 0.4-0.7°C (0.72-1.26°F). Although Tambora's eruption has not been conclusively tied to the cold summer in 1816, the impact of the eruption appears so large that, despite measurement uncertainties, it remains as powerful evidence of global circulation.

2. Explain the statement, "the atmosphere socializes humanity, making all the world a spatially linked society." Illustrate your answer with some examples.

Tambora, Chernobyl, the Persian Gulf War, the Mount Pinatubo eruption in 1991, cross-boundary pollution between the United States and Canada, global impact on atmospheric carbon dioxide caused by equatorial and tropical deforestation, and the spread of acid deposition are all examples of atmospheric redistribution of materials affecting society as well as a good basis for a philosophical discussion of human-Earth interdependence.

3. Define wind. How is it measured? How is its direction determined?

Wind is the horizontal motion of air relative to Earth's surface. It is produced by differences in air pressure from one location to another and is influenced by several variables. The two principal variables are speed and direction. Wind speed is measured with an anemometer, and may be expressed in kilometers per hour (kmph), miles per hour (mph), meters per second (m/sec), or knots. (A knot is a nautical mile per hour, covering 1 minute of Earth's arc in an hour, equivalent to 1.85 kmph or 1.15 mph.) Wind direction is determined with a wind vane; the standard measurement is taken 10 m (33 ft) above the ground to avoid, as much as possible, local effects of topography upon wind direction.

4. Distinguish among primary, secondary, and tertiary classifications of global atmospheric circulation.

Primary (general) circulation, secondary circulation of migratory high-pressure and low-pressure systems, and tertiary circulation that includes local winds and temporal weather patterns are three classes of global winds. Winds that move principally north or south along meridians are known as meridianal flow or meridianal circulations. Winds moving east or west along parallels of latitude are called zonal flows, or zonal circulations.

5. What is the purpose of the Beaufort wind scale? Characterize winds given Beaufort numbers of 4, 8, and 12, giving effects over both water and land.

A descriptive scale useful in visually estimating winds is the traditional Beaufort wind scale. Originally established in 1806 by Admiral Beaufort of the British Navy for use at sea, the scale was expanded to include wind speeds by G. C. Simpson in 1926 and standardized by the National Weather Service (formerly the Weather Bureau) in 1955. The scale presented in Table 6-1 is modernized and adapted from these earlier versions. It still is referenced on ocean charts and is presented here with descriptions of visual wind effects on land and sea. This observational scale makes possible the estimation of wind speed without instruments, although most modern ships use sophisticated equipment to perform such measurements.

6. How does air exert pressure? Describe the basic instrument used to measure air pressure. Compare the two different types of instruments discussed.

Air molecules–through their motion, size, and number–produce pressure that is exerted on all surfaces in contact with the air. The weight of the atmosphere, or air pressure, crushes in on all of us; fortunately, that same pressure is also inside us, pushing out. The atmosphere exerts an average force of 14.7 pounds per square inch (approximately 1 kg/cm^2) at sea level. Under the acceleration of gravity, air is compressed and is more dense near Earth's surface, rapidly thinning out with increased altitude (Figure 6-5). This decrease is measurable since air exerts its weight as a pressure.

Any instrument that measures air pressure is called a barometer. Torricelli developed a mercury barometer and established the average height of the column of mercury in the barometric tube. The column of mercury was counterbalanced by the mass of surrounding air exerting an equivalent pressure on the mercury in the vessel. A more compact design that works without a meter tube of mercury is called an aneroid barometer (Figure 6-7).

7. What is normal sea level pressure? In mm? mb? in.? kPa?

Using such instruments, normal sea level pressure is expressed as 1013.2 millibars (a way of expressing force per square meter of surface area), or 29.92 in., or 760 mm, of mercury; or 101.32 kPa. (10 mb = 1 kPa.)

8. What does an isobaric map of surface air pressure portray? Contrast pressures over North America for January and July.

An isobaric map portrays the weight of the atmosphere in specific locations. By delineating high and low pressure air masses, geographers are able to determine the movement of air caused by the pressure gradient force, and determine the stability of an air mass, predicting patterns of precipitation and aridity. Figure 6-8 illustrates the extremes in pressure and how they relate to predictable weather patterns.

Pressures over North America during January reflect the heat capacity of water. Low pressures, such as the Aleutian Low and the Icelandic Low are found over the Pacific and Atlantic Oceans, respectively. During January, high pressures dominate the North American land mass due to extreme low temperatures, correlate with Figure 5-11. In July these pressures switch locations, due to the low specific heat of land, the North American landmass heats rapidly, causing low pressures to be located over the North American continent and high pressures over the oceans. The stable, high pressure air limits much precipitation on the West coast, while warm, low pressure air dominates the Eastern Seaboard causing greater evaporation rates, higher humidity, and summer showers.

9. What wind speed effects do the spacing of isobars produce?

Isobars allow us to see the pressure gradient that exists in specific locations in the atmosphere. When the isobars are close together, this shows a steep pressure gradient, or difference, causing greater wind speeds. Similar to a topographic map, where closely clustered lines of elevation show a steep slope, we expect to find strongwinds in areas of closely clustered isobars. If isobars are spaced far apart, this reflects a gentle pressure gradient resulting in a slow wind speed, or a mild breeze. See Figure 6-9.

10. Describe the effect of the Coriolis force. How does it apparently deflect atmospheric and ocanic circulations? Explain.

The Coriolis force applies to objects moving across Earth's surface that appear to deflect from a straight path. Because the physicist Sir Isaac Newton stated that, if something is accelerating over a space, a force is in operation, the label *force* is appropriate here. The deflection produced by the Coriolis force is caused by the fact that Earth's rotational speed varies with latitude, decreasing from 1675 kmph (1041 mph) at the equator to 0 at the poles (see Table 2-3). The Coriolis effect increases as the speed of moving objects increases; thus, the faster the movement, the greater the apparent deflection. And, because Earth rotates eastward, objects that move in an absolute straight line over a distance for some time (such as winds and ocean currents) appear to curve to the right in the Northern Hemisphere and to the left in the Southern Hemisphere.

11. What are geostrophic winds, and where are they encountered in the atmosphere?

Figure 6-10 (b) illustrates the combined effect of the pressure gradient force and the Coriolis force on the atmosphere, producing geostrophic winds. Geostrophic winds are characteristic of upper tropospheric circulation. The air does not flow directly from high to low, but around the pressure areas instead, remaining parallel to the isobars and producing the characteristic pattern shown on the upper-air weather chart.

12. Describe the horizontal and vertical air motions in a high-pressure cell and in a low-pressure cell.

The pressure gradient force acting alone is shown in Figure 6-6 (a) from two perspectives. As air descends in the high-pressure area, a field of subsiding, or sinking, air develops. Air moves out of the high-pressure area in a flow described as diverging. High-pressure areas feature descending, diverging air flows. On the other hand, air moving into a low-pressure area does so with a converging flow. Thus, low-pressure areas feature converging, ascending air flows.

13. Construct a simple diagram of Earth's general circulation, including the four principal pressure belts or zones and the three principal wind systems.

For the student to complete (Figures 6-15, and Table 6-2).

14. How does the intertropical convergence zone (ITCZ) relate to the equatorial low-pressure trough? How might it appear on a satellite image?

The equatorial low-pressure trough is an elongated, narrow band of low pressure that nearly girdles Earth, following an undulating linear axis. Constant high Sun altitude and consistent daylength make large amounts of energy available at this region of Earth's surface throughout the year. The warming creates lighter, less-dense, ascending air, with winds converging all along the extent of the trough. The combination of heating and convergence forces air aloft and forms the intertropical convergence zone (ITCZ). This converging air is extremely moist and full of latent heat energy. Vertical cloud columns frequently reach the tropopause, and precipitation is heavy throughout this zone. On a satellite image it is identified by bands of clouds along the equator.

15. Characterize the belt of subtropical high pressure on Earth: names of specific cells? generation of westerlies and trade winds? sailing conditions?

Subtropical high pressure consists of hot, dry air resulting from air diverging from the Hadley cell deflected to the poles by the Coriolis Force. This air is associated with cloudless, desert regions, such as the Sahara and the Arabian desert. Examples of subtropical high pressure cells are; the Azores high, which dominates the west coast of Africa in the Atlantic Ocean, the Bermuda High, located in the western Atlantic, and the Pacific High which dominates the Pacific Ocean. The divergence of this dry air creates the ocean gyres, spinning in a clockwise direction in the North Hemisphere due to the Coriolis Force. The ocean currents stimulated by the divergence of high pressure create two regions of predictable currents, the tradewinds, which migrate in an easterly direction, creating currents such as the Canary current and the Equatorial Counter Current between Europe and North America. And the westerlies, which move in a westerly direction causing such currents as the Gulf Stream in the Atlantic and the Kurioshio Current in the Pacific. The combination of tradewinds, or easterlies, and westerlies enabled European sailors to travel to North America and return to Europe using the currents created by the subtropical highs. If ships were not geographically accurate, they may have gotten caught in the center of the subtropical highs, which are characterized by little wind. These areas are called the horse latitudes, where ships often killed livestock in order to ration water and food, fearing they would not locate the tradewinds.

16. What is the relation among the Aleutian low, the Icelandic low and migratory low-pressure cyclonic storms in North America? In Europe?

The Aleutian low and Icelandic low are migratory pressure cells. During the summer, these low pressure cells are found in high latitudes over the Pacific and Atlantic Oceans, bring precipitation to Ireland and the Pacific Northwest, such as Seattle and Alaska. In January, as the continent of North America becomes dominated by high pressure, these subpolar lows migrate to lower latitudes (due to their relative temperature), causing cyclonic storms of the west coast of North America and Europe, such as California and Spain. The contrasts that these areas experience, variations between high pressure in the winter and low pressures in the summer, cause frontal interaction and predictable patterns of precipitation.

17. How does the isobaric surface (ridges and troughs) relate to surface pressure systems? To divergence aloft and surface lows? Convergence aloft and surface highs?

As on surface maps, closer spacing of the isobars indicates faster winds; wider spacing indicates slower winds. On this isobaric pressure surface, altitude variations from the reference datum are called ridges for high pressure, and troughs for low pressure. The patterns of ridges and troughs in the upper-wind flow is important in sustaining surface cyclonic (low) and anticyclonic (high) circulation. Frequently surface pressure systems are generated by the upper-air wind flow. Near ridges in the isobaric surface, winds slow and converge. Conversely, near the area of maximum wind speeds in the isobaric surface, winds accelerate and diverge. This divergence in the upper-air flow is important to cyclonic circulation at the surface because it creates an outflow of air aloft that stimulates an inflow of air into the low-pressure cyclone.

18. Relate the jet-stream phenomenon to general upper-air circulation. How does the presence of this circulation relate to airline schedules from New York to San Francisco and the return trip to New York?

Within the westerly flow of geostrophic winds are great waving undulations called Rossby waves. These Rossby waves develop along the flow axis of a jet stream. The most prominent movement in these upper-level westerly wind flows is the jet stream, an irregular, concentrated band of wind occurring at several different locations. Airline schedules reflect the presence of these upper-level winds, for they allot shorter flight times from west to east and longer flight times from east to west.

19. People living along coastlines generally experience variations in day-night winds. Explain the factors that produce these changing wind patterns.

The differential heating characteristics of land and water surfaces create these winds. During the day, land heats faster and becomes warmer than does the water offshore. Because warm air is less dense, it rises and triggers an onshore flow of cooler marine air, which is usually stronger in the afternoon. At night, inland areas cool radiatively faster than do offshore waters. As a result, the cooler air over the land subsides and flows offshore over the warmer water, where the air is lifted. This night pattern reverses the process that developed during the day.

20. The arrangement of mountains and nearby valleys produces local wind patterns. Explain the day and night winds that might develop.

Mountain air cools rapidly at night, and valley air gains heat rapidly during the day (Figure 6-22). Thus, warm air rises upslope during the day, particularly in the afternoon; at night, cooler air subsides downslope into the valleys.

21. Describe the seasonal pressure patterns that produce the Asian monsoonal wind and precipitation patterns. Contrast January and July conditions.

The monsoons of southern and eastern Asia are thought to be driven by the location and size of the Asian landmass, and its proximity to the Indian Ocean and the seasonally shifting ITCZ. Also important to the generation of monsoonal flows are wind and pressure patterns in the upper-air circulation. The extreme temperature range from summer to winter over the Asian landmass is due to its continentality, reflecting its isolation from the modifying effects of the ocean. Resultant cold, dry winds blow from the Asian interior over the Himalayas, downslope, and across India, producing average temperatures of between 15° and 20°C (60° and 68°F) at lower elevations. These dry winds desiccate the landscape. During the June-September wet period, the Sun shifts northward to the Tropic of Cancer, near the mouths of the Indus and Ganges Rivers. The Asian continental interior develops a thermal low pressure, associated with higher average temperatures, and the intertropical convergence zone shifts northward over southern Asia. Meanwhile, the Indian Ocean, with surface temperatures of 30°C (86°F), is under the influence of subtropical high pressure. These conditions produce the wet monsoon of India, where world-record rainfalls have occurred: both the second-highest average rainfall (1143 cm or 450 in.) and the highest single-year rainfall total (2647 cm or 1042 in.), were measured at Cherrapungi, India.

22. What is the relationship between global atmospheric circulation and ocean currents? Relate oceanic gyres to patterns of subtropical high pressure.

Earth's atmospheric and oceanic circulations are interrelated. The driving force for ocean currents is the frictional drag of the winds. Also important in shaping these currents is the interplay of the Coriolis Force, density differences associated with temperature and salinity, the configuration of the continents and ocean floor, and astronomical forces (the tides). Ocean currents are driven by atmospheric circulation around subtropical high-pressure cells in both hemispheres. These circulation systems are known as gyres and generally appear offset toward the western side of each ocean basin.

23. Define the western intensification. How is it related to the Gulf Stream and Kuroshio currents?

Gyres and their western margins feature slightly stronger currents than do the eastern portions. Along the full extent of areas adjoining the equator, trade winds drive the oceans westward in a concentrated channel. These currents are kept near the equator by the weaker Coriolis Force influence. As the surface current approaches the western margins of the oceans, the water actually piles up an average of 15 cm (6 in.). From this western edge, ocean water then spills northward and southward in strong currents, flowing in tight channels along the western edges of the ocean basins (eastern shorelines of continents). This is the process known as western intensification.

In the Northern Hemisphere, the Gulf Stream and the Kuroshio move forcefully northward as a result of western intensification, with their speed and depth increasing with the constriction of the area they occupy. The warm, deep-blue water of the ribbonlike Gulf Stream (Figure 5-7) usually is 50-80 km wide and 1.5-2.0 km deep (30-50 mi wide and 0.9-1.2 mi deep), moving at 3-10 kmph (1.8-6.2 mph). In 24 hours, ocean water can move 70-240 km (40-150 mi) in the Gulf Stream, although a complete circuit around an entire gyre may take a year.

24. Where on Earth are upwelling currents experienced? What is the nature of these currents?

Where surface water is swept away from a coast, either by surface divergence (induced by the Coriolis Force) or by offshore winds, an upwelling current occurs. This cool water generally is nutrient-rich and rises from great depth to replace the vacating water. Such cold upwelling currents occur off the Pacific coasts of North and South America and the subtropical and midlatitude west coast of Africa.

25. What is meant by deep-ocean circulation? At what rates do these currents flow? How might this relate to the Gulf Stream in the western Atlantic Ocean?

Important mixing currents along the ocean floor are generated from such downwelling zones and travel the full extent of the ocean basins, carrying heat energy and salinity.

Imagine a continuous channel of water beginning with cold water downwelling in the North Atlantic, flowing deep and strong to upwellings in the Indian Ocean and North Pacific. Here it warms and then is carried in surface currents back to the North Atlantic. A complete circuit may require 1000 years from downwelling in the Labrador Sea off Greenland to its reemergence in the southern Indian Ocean and return. Even deeper Antarctic bottom water flows northward in the Atlantic Basin beneath these currents.

Overhead Transparencies

As an adopter you are provided with the following figures for overhead projector use.

- Figure 6-1: Volcanic eruption effects spread worldwide by winds
- Figure 6-6: Atmospheric pressure profile
- Figure 6-8: Air pressure readings

and conversions
- Figure 6-10 a, b, c : Three physical forces that produce winds
- Figure 6-13: Global barometric pressures for January and July
- Figure 6-15: General atmospheric model
- Figure 6-18: Analysis of constant isobaric surface
- Mounted together: Figure 6-24- Major ocean currents (top); and, Figure 6-25: Deep ocean circulation (bottom)

Mean sea-level pressure for January and July based on long-term observations. National Climatic Data Center, *Monthly Climatic Data for the World*, 46, no. 1 (January and July 1993). Prepared in cooperation with the World Meteorological Organization. Washington, DC: National Oceanic and Atmospheric Administration.

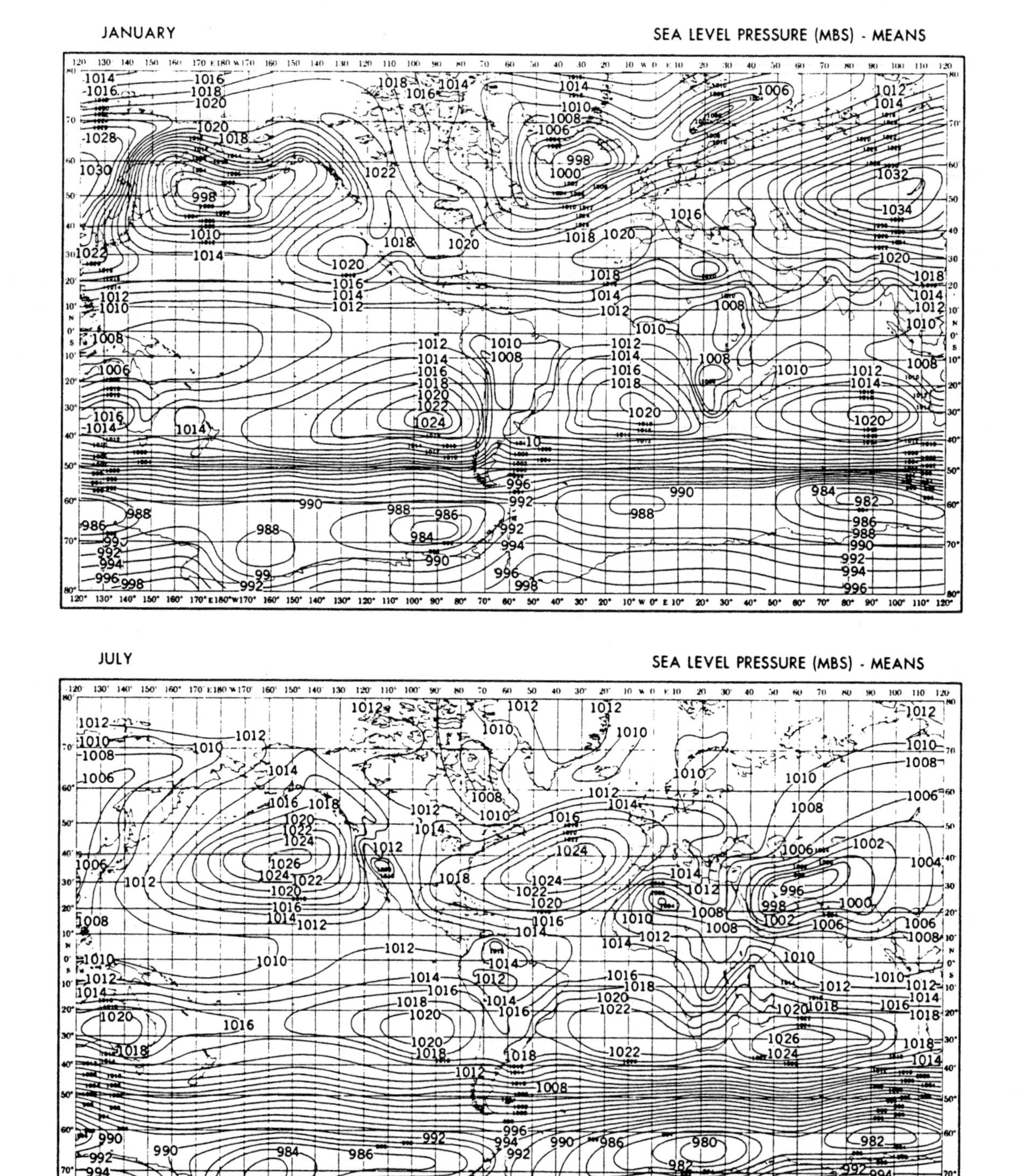

PART TWO:
The Water, Weather, and Climate System

Overview–Part Two

Part Two presents aspects of hydrology, meteorology and weather, oceanography, and climate, in a flowing sequence in Chapters 7 through 10. We begin with water itself–its origin, location, properties, and dynamic circulation in the hydrologic cycle. The global oceans and seas are identified as the greatest repository of water on Earth. The dynamics of daily weather phenomena follow: the effects of moisture and energy in the atmosphere, the interpretation of cloud forms, conditions of stability or instability, the interaction of air masses, and the occurrence of violent weather. The specifics of the hydrologic cycle are explained through the water-balance concept, which is useful in understanding water-resource relationships, whether global, regional, or local. Important water resources include rivers, lakes, groundwater, and oceans. The spatial implications over time of this water-weather system lead to the final topic in Part Two: world climate patterns.

7

Water and Atmospheric Moisture

Overview–Chapter 7

Water is the essential medium of our daily lives and a principal compound in nature. It covers 71% of Earth (by area) and in the solar system occurs in such significant quantities only on our planet. It constitutes nearly 70% of our bodies by weight and is the major component in plants and animals. The water we use must be adequate in quantity as well as quality for its many tasks–everything from personal hygiene to vast national water projects. Indeed water occupies the place between land and sky, mediating energy and shaping both the lithosphere and atmosphere, as stated in the text.

This chapter examines the dynamics of water and atmospheric moisture in particular. Discussion of the water, weather, and climate system, Part 2, Chapters 7 through 10, starts with the beginning of Earth's water. The text then follows the accumulation of water, and describes the unique properties of water. A description of adiabatic processes ends the chapter, setting the foundation to weather phenomena described in Chapter 8.

The *Student Study Guide* presents 20 "Learning Objectives" to guide the student in reading the chapter. The *Applied Physical Geography* lab manual has portions of one exercise with several steps that involve aspects of this chapter.

New to the Third Edition

(Note: This section highlights major changes, new features, and additions in the third edition. This does not describe all the rewrite and recast of the text.)

1. A list of key learning concepts begins the chapter.

2. A visual representation of the difference between specific humidity and relative humidity. Figure 7-8 shows the relative and specific humidity levels in air masses of varying temperature.

3. Figure 7-9 illustrates how dew point temperatures can be reached by adding ice to a drink, causing condensation. This process is further illustrated global examples of condensation.

4. An illustration of a hair hygrometer, has been added to Figure 7-13.

5. The ability of ice to cause damage to roads, bridges, automobiles, water pipes and ships is discussed in News Report #1: "Breaking Roads and Pipes and Sinking Ships." The report summarizes how people living in cold climates attempt to avoid damage caused by ice, as well as how people have attempted to utilize the expansion of ice for quarrying and other activities.

6. A new satellite image of the water vapor content in the atmosphere, Figure 7-15.

7. News Report #2: "The Power of Clouds," describes the amount of energy (latent heat of evaporation and condensation) found in clouds, using hurricanes as a primary example. The article estimates that the latent heat energy liberated from Hurricane Andrew approximates the total energy consumption of the U.S. for 6 months!

8. A discussion of adiabatic processes, stable and unstable air masses, connects to the concept of relative humidity. This includes new figures to add greater illustration of these processes.

- Figures 7-16 depicts the forces acting on an air parcel causing upward or downward movement.
- Figure 7-17 illustrates the forces causing atmospheric movement by examining how hot air balloons rise and fall.
- Figure 7-18: an image of air heating by compression and cooling by expansion.
- New examples of the Dry Adiabatic Rate (DAR), Figure 7-19 a and b.

9. New data concerning the increase of iceberg calving caused by global warming.

10. A new summary and review section ends the chapter.

Key Learning Concepts

1. *Describe* the origin of Earth's waters, relate the quantity of water that exists today, and *list* the locations of Earth's freshwater supply.

2. *Describe* the heat properties of water and *identify* the traits of its three phases: solid, liquid, and gas.

3. *Define* humidity and the expressions of the relative humidity concept; *explain* dew-point temperature and saturated conditions in the atmosphere.

4. *Define* atmospheric stability and *relate* it to a parcel of air that is ascending or descending.

5. *Illustrate* three atmospheric conditions-unstable, conditionally unstable, and stable-with a simple graph that relates the environmental lapse rate to the dry adiabatic rate (DAR) and moist adiabatic rate (MAR).

Expanded Outline Discussion

The following headings (boldfaced) match some of the first, second, and third order headings in Chapter 7. The narrative under each heading contains information, sources, and anecdotal facts relating to portions of the chapter. Not all text headings are discussed.

WATER ON EARTH

Water initially was imparted to the planet from the accretion of icy comets as Earth formed and condensed. Water formed from subsurface combinations of the elements hydrogen and oxygen. Water subsequently outgassed from the planet, forming beneath the crust.

Water Quantity: In Equilibrium Worldwide

Figure 7-3 illustrates the basic distribution of water–ocean and freshwater. Table 7-1 then locates just the freshwater in subsurface and surface categories. Table 7-4 portrays freshwater lakes alone, showing that over a third of this water is in only 3 lakes: Baykal, Tanganyika, and Superior. The science of limnology studies the physical, biological, and chemical nature of lakes. Very few lakes comprise most of the water volume.

GLOBAL OCEANS AND SEAS

Oceanography is the integrative discipline that studies the ocean. Oceanography, much like geography, draws on many academic disciplines: geology, geophysics, geochemistry, physics, biology, meteorology, and climatology. From these fields, specialties in chemical, biological and physical oceanography are derived; also, marine geology, marine engineering, and public policy disciplines involving the law of the sea are important today. Figure 7-5 lists with reference numbers each of Earth's principal oceans and seas. The physical structure of the oceans is portrayed in an integrative illustration in Figure 7-6 (p. 184) with plots of temperature, salinity, dissolved carbon dioxide, and oxygen levels, with increasing depth.

The comparison of land and ocean hemispheres in Figure 7-2 can be demonstrated to class using a globe.

Figure 7-4 compares the four oceans as to area, volume, depth, and deepest point. The southern portions of the Indian, Atlantic, and Pacific is sometimes called the Southern Ocean. See: Institute of Oceanographic Sciences, Deacon Laboratory. *The FRAM (Fine Resolution Antarctic Model) Atlas of the Southern Ocean.* Wormley, Surrey: Natural Environment Research Council, 1991, for detailed description of currents and extent of the Southern Ocean.

UNIQUE PROPERTIES OF WATER

The text briefly establishes the unique properties of water: stability of the molecule, solvency ability, and extraordinary heat properties. The explanation for these remarkable properties resides at the molecular level with covalent bonding, polarity of the water molecule, and hydrogen bonding, respectively. Hydrogen bonding between water molecules establishes the heat characteristics of the changes in state between ice, water, and water vapor–that is, the amount of latent heat required to alter the hydrogen bonding. These states are portrayed in Figure 7-5. I have included three inset drawings to suggest what is going on at the molecular level with each phase. The properties we experience are dictated by this structure at the molecular level.

The concept of surface tension (created by hydrogen bonding) is driven home when you use a paper towel to blot up a spill, capillary action is visible as liquid creeps through the towel. Farmers practicing dry farming leave land fallow, or out-of-production, which allows the water content to build up. They lightly disk plow the soil surface to break up "capillary straws" in the soil, thus trapping the moisture in the soil for later planting.

Heat Properties

Ice, the Solid Phase

A soft drink forgotten in the freezer will eventually explode because the water in it expands with freezing (Figures 7-5 and 7-6). Students from lower latitudes may not know that a campus parking space in the northern states and Canadian provinces comes with an electrical outlet to run the engine-block heaters and that the parking fee comes with a monthly electric bill. The power of ice is such that an engine block is easily cracked by freezing water. Those of you who live at high latitudes or at high elevation are well aware of the need for insulation around water pipes in the winter to prevent damage. In the far north, landscapes are marked by patterns of heaving soil masses caused by freezing water. Frozen ground phenomena are introduced in Chapter 13 under physical weathering and detailed in the Focus Study of Chapter 17 titled "Periglacial Landscapes." This freeze-thaw action is an important physical weathering process in the breaking up of rock and its effects on soils.

The story of the Titanic's tragic loss and the hidden hazard of icebergs is evidence of the 0.91 density of ice compared with water. (Aside: The fateful trip of the Titanic is a powerful metaphor for this era as our society plunges through a sea of icebergs representing hidden risks, with passengers and crew expected to concern themselves with the arrangement of deck chairs and the songs the band is playing. Some minor corrections of speed and course are all that are needed for the successful continuation of our journey.)

In Figure 7-7 students can follow a gram of water across from the ice phase to liquid to the vapor phase, totaling up the calories needed at each step. Then a gram of vapor can be taken back through the phases to ice, thus releasing that same amount of energy to the environment.

HUMIDITY

Relative Humidity

It has been my experience that to begin this discussion with the indirect measure of humidity, that is, relative humidity, and then progress through the several ways of expressing water vapor content simplifies the concept of humidity for students. The saturation concept then forms the basis for understanding clouds and atmospheric stability.

In laboratory experiments, filtered air with no impurities can be cooled to a point where it reaches more than 400% relative humidity before water vapor begins to sublimate directly to ice crystals, a condition called supersaturation, whereas some hygroscopic surfaces will produce condensation at relative humidity as low as 65%.

Expressions of Relative Humidity: Vapor Pressure

Since boiling is simply rapid, or explosive, vaporization of water, how would the concept of vapor pressure and saturation vapor pressure help determine at what temperature water boils? At Yellowstone National Park in the United States, elevation 2350m (7700 ft), water boils at 92°C (199°F). Why? Lower air pressure with altitude creates a lower boiling temperature.

Specific Humidity

Another expression that is equivalent for all practical purposes to specific humidity is the water-vapor mixing ratio, which is the weight in grams of water vapor per kilogram of dry air. Since air weighs so much more than water vapor, this term is about the same as specific humidity. To derive relative humidity, you would compare the water-vapor mixing ratio (content) with the saturation water vapor mixing ratio (capacity) to determine the percentage of saturation.

Instruments for Measurement

A classroom demonstration of a sling psychrometer and psychrometric tables works well to illustrate these concepts. An enlarged psychrometric table on the overhead projector will assist the presentation if supplies are short for each student to have a booklet. The *Applied Physical Geography* lab manual presents these tables and offers some exercises requiring calculations of relative humidity.

If time permits, several measurements can be made on different lecture days to illustrate changes in relative humidity. If you have a recording hair hygrometer in the classroom changes can be noted over time. We have all noticed how relative humidity readings in the classroom rise during exams!

Global Distribution of Relative Humidity

Compare the graphs in Figure 7-14. Establish the low relative humidity readings (a) around 30° latitude (very warm air) yet note the higher water vapor content than air at 50° latitude (b).

ATMOSPHERIC STABILITY: Adiabatic Processes, Dry Adiabatic Rate (DAR), and Moist Adiabatic Rate (MAR)

Figure 7-20, the relationship between dry and moist adiabatic rates and environmental lapse rates, provides a basis for discussion. Note that the greater the slope of the lines, the greater the temperature decrease.

Remember that it is not proper to use the term *lapse* with the DAR and the MAR, as is sometimes done. The normal and environmental rates are the true lapse rates: rates of temperature decrease with increasing altitude. I hope the acronyms (DAR, MAR) prove to be helpful.

Stable and Unstable Atmospheric Conditions

Figure 7-21 develops three states of stability in three pairs of graphs. The column to the left has the three graphs in the style of Figure 7-20, tying together this section of the text. The three graphs in the column to the right display these conditions more pictorially. I used the same beginning conditions at ground level for easier comparison.

Often, orographic conditions will demonstrate conditionally unstable relationships. The air is stable as it moves across a valley and then as it is lifted orographically it adiabatically cools and achieves saturation when it becomes unstable. A visual line of cloud development along the mountains often marks the saturated conditions that occur with uplift.

Glossary Review for Chapter 7
(in alphabetical order)

adiabatic
dew-point temperature
dry adiabatic rate (DAR)
eustasy
glacio-eustatic
hair hygrometer
humidity
latent heat
latent heat of condensation

latent heat of vaporization
moist adiabatic rate (MAR)
outgassing
phase change
relative humidity
saturated
sling psychrometer
specific humidity
stability
sublimation
vapor pressure

Annotated Chapter Review Questions

1. Approximately where and when did Earth's waters originate?

All water on Earth was formed within the planet, reaching Earth's surface in an on-going process called outgassing, by which water and water vapor emerge from layers deep within and below the crust, as much as 25 km (15.5 mi) or more below Earth's surface (Figure 7-1). Various geophysical factors explain the timing of this outgassing over the past 4 billion years. In the early atmosphere, massive quantities of outgassed water vapor condensed and fell in torrents, only to vaporize again because of high temperature at Earth's surface. For water to remain on Earth's surface, land temperatures had to drop below the boiling point of 100°C (212°F), something that occurred about 3.8 billion years ago.

2. If the quantity of water on Earth has been quite constant in volume for at least 2 billion years, how can sea level have fluctuated? Explain.

Today, water is the most common compound on Earth, having achieved the present volume of 1.36 billion km^3 (326 million mi^3) approximately 2 billion years ago. This quantity has remained relatively constant, even though water is continuously being lost from the system, escaping to space or breaking down and forming new compounds with other elements. Lost water is replaced by pristine water not previously at the surface, water that emerges from within Earth. The net result of these inputs and outputs to water quantity is a steady-state equilibrium in Earth's hydrosphere. Despite this overall net balance in quantity, worldwide changes in sea level do occur and are called eustasy, which is specifically related to changes in volume of water and not movement of land. These changes are explained by glacio-eustatic factors (see Chapter 17). Glacio-eustatic factors are based on the amount of water stored on Earth as ice. Over the past 100 years, mean sea level has steadily risen and is still rising worldwide at this time. Apparent changes in sea level also are related to actual physical changes in landmasses called isostatic change, such as continental uplift or subsidence (isostasy is discussed in Chapter 11).

3. Describe the location of Earth's water, both oceanic and fresh. What is the largest repository of freshwater at this time? In what ways is this significant to modern society?

See Table 7-1. The greatest single repository of surface freshwater is ice. Ice sheets and glaciers account for 77.78% of all freshwater on Earth. Add to this the subsurface groundwater, and that accounts for 99.36% of all freshwater. The remaining freshwater, although very familiar to us, and present in seemingly huge amounts in lakes, rivers, and streams, actually represents but a small quantity, less than 1%.

4. Why might you describe Earth as the water planet? Explain.

Earth's waters exist in the atmosphere, on the surface, and in the crust near the surface, in liquid, solid, and gaseous forms. Water occurs as fresh and saline, and exhibits important heat properties as well as an extraordinary role as a solvent. Among the planets in the solar system, only Earth possesses water in any quantity.

5. Describe the three states of matter as they apply to ice, water, and water vapor.

Earth's distance from the Sun places it within a most remarkable temperate zone compared with the other planets. This temperate location allows all states of water to occur naturally on Earth: ice, in the solid phase; water, in the liquid phase; and water vapor, in the gas phase.

6. What happens to the physical structure of water as it cools below 4°C (39°F)? What are some of the visible indications of these physical changes?

As water cools, it behaves like most compounds and contracts in volume, reaching its greatest density at 4°C (39°F). But below that temperature, water behaves very differently from most compounds, and begins to expand as more hydrogen bonds form among the slower-moving molecules, creating the hexagonal structures shown in Figure 7-5. This expansion continues to a temperature of –29°C (–20°F), with up to a 9% increase in volume possible. As shown in Figure 7-6a, the rigid internal structure of ice dictates the six-sided appearance of all ice crystals, which can loosely combine to form snowflakes. The expansion in volume that accompanies the freezing process results in a decrease in density. Specifically, ice has 0.91 times the density of water, and so it floats. The freezing action of ice as an important physical weathering process is discussed in Chapters 13 and 17.

7. What is latent heat? How is it involved in the phase changes of water? What amounts of energy are involved?

For water to change from one state to another, heat energy must be added to it or released from it. The amount of heat energy must be sufficient to affect the hydrogen bonds between the molecules. For ice to melt, heat energy must increase the motion of the water molecules to break some of the hydrogen bonds. Despite the fact that there is no change in sensible temperature between ice at 0°C and water at 0°C, 80 calories are required for the phase change of 1 gram of ice to 1 gram of water. This heat, called latent heat, is hidden within the water and is liberated whenever a gram of water freezes. To accomplish the phase change of liquid to vapor at boiling, under normal sea-level pressure, 540 calories must be added to 1 gram of boiling water to achieve a phase change to water vapor. Those calories are the latent heat of vaporization. To summarize, taking 1 gram of ice at 0°C and raising it to water vapor at 100°C–changing it from a solid, to a liquid, to a gas–absorbs 720 calories (80 cal + 100 cal + 540 cal). Or, reversing the process, changing from 1 gram of water vapor at 100°C to ice at 0°C liberates 720 calories.

8. Take one gram of water at 0°C and follow it through to one gram of water vapor at 100°C, describing what happens along the way. What amounts of energy are involved in the changes that take place?

If we take a gram of water at 0°C, it will require 100 calories to raise the temperature of water to 100°C, one calorie for each degree of change in temperature. At 100°C, water is at its boiling point. To change water's phase from liquid to vapor, will require another 540 calories. This energy is called the latent heat of vaporization. The total amount of energy absorbed by water in this process was 640 calories, 100 cal + 540 cal = 640 cal. To take a gram of ice at 0°C and turn it into vapor at 100°C, would require 80 additional calories, this energy absorbed by the water is called the latent heat of melting, or the latent heat of fusion. Thus, to transform ice at 0°C to vapor at 100°C would require a total of 720 calories, 80 calories to change from ice at 0°C to water at 0°C, plus 100 calories to heat water from 0°C to 100°C, and finally the additional 540 calories required to heat water at 100°C to vapor at 100°C, 80 cal + 100 cal + 540 cal = 720 cal.

9. What is humidity? How does it relate to the energy present in the atmosphere? To our personal comfort and how we perceive apparent temperatures.

Humidity is the vapor content of air. The vapor content of air changes according to the temperature of air and the temperature of water vapor. Warm air has a greater capacity to hold water than does cold air. Warm air stores a greater amount of energy in the atmosphere as the latent heat of evaporation, than cold air. Humidity affects the apparent temperature which we experience. For example, air with very little humidity, while it may be hot, is much more comfortable for humans. The warmth of the air enables the atmosphere to hold more moisture, and as we perspire, our bodies are cooled by this heat transfer. If we live in areas that are hot and humid, this means that the air is close to being saturated, and may not be able to hold more moisture. Our perspiration then may not be absorbed by the atmosphere, making the apparent heat seem much higher.

Humans attempt to adjust the humidity levels in our homes through humidifiers, which add vapor to the air, increasing apparent temperatures in the winter, and through the use of air conditioners, we can cool the air, extracting the vapor from the atmosphere, reducing the apparent temperatures.

10. Define relative humidity. What does the concept represent? What is meant by saturation and dew-point temperature?

The water vapor content of air is termed humidity. The capacity of air to hold water vapor is primarily a function of temperature: warmer air has a greater capacity for water vapor, whereas cooler air has a lesser capacity. Relative humidity is not a direct measurement of water vapor; rather, it is expressed as a percentage of the amount of water vapor that is actually in the air (content), compared with the maximum water vapor the air could hold at a given temperature (capacity). Air is said to be saturated, or full, if it is holding all the water vapor that it can hold at a given temperature; under such conditions, the net transfer of water molecules between surface and air achieves a saturation equilibrium. Saturation indicates that any further addition of water vapor (change in content) or any decrease in temperature (change in capacity) will result in active condensation.

The temperature at which a given mass of air becomes saturated is termed the dew-point temperature. In other words, air is saturated when the dew-point temperature and the air temperature are the same.

11. Using several different measures of humidity in the air, derive relative humidity values (vapor pressure/saturation vapor pressure; specific humidity/maximum specific humidity).

As the Figure 7-11 graph shows, air at 20°C (68°F) has a saturation vapor pressure of 24 mb; that is, the air is saturated if the water vapor portion of the air pressure is at 24 mb. Thus, if the water vapor content actually present is exerting a vapor pressure of only 12 mb in 20°C air, the relative humidity is only 50% (12 mb + 24 mb = 0.50 x 100 = 50%). The graph illustrates that, for every temperature increase of 10C° (18F°), the vapor pressure capacity of air nearly doubles.

The maximum mass of water vapor that a kilogram of air can hold at any specified temperature is termed the maximum specific humidity and is plotted in Figure 7-12 (p. 194). This graph shows that a kilogram of air could hold a maximum specific humidity of 47 g of water vapor at 40°C (104°F), 15 g at 20°C (68°F), and 4 g at 0°C (32°F). Therefore, if a kilogram of air at 40°C has a specific humidity of 12 g, its relative humidity is 25.5% (12 g + 47 g = 0.255 x 100 = 25.5%). Specific humidity is useful in describing the moisture content of large air masses that are interacting in a weather system.

12. How do the two instruments described in this chapter measure relative humidity?

The hair hygrometer uses the principle that human hair changes as much as 4% in length between 0 and 100% relative humidity. The instrument connects a standardized bundle of human hair through a mechanism to a gauge and a graph to indicate relative humidity. Another instrument used to measure relative humidity is a sling psychrometer. Figure 7-13 shows this device, which has two thermometers mounted side-by-side on a metal holder. One is called the dry-bulb thermometer; it simply records the ambient air temperature. The other thermometer is called the wet-bulb thermometer; it is set lower in the holder and has a cloth wick over the bulb, which is moistened with distilled water. The psychrometer is then spun by its handle for a minute or two. The readings on the two thermometers can then be checked on a psychrometric table to determine relative humidity. The greater the difference between the two thermometers, the lower the relative humidity.

13. How does this latitudinal distribution of relative humidity compare with the actual water vapor content of air?

In terms of vapor pressure, the water vapor actually in the air above subtropical deserts at 30° latitude is about double the amount at higher latitudes such as 50°, yet the relative humidity in the subtropical regions is lower because the warmer air has a greater capacity to hold water vapor. Thus, the relatively dry air over the cloudless Sahara (warmer air) actually contains more water vapor than does the relatively moist air in the midlatitudes (cooler air).

14. Differentiate between stability and instability relative to a parcel of air rising vertically in the atmosphere.

Stability refers to the tendency of a parcel of air to either remain as it is or change its initial position by lifting or falling. An air parcel is termed stable when it resists displacement upward or, if disturbed, it tends to return to its starting place. On the other hand, an air parcel is considered unstable when it continues to rise until it reaches an altitude where the surrounding air has a density similar to its own. Determining the degree of stability or instability involves measuring simple temperature relationships between the air parcel and the surrounding air.

15. What are the forces acting on a vertically moving parcel of air? How are they affected by the density of the air parcel?

There are two opposing forces which act on a vertically moving parcel of air. These forces are an upward bouyant force and a downward gravitational force. Temperature and density characteristics of the air mass determines which force will be more influential. If an air mass is higher in temperature than the surrounding atmosphere or if an air mass is less dense than the surrounding air, it will continue to rise vertically. Similar to any gas, as it rises in the air, the air mass begins to expand due to decreasing pressure in higher altitudes of the atmosphere. An unstable air mass will continue to rise until the surrounding air has similar density and temperature characteristics. An air mass will become dominated by the downward pull of gravity as it cools in the atmosphere. The temperature and density of the air mass may be lower than the surrounding air. This will cause the air mass to descend. As the parcel falls, external pressure in the atmosphere increases causing the air mass to compress.

16. How do the adiabatic rates of heating or cooling differ from the normal lapse rate and environmental lapse rate?

The normal lapse rate, as introduced in Chapter 3, is the average decrease in temperature with increasing altitude, a value of 6.4°C per 1000 m

(3.5°F per 1000 ft). This rate of temperature change is for still, calm air, but it can differ greatly under varying weather conditions, and so the actual lapse rate at a particular place and time is labeled the environmental lapse rate. Within this environment, an ascending (lifting) parcel of air tends to cool by expansion, responding to the reduced pressure at higher altitudes. Descending (subsiding) air tends to heat by compression.

The dry adiabatic rate is the rate at which dry air cools by expansion (if ascending) or heats by compression (if descending). Dry air is less than saturated, with a relative humidity of less than 100%. The DAR is 10°C per 1000 m (5.5°F per 1000 ft), as illustrated in Figure 7-19. The moist adiabatic rate is the average rate at which air that is moist (saturated) cools by expansion (if ascending) or heats by compression (if descending). The average MAR is 6°C per 1000 m (3.3°F per 1000 ft), or roughly 4°C (2°F) less than the DAR. However, the MAR varies with moisture content and temperature, and can range from 4°C to 10°C per 1000 m (2°F to 6°F per 1000 ft).

17. Why is there a difference between the dry adiabatic rate (DAR) and the moist adiabatic rate (MAR)?

In a saturated air parcel, latent heat of condensation is liberated as sensible heat, which reduces the adiabatic rate of cooling. The release of latent heat may vary, which affects the MAR. The MAR is much lower than the DAR in warm air, whereas the two rates are more similar in cold air.

18. What would atmospheric temperature and moisture conditions be on a day when the weather is unstable? Stable? Relate in your answer what you would experience if you were outside watching.

On a day when the weather is unstable, the atmosphere is dominated by warm air which may absorb available moisture and reach the level of saturation. These warm air parcels may rise vertically causing them to cool, condense and precipitate. You would experience evaporation during the warm period of the day, perhaps beginning the morning with a cloudless sky and seeing clouds form throughout the day as the air warms and the humid parcels of air rise in the atmosphere, cooled by the environmental lapse rate. As the parcels cool in the atmosphere, or after 4 pm, when the atmosphere begins to cool, precipitation may occur.

On a day when the weather is stable, there is no vertical movement of air. The atmosphere remains cool and dry, due to the lower capacity of cool air to hold moisture. You would experience a cloudless sky, or perhaps a greenhouse effect caused by pollution or clouds which may cool the lower atmosphere.

Overhead Transparencies

As an adopter you are provided with the following figures for overhead projector use.

- Figure 7-5: The three states of water and water's phase changes
- Figure 7-7: Water's heat-energy characteristics
- Figure 7-20: Temperature relationships and atmospheric stability
- Figure 7-21: Stability-three examples

8

Weather

Overview

We begin our study of weather with a discussion of atmospheric moisture and clouds. Clouds form as air becomes water-saturated and are more than whimsical, beautiful configurations; they are important indicators of overall atmospheric conditions. We follow huge air masses across North America, observe powerful lifting mechanisms in the atmosphere, examine cyclonic systems, and conclude with a portrait of the violent and dramatic weather that occurs in the atmosphere. Temperature, air pressure, relative humidity, wind speed and direction, daylength, and Sun angle are important measurable elements that contribute to the weather. We tune to local stations for the day's weather report from the National Weather Service (in the United States) or the Atmospheric Environment Service (in Canada) to see the current satellite images and to hear tomorrow's forecast.

The *Student Study Guide* presents 20 "Learning Objectives" to guide the student in reading the chapter. The *Applied Physical Geography* lab manual has portions of two exercises with several steps that involve aspects of this chapter.

New to the Third Edition

(Note: This section highlights major changes, new features, and additions in the third edition. This does not describe all the rewrite and recast of the text.)

1. A list of key learning ideas begins the chapter.

2. Discussion of clouds and fog have been moved from Chapter 7.

3. News Report #1: "Harvesting Fog," examines the adaptation of plants and animals to fog environments. Human technologies to collect condensation and increase local fresh water supplies are also addressed.

4. Figure 8-2 illustrates the scale and size of moisture droplets and raindrops.

5. Convergent lifting is described and illustrated in Figure 8-14.

6. News Report #2: "Mountains Set Precipitation Records," describes the ability of orographic precipitation processes to yield some of the highest recorded statistics for precipitation.

7. The rainshadow effect created by orographic lifting in Washington state is illustrated in Figure 8-18.

8. A more elaborate discussion of air mass interaction explains the maritime tropical air masses which dominate the southeastern region of the United States.

9. The structure and development of wave cyclones are given greater attention, describing easterly waves and connecting to rainy seasons in West Africa.

10. Figures 8-13, 8-22, and 8-31 illustrate the development and location of cyclone activity.

11. Focus Study 1 summarizes the 1995 hurricane season using recent statistics and satellite images (Focus Study, Figures 1 and 2).

12. New illustrations of tornado formation and damage are found in Figures 8-27, 8-28, and 8-29.

13. Recent statistics concerning tornadoes, hailstorms and lightning strikes.

14. The lake-effect of cP and cA air masses moving over the Great Lakes is fully discussed. A snowfall map reflecting this process is found on Figure 8-13.

15. A new summary and review section ends the chapter.

Key Learning Concepts

1. *Identify* the necessary requirements for cloud formation and *explain* the major cloud classes and types.

2. *Identify* the basic types of fog and *explain* the conditions that lead to their formation.

3. *Describe* air masses that affect North America and *relate* their qualities to source regions.

4. *Identify* types of atmospheric lifting mechanisms and *describe* four principal examples.

5. *List* measurable elements that contribute to weather.

6. *Analyze* various types of violent weather and the characteristics of each.

Expanded Outline Discussion

The following headings (boldfaced) match some of the first, second, and third order headings in Chapter 8. The narrative under each heading contains information, sources, and anecdotal facts relating to portions of the chapter. Not all text headings are discussed.

CLOUDS AND FOG

Slides are invaluable to identify the types and classes of clouds. The color photos in the text are a starting point, but your slides will be the thing that works, with your experiences and insights added. We have all let some beautiful cloud examples float by uncaptured on film–so keep a camera handy for shots, and after a few years, you will have an inventory.

Cloud Formation Processes

Ask your students if they have ever noticed that when they open the freezer door, moisture droplets form in the kitchen air. Our houses always contain condensation nuclei. The cold air from the freezer chills the kitchen air to the dew point, causing water vapor to condense on each available nucleus. If you wave your hand through the "cloud" in the kitchen, you cannot feel or see any individual moisture droplets. This should score the point that clouds basically are composed of microscopic droplets.

To form a raindrop, a cloud must be of a height (at least a kilometer or two) to allow the growing droplet to circulate for half an hour or so. As the cloud of droplets develop, simple condensation is replaced by coalescence as the principal process of raindrop formation.

Cloud Types and Identification

Figure 8-4 is in the overhead transparency packet. This figure presents eight representative cloud photographs. This figure relates to the descriptive material on clouds in Table 8-1, which includes the weather map symbols for each cloud type.

Fog

Fog is quite rare within twenty or so degrees on either side of the equator, where temperature inversions are infrequent. Fog layers will usually not clear away until the responsible inversion layer dissipates or is swept away. The phrase often used for this is, "the fog burns off." Obviously, fog does not burn off; rather, the temperature of the air warms as the morning progresses. Warmer air has a greater capacity to hold water vapor than does cooler air, allowing the fog to evaporate.

Figure 8-10 presents the mean annual number of days with heavy fog for the United States and Canada. Note the long caption giving details where the maximum fogs are experienced. Photos of areas which experience evaporation fog, valley fog and radiation fog are included in Figures 8-7, 8-8, and 8-9 respectively.

If the area where you live experiences fog during some time of the year and if there are any tall television broadcasting towers near your town, you can check this out. The engineers who service the antenna might let visitors go up the elevator in the tower. If not, they can be interviewed and they might place a min/max thermometer at the top of the tower, which you can compare with one placed at the bottom. Your local broadcasting "weather personality" might be interested in having such an installation.

AIR MASSES

Bjerknes, and others, developed the "Norwegian methods" for examining air mass properties: their characteristics, the nature of their movements, and the development of "fronts" and "frontal-wave" disturbances. Air masses provide an effective basis for weather analysis and for airmass climatology.

Air Masses Affecting North America

Figure 8-11 showing air masses that influence North America in a winter pattern and a summer pattern is in the overhead-transparency packet. Figure 8-11 provides sea-surface temperature patterns in C, details of each air mass source region and their general characteristics.

ATMOSPHERIC LIFTING MECHANISMS:
Convectional Lifting

A film or tape loop of satellite images spanning several days will demonstrate the afternoon and early evening build-up of convectional activity over land. This is when a classroom video capability can be invaluable. A bit later I mention the PBS weather show called "A. M. Weather" produced by Maryland Public Television. They present complete *GOES* satellite loops each morning that can be taped and used in class.

Orographic Lifting

The important role that mountain ranges play in producing precipitation is demonstrated in Figure 9-6 showing the annual precipitation for the United States and Canada, and Figure 10-3, showing world average annual precipitation.

Orographic Precipitation World Records

Relative to the precipitation record on Mount Waialeale, Kauai mentioned in the text (annual average of 1234 cm, 486 in. or 40.5 ft), the rain gauge is at the 1547m (5075.5 ft) elevation and has been for the period of 1941 to the present. The previous record, one that is often cited in texts and articles, is for the 1931 to 1960 period–for a total of 1168 cm (460 in.) a year. The 1981-82 season received the highest precipitation for this station for a single twelve month period, a whopping 1585 cm (624 in.)! Check the updated publication about these and other records: Pauline Riordan and Paul G. Bourget, *World Weather Extremes*, ETL-0416. Fort Belvoir, VA: U.S. Army Corps of Engineers, Engineering Topographic Laboratories, December 1985 (updated previous edition dated January 1970.)

You might want to obtain a copy of *Climate of the States–Hawaii*, Environmental Data Service, Washington: U.S. Government Printing Office; it includes a map of the islands, isohyet maps of each island, tables of data, and bibliography.

The record precipitation for contiguous North America falls to Canada, where the focusing effects of local topography and orographic lifting on the west side of Vancouver Island produced an average annual amount at Henderson Lake, British Columbia of 650 cm (256 in.).

MIDLATITUDE CYCLONIC SYSTEMS

(including weather information sources)

Many sources and materials are available for this section on weather. Of course, the best is to have a facsimile machine or on-line computer and small satellite dish or dedicated phone line so that you can receive up-to-the-minute NAFAX Service transmissions of images, weather maps, radar soundings, upper air charts, ocean temperatures, and Palmer Moisture Index maps for the United States from the National Weather Service. Or, a computer modem, such as Accu-Weather®, with data downloading into a computer for later playback and use, although a luxury, is quite effective in class as you can have current images and maps as weather events occur.

An excellent book on global weather, satellite meteorology; data handling, interpreting images; measuring rainfall, the oceans, snow and ice; weather forecasting; climate change; and what the future holds is W.J. Burroughs', *Watching the World's Weather*, New York (London): Cambridge University Press, 1991.

Satellite images of weather phenomena and weather movies are available from the Internet at:

http://rs560.cl.msu.edu/weather/ as well as local television stations Web sites or USA Today's Web site. The Defense Meteorological Satellite Program archives information related to weather, ocean, and Earth. Contact them at:

http://web.ngdc.noaa.gov/dmsp/dmsp.html. Past hurricane radar images and maps that tract the path of tropical storms are available at:

http://thunder.atms.purdue.edu/hurricane.html. The National Center for Atmospheric Research is on the net. Contact them at:

http://http.ucar.edu/metapage/html. And the University of Illinois is making itself the mecca of weather information, plug into:

http://www.atmos.uiuc.edu/wxworld/html/general.html.

"A.M. Weather," originating from Maryland Public Television, co-produced by NOAA and AOPA, as presented on most Public Broadcasting System (PBS) stations offers a 15-minute, detailed weather analysis with satellite loops and five-day forecasts. The service is usually broadcast on most affiliates in the morning (6:00 A.M.). "The Weather Channel" is on most cable systems and is a commercial outlet for local, national, and international coverage, including *GMS* and *METEOSAT* images. They publish an annual "Weather Guide Calendar" that highlights specific weather data and events for the year. The Weather Channel also creates 15 minute educational segments about weather phenomenon, such as easterly waves, cyclones, tornadoes, etc. The Weather Channel runs these series every day at a set time, so that you could program your VCR and have a number of weather videos correlating to your lectures.

The new weather satellites. During the lifetime of this edition of *GEOSYSTEMS*, it is entirely possible, if there are any further delays in satellite development and launches, that the Western Hemisphere could have a period without a U.S.-made weather satellite in operation! The present *GOES-7* (Geostationary Operational Environmental Satellite-East), with thruster-fuel conservation measures presently being taken, can operate only for a year or so. The western satellite finished its lifecycle several years ago, necessitating moving the remaining satellite westward to maintain coverage of the Pacific and Atlantic coasts. In the past few years, with increased hurricane strength, record numbers of tornadoes in 1991, and increasingly complex weather patterns emerging, the loss of this visible- and infrared-surveillance is very serious. To take up some of the observational void the U. S. borrowed *METEOSAT-3* from the European Space Agency for Atlantic surveillance.

The launch of *GOES-NEXT*, the first of a new generation of weather satellites, has been chronically delayed with a present target launch date in April 1994. The innovations in the new system are causing the difficulty, such as the major switch from spinning satellites to three axis-stabilized satellites that remain oriented with the same side facing Earth. A possible fall-back position could be to resuscitate the old GOES-design assembly-line, but the gap of operational-satellite coverage probably would still happen. (See: David Hamilton, "Will GOES-NEXT Go Next?" *Science* Vol. 253 No. 5016 (12 July 1991): 133. And, R. Cowen, "Launch Delays Jeopardize Weather Forecasts," *Science News* Vol. 140 No. 1 (6 July 1991): 5. Also, for additional background, see W.L. Smith, *et.al.*, "The Meteorological Satellite: Overview of 25 Years of Operation," *Science* Vol. 231, January 31, 1986: 455-462.)

If the April 1994 rocket launch of *GOES-NEXT* is successful it will assume the eastern position and *GOES-7* will be moved to the western position for its short remaining life. The next launch of the new design satellite is *GOES-J* sometime in 1995.

Images from the *GMS* satellite are obtained through the Meteorological Information Center/ Japan Weather Association/ Kaiji, Centre Building/ 5, 4-Chome, Kojimachi, Chiyoda-Ku/ Tokyo 102 Japan. For *METEOSAT* images and information contact METEOSAT Data Services, METEOSAT Exploitation Project, European Space Operations Centre, Robert-Bosch-Strasse 5, 6100 Darmstadt, Germany. Both sources are very helpful with filling orders and information requests in a timely manner.

The National Climate Data Center, of the National Environmental Satellite, Data, and Information Service, which is part of the National Oceanic and Atmospheric Administration, Depart-ment of Commerce, is located in Asheville, North Carolina 28801, and publishes a "Daily Weather Map" series on a weekly basis. This features a detailed surface map, 500-millibar chart, highest and lowest temperatures, and precipitation areas and amounts for the week. From the same data center, you can obtain a poster entitled "Explanation of the Daily Weather Map" which is a guide showing all the standard symbols presently used. These poster-charts are available in single copies (free) or in lots of 50 at a very low price.

A subscription to *Weatherwise* ("The Magazine About the Weather"), is helpful, for it contains interesting articles, annual reviews of hurricanes and tornadoes, an annual weather photo contest, and historical information. It is published 6 times a year by the Helen Dwight Reid Foundation in association with the American Meteorological Society, Heldref Publications, Washington, DC, 1-800-365-9753). Other periodicals of interest are the *Bulletin of the American Meteorological Society* (monthly), *Journal of Atmospheric Science*, and *Monthly Weather Review* from the American Meteorological Society; *Weather* (monthly), from the Royal Meteorological Society; and NOAA (bimonthly), from the Office of Public Affairs NOAA. (See Appendix B in the text for addresses.)

An idealized cycle of a midlatitude wave cyclone is presented in Figure 8-22 and its characteristic storm tracks in

Figure 8-23. Note the various regions of cyclogenesis on the latter map.

Cyclogenesis: The Birth of a Cyclone

A generalized map of cyclonic storm tracks for North America is presented in Figure 8-23. During Geography Awareness Week in November 1987, thousands of geography students throughout the United States launched helium-filled balloons. They generally followed the same upper-air winds that direct storm systems across the country–in other words, the balloon tracks shown in the map below depict the prevailing storm tracks for those days. These balloon tracks follow the prevailing surface and tropospheric air patterns. The data used to prepare this map were provided by the Geography Education Program, National Geographic Society Alliances Program.

The National Climatic Data Center (NCDC), National Environmental Satellite Data, and Information Service (NESDIS), NOAA, publishes a monthly report titled *Storm Data–and Unusual Weather Phenomena with Late Reports and Corrections* (ISSN 0039-1972), edited by Grant W. Goodge. (NCDC is at 37 Battery Park Avenue, Asheville, NC 28801-2733, 704-271-4800.) In this publication a map is published showing cyclogenesis and cyclonic storm tracks during the month. The cyclonic track maps for December 1990 (v. 32, no. 12) and July 1992 (v. 34, no. 7) are presented on the next page. You may want to make overheads of these for comparison to Figure 8-23.

Daily Weather Map and the Midlatitude Cyclone

The weather map presentation in Figure 8-24 gives you three daily weather maps and three matching *GOES-7* satellite images (in enhanced infrared wavelengths), with standard weather station symbols. I hope this compact presentation will provide you with the basis for class discussions.

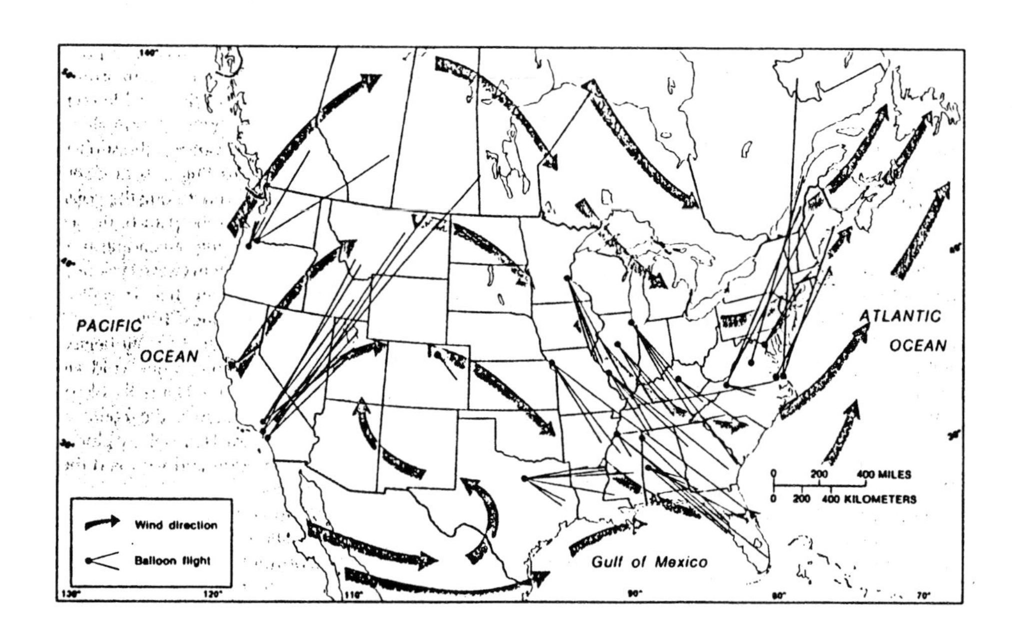

From the National Climatic Data Center (NCDC), National Environmental Satellite Data, and Information Service (NESDIS), NOAA, monthly report*Storm Data–and Unusual Weather Phenomena with Late Reports and Corrections* , edited by Grant W. Goodge, December 1990 (v. 32, no. 12) and July 1992 (v. 34, no 7).

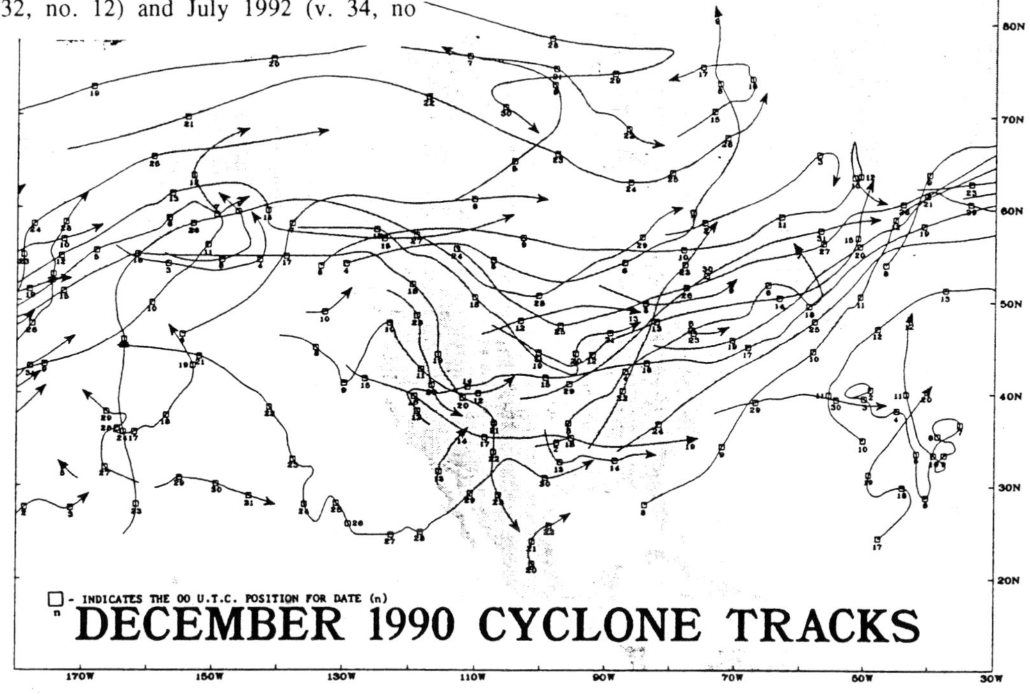

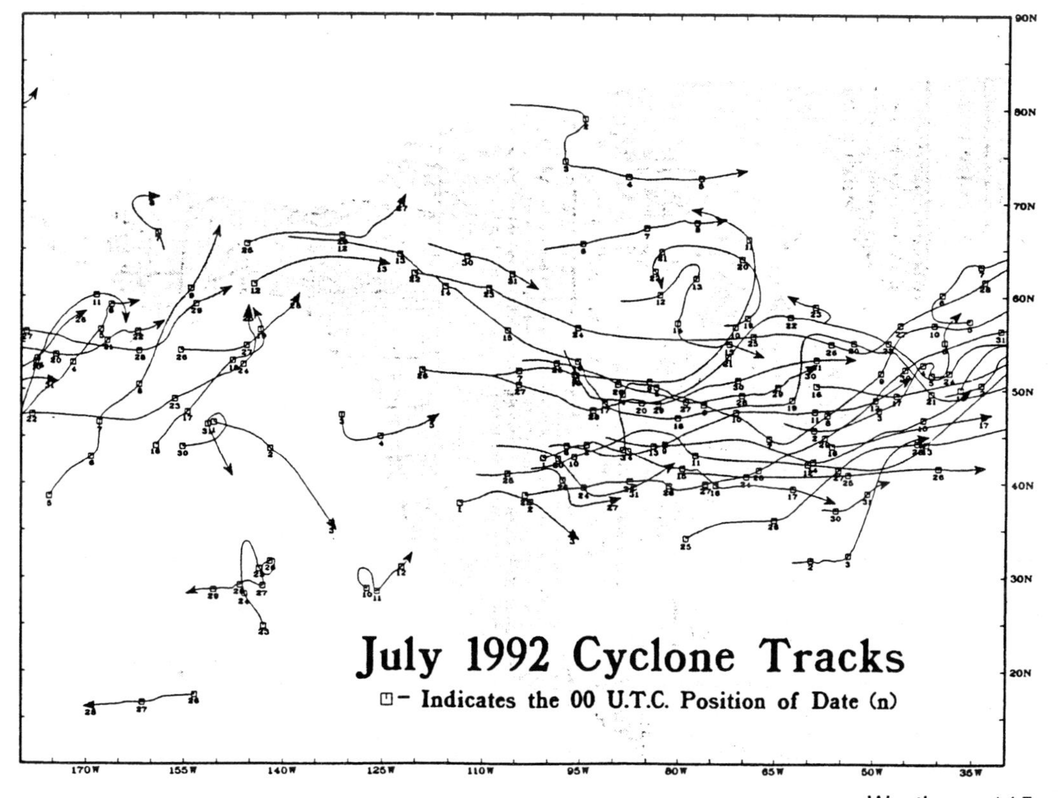

I obtained these maps and images from National Climate Data Center and the Satellite Data Services Division of the NESDIS of NOAA. In the *Student Study Guide* and the *Applied Physical Geography* lab manual I present the students with a weather map for April 1, 1971 with air and dew-point temperatures, air pressure, wind direction, and state of the sky. They are instructed to analyze and label the air masses, sketch the isobars at 4 mb intervals, and add cold and warm fronts.

To forecast what will occur in the future is a challenge. Weather forecasters rely on several methods all facilitated by modern computers. Today's forecasts are termed *numerical weather predictions* reflective of the role of computers. A *trend* forecast examines the tendencies of the weather up to the moment in time and space and assumes a continuation of the pattern. An *analog* forecast method examines the times in the past when similar conditions existed and attempts were made to match the developing weather with these past average events. Computers allow a synthesis of aspects from various techniques, augmented by complete daily and hourly orbital satellite surveillance of Earth. See Richard Monastersky, "Forecasting Into Chaos–Meteorologists seek to foresee unpredictability," *Science News* Vol. 137 No.18 (5 May 1990): 280-282; and, Richard A. Kerr, "Forecasting the Weather a Bit Better–A new computer model," *Science* Vol. 228 (5 April 1985): 40-41.

VIOLENT WEATHER: Thunderstorms, Lightning, and Thunder

A rule of thumb to use in determining the distance a lightning strike is from your location assumes that the flash arrived instantaneously at the speed of light, whereas the sound traveled at the speed of sound, some *3 seconds per km (1090 ft per sec*, or *5 sec per mile*). Simply begin counting at the moment of the flash to determine the elapsed time before you hear the thunder. Given the number of seconds elapsed you will know the distance you are from the lightning in km, feet, or mile units given above speeds. Suffice it to say that if you experience no delay and witness a simultaneous flash and crack of thunder then you are in the wrong place at the wrong time!

Thunder is enhanced by greater moisture density within the cloud and by topography which can act further to reverberate the sound waves.

Hail

The largest authenticated hailstone in the world landed in Coffeyville, Kansas in 1970, measuring in circumference 44.5 cm (17.5 in.), and weighing 758 gm (1.67 lbs). Cedoux, Saskatchewan, holds the official record for Canada at 10.2 cm (4 in.), but it is believed that hailstones larger than this have fallen in Canada. Hail exceeding these examples has no doubt fallen, but worldwide observation records are just too inadequate for us to know for certain.

Tornadoes

For an overview and update (of tornado formation and evolution, thunderstorms, mesocyclones, why this geographic occurrence, multiple-vortex structure, and Doppler radar systems), see John T. Snow, "The Tornado," *Scientific American* (April 1984): 86-96. The Fujita scale of damaging winds is presented on the last page of this chapter.

A large tornado can contain several secondary vortices rotating individually as well as rotating within the larger tornado. A tornado actually forms suction prints on the ground, which afterward are seen as sequences of semicircular marks across a field.

Even though this text is being written in central California, far from the "tornado alley" of Oklahoma and Kansas, a tornado touched down just a block from this word processor on March 22, 1983. It was related to the incredible weather of the last intense "El Niño" phenomena–discussed in a focus study with Chapter 10. We were not at home and so no photos are included. It hit about 2 P.M., was moderate on the Fujita Scale, moved northeastward, hopping along and damaging about 30 homes and several businesses.

You may have noticed that the tornado presentation in Figure 8-30 is up-to-date through 1993 and the text is up-to-date through 1995. This was accomplished with the help of the staff, especially Mr. Leo A. Grenier, at the National Severe Storms Forecast Center, National Weather Service, Room 1728 Federal Building, 601 E. 12th Street, Kansas City, Missouri 64106. (For the reference to the *United States Government Manual* publication where this and every other government bureau and agency is listed, see Appendix B.)

For an analysis of tornadoes in Canada see M.J. Newark, "Tornadoes in Canada for the Period 1950 to 1979," CL 1-2-83, Downsview, Ontario: Analysis and Impact Division, Canadian Climate Centre, Atmospheric Environment Service, Environment Canada, 1983.

Tropical Cyclones

Much is still not known about the formation of the tropical cyclone, so that what is presented here represents a brief overview of present best thinking. Tropical cyclones require warm air and ocean surface temperatures. Lower water temperature thresholds range near 27°C (80.6°F). Tropical oceans often exceed 30°C (86°F). Temperature profiles in the atmosphere must be unstable to accelerate the necessary uplift as the system develops. Tropical cyclones do not form any closer than 5° latitude on either side of the equator. This indicates that the complete lack of the Coriolis (deflective) force at the equator inhibits cyclonic development. The tropical cyclone requires the Coriolis Force to deflect converging winds and initiate the inward spiraling along the easterly tropical wave. Tropical cyclones were once thought to simply represent a deepening and intensification of a migrating easterly wave with convergence and cumulonimbus cloud development on the back of the wave, diverging and clearing air in front of the wave. However, this simplistic view is not regarded as complete today. The cyclonic circulation must be imparted to the converging mass through several other factors, such as an interaction with residual cyclonic circulation migrating southward from higher latitudes and the presence of high-pressure divergence in the upper-air circulation that is necessary for ventilation ("a chimney") of the forming system.

Hurricanes, Typhoons, and Cyclones

The key federal agency dealing with hurricanes is the National Hurricane Center of the National Weather Service, NOAA, Gables One Tower, Room 631, 1320 S. Dixie Hwy, Coral Gables, FL 33146. They have data on tropical cyclones dating from 1886 and are helpful when severe events occur issuing press kits, news releases, etc.

The hurricane's eye is created by a column of air which heats by compression as it descends within the surrounding eyewall. The eye usually represents an area about 10% of the overall diameter of the hurricane. Citizens are warned that they should not go outside should they suddenly hear the intense noise stop, for their curiosity will surely lead to death. As quickly as the winds and rains ceased, they will begin again with little or no warning as the eyewall passes.

See Table 8-3 the Saffir-Simpson hurricane damage scale. Hurricane winds ranging from 64 to 113 knots (74 to 130 mph) cause damage to trees, roofs, and mobile homes in particular; from 114 to 135 knots (131 to 155 mph) damage to roofs, some buildings, mobile homes, and inland flooding occurs; and from 136 knots (155 mph) on up there is major severe damage to all structures. Average hurricanes range from 100 to 150 knots

(115 to 172 mph). Hurricane Andrew (featured in Chapter 8) was one of the few category five hurricanes this century. The damage estimates now exceed $20 billion making it the second most costly disaster in the history of the U.S.– the 1994 Northridge earthquake now estimated at $30 billion in damage.

Two excellent color maps showing first-wind and second-wind patterns, microbursts, and mini-swirl patterns for Hurricane Iniki and Hurricane Andrew are available, published as supplements in *NOAA Strom Data*, September 1992 (34, no. 9) and August 1992 (34, no. 8) respectively. The articles that accompany these maps were prepared by Theodore Fujita and contains photographs.

For further analysis see: Roger M. Wakimoto and Peter G. Black, "Damage Survey of Hurricane Andrew and Its Relationship to the Eyewall," *Bulletin of the American Meteorological Society*, 75, no. 2, February 1994: 189-200.

Glossary Review for Chapter 8 (in alphabetical order)

advection fog
air mass
altocumulus
Bergeron ice-crystal process
chinook winds
cirrus
cloud
cloud-condensation nuclei
cold front
collision-coalescence process
convectional lifting
convergent lifting
cumulus
cumulonimbus
cyclogenesis
evaporation fog
fog
funnel cloud
hail
hurricane
landfall
lightning
mesocyclone
meteorology
midlatitude cyclone
moisture droplet
nimbostratus
occluded front
orographic lifting
radiation fog
rain shadow
squall line
stationary front
stratocumulus
stratus
storm surge
storm track
thunder
tornado
tropical cyclone
typhoon
upslope fog
valley fog
warm front
waterspout
wave cyclone
weather

Annotated Chapter Review Questions

1. Specifically, what is a cloud? Describe the droplets that form a cloud.

Clouds are not initially composed of raindrops. Instead, they are made up of a multitude of moisture droplets, each individually invisible to the human eye without magnification. An average raindrop, at 2000 µm diameter (0.2 cm or 0.078 in.), is made up of a million or more moisture droplets, each approximately 20 µm diameter (0.002 cm or 0.0008 in.).

2. Explain the condensation process: What are the necessary requirements? What two principal processes are discussed in this chapter?

Under unstable conditions, a parcel of air may rise to where it becomes saturated, i.e., the air cools to the dew-point temperature and 100% relative humidity. Further cooling (lifting) of the air parcel produces active condensation of water vapor to water. This condensation requires microscopic particles called condensation nuclei, which always are present in the atmosphere. In continental air masses, which average 10 billion nuclei per cubic meter, condensation nuclei are derived from ordinary dust, volcanic and forest-fire soot and ash, and particles from combustion.

The collision-coalescence process predominates in clouds that form at above-freezing temperatures, principally in the warm clouds of the tropics. Initially, simple condensation takes place on small nuclei, some of which are larger than others and produce larger water droplets. As those larger droplets fall through a cloud, they combine with smaller droplets, gradually coalescing in size until their weight is beyond the ability of the cloud circulation to hold them aloft. The existence of ice crystals and supercooled water droplets is also a significant mechanism for raindrop production. Figure 7-11 shows that the saturation vapor pressure near an ice surface is lower than that near a water surface; therefore, supercooled water droplets will evaporate more rapidly near ice crystals, which then absorb the vapor. In this ice-crystal process, first proposed in 1928 by Swedish scientist Tor Bergeron, the ice crystals feed on the supercooled cloud droplets, grow in size, and eventually fall as snow or rain (Figure 8-3b). Precipitation in middle and high latitudes begins as ice and snow high in the clouds, then melts and gathers moisture as it falls through the warmer portions of the clouds.

3. What are the basic forms of clouds? Using Table 8-1, describe how the basic cloud forms vary with altitude.

See Table 8-1 and Figure 8-4. Clouds usually are classified by altitude and by shape. They come in three basic forms–flat, puffy, and wispy–which are found in four primary classes and ten basic types. Clouds that are developed horizontally and are flat and layered are called stratiform clouds. Those that are developed vertically and are puffy and globular are termed cumuliform. Wispy clouds usually are quite high, composed of ice crystals, and are labeled cirroform. These three basic forms occur in four altitudinal classes: low, middle, high, and those that are vertically developed occur across all altitudes.

4. Explain how clouds might be used as indicators of the conditions of the atmosphere? Of expected weather?

Clouds can be used to indicate conditions of the atmosphere and expected weather due to their ability to demonstrate the atmospheric stability, atmospheric temperature and moisture level, or relative humidity. The altitude at which clouds develop are good indicators of atmospheric temperatures, see Figure 8-4 and Table 8-1. Clouds low in elevation, such as stratus or cumulus clouds, reflect a cool atmosphere, which allows clouds to form at low altitudes in the troposphere. This may lead to condensation and precipitation of a saturated air mass. Cirrus clouds are a good example of this. As they thicken and sink to low elevations, they are often associated with oncoming storms. Vertically developed clouds, such as cumulus or cumulonimbus clouds, demonstrate air masses laden with moisture, and one could expect to see precipitation depending on their height and mass. Other cloud formations illustrate scarcity of moisture in the atmosphere, or a warm atmosphere that limits condensation in the lower atmosphere. Good examples of these clouds could be stratoculumous clouds often associated with clearing weather and wispy cirrus clouds high in the upper troposphere.

5. What type of cloud is fog? List and define the principal types of fog.

A cloud in contact with the ground is commonly referred to as fog. The presence of fog tells us that the air temperature and the dew-point temperature at ground level are nearly identical, producing saturated conditions. Generally, fog is capped by an inversion layer, with as much as 30C° (50F°) difference in air temperature between the ground under the fog and the clear, sunny skies above. By international agreement, fog is officially described as a cloud layer on the ground, with visibility restricted to less than 1 km (3300 ft).

As the name implies, advection fog forms when air in one place migrates to another place where saturated conditions exist. Another type of fog forms when cold air flows over the warm water of a lake, ocean surface, or even a swimming pool. An evaporation fog, or steam fog, may form as the water molecules evaporate from the water surface into the cold overlying air. A type of advection fog, involving the movement of air, forms when moist air is forced to higher elevations along a hill or mountain. This upslope lifting leads to cooling by expansion as the air rises. Radiation fog forms when radiative cooling of a surface chills the air layer directly above that surface to the dew-point temperature, creating saturated conditions and fog. This fog occurs especially on clear nights over moist ground.

6. Describe the occurrence of fog in the United States and Canada. Where are the regions of highest incidence?

According to Figure 8-10, fog occurs in greatest frequency along the coastlines of the Pacific, the North Atlantic and the Labrador Sea, with over 80, 80-180, and 90-150 days of fog per year in these respective locations. This is due to cold air temperatures causing evaporation fog in the Atlantic and Labrador regions. While advection fog, the movement of warm air along a cold body of water, occurs along the Pacific, due to the cool temperature of the California current. Advection fog also occurs along the Great Lakes, as warm air from the tropics travels over the Great Lakes giving regions such as Montreal more than 60 days of fog. Other inland locations receive a high number of days with fog, mostly associated with orographic or upslope fog where the warm, moist air is forced upslope and cools to create fog. Good examples of this are in the Cascades causing 100-200 days of fog in Oregon and Washington states, and causing more than 80 days of fog along the Appalachian Mountains.

7. How does a source region influence the type of air mass that forms over it? Give specific examples of each basic classification.

See Table 8-11 for details. Earth's surface imparts its varying characteristics to the air it touches. Such a distinctive body of air is called an air mass, and it initially reflects the characteristics of its source region. The longer an air mass remains stationary over a region, the more definite its physical attributes become. Within each air mass there is a homogeneity of temperature and humidity that sometimes extends through the lower half of the troposphere.

8. Of all the air masses, which are of greatest significance to the United States and Canada? What happens to them as they migrate to locations different from their source regions? Give an example of air-mass modification.

See Figure 8-11. As air masses migrate from their source regions, their temperature and moisture characteristics are modified. For example, a mT Gulf/Atlantic air mass may carry humidity to Chicago and on to Winnipeg, but gradually will become more stable with each day's passage northward. Similarly, temperatures below freezing occasionally reach into southern Texas and Florida, influenced by an invading winter cP air mass from the north. However, that air mass will have warmed from the –50°C (–58°F) of its source region in central Canada. In winter, as a cP air mass moves southward over warmer land, it is modified, warming especially after it crosses the snowline. If it passes over the warmer Great Lakes, it will absorb heat and moisture and produce heavy lake-effect snowfall downwind as far east as New York, Pennsylvania, New England, and the Maritime Provinces of Canada (see Figure 8-13).

9. Explain why it is necessary for an air mass to rise if thereis to be precipitation. What are the three main lifting mechanisms described in this chapter?

If air masses are to cool adiabatically (by expansion), and if they are to reach the dew-point temperature and saturate, condense, form clouds, and perhaps precipitate, then they must be lifted. Three principal lifting mechanisms operate in the atmosphere: convectional lifting (stimulated by local surface heating), orographic lifting (when air is forced over a barrier such as a mountain range), and frontal lifting (along the leading edges of contrasting air masses).

10. When an air mass passes across a mountain range, many things happen to it. Describe each aspect of a mountain crossing by a moist air mass. What is the pattern of precipitation that results?

See Figure 8-18. The physical presence of a mountain acts as a topographic barrier to migrating air masses. Orographic lifting occurs when air is pushed upslope against a mountain and cools adiabatically. Stable air that is forced upward in this manner may produce stratiform clouds, whereas unstable or conditionally unstable air usually forms a line of cumulus and cumulonimbus clouds. An orographic barrier enhances convectional activity and causes additional lifting during the passage of weather fronts and cyclonic systems, thereby extracting more moisture from passing air masses. The wetter intercepting slope is termed the windward slope, as opposed to the drier farside slope, known as the leeward slope, which is the location of a rain shadow. Moisture is condensed from the lifting air mass on the windward side of the mountain; on the leeward side the descending air mass is heated by compression, causing evaporation of any remaining water in the air. Thus, air can begin its ascent up a mountain warm and moist but finish its descent hot and dry.

11. What are the four principal lifting mechanisms that cause air masses to ascend, cool, condense, form clouds, and perhaps produce precipitation?

The four principal lifting mechanisms causing adiabatic cooling are convergent lifting, convectional lifting, orographic lifting and frontal lifting. Convergent lifting occurs along the ITCZ where air flows from areas of higher pressure towards areas of low pressure (pressure gradient force). The tradewinds and easterlies form this type of convergence, warming air masses which create high vertical cumulonimbus clouds and high amounts of precipitation. Convectional lifting is stimulated by local surface heating, often associated with differential heating between land and water surfaces. Maritime tropical air masses migrate out from the ITCZ transporting warm, moist unstable air to continental locations causing precipitation. Adiabatic lifting occurs in both convergent and convectional lifting as warm, moist air masses rise vertically in the atmosphere, and are cooled by the surrounding air. Orographic lifting occurs when air masses are forced over barriers, such as mountains. Such movement causes the vertical movement of air masses, which are cooled by the cold air at higher elevations. And frontal lifting takes place when air masses of differing temperature and humidity interact. Such collision is caused by the jet stream, affecting mostly midlatitude locations, as the jet stream pulls air masses from their source region into contact with other air masses. Adiabatic lifting occurs in frontal lifting when a cold air mass, or cold front, displaces warm air along the surface, or when a warm air mass or warm front slides over cooler air and is forced to rise vertically in the atmosphere.

12. Explain how the precipitation distribution in the state of Washington is influenced by the principles of orographic lifting.

Figure 8-18 illustrates the rain-shadow effect created by orographic lifting as it affects four locations in Washington state. The illustration examines Quinalt, Sequim, Rainier Paradise and Yakima, in order to illustrate the influence of mountain barriers upon the progression of air masses that migrate eastward across the state. Two of these locations, Quinalt and Rainier Paradise, are examples of the effects of orographic lifting on the windward side of mountain barriers. Quinalt, located on the western slope of the Olympic National Forest, receives 122.2 inches of precipitation each year, despite a relatively low elevation of 219 feet above sea level. Rainier Paradise, located on the Western slope of the Cascade Mountains, records 103.7 inches of rain per year at an elevation of 5550 feet. Sequim and Yakima are both located on the leeward slope of these mountain systems, where dry, descending air often desiccates these environments, and does not yield precipitation. Sequim, located in the Puget Trough at 180 feet above sea level, has an annual precipitation of 16.2 inches, while Yakima, bears the majority of the rain-shadow effect, receiving only 8.4 inches of precipitation a year, despite its elevation of 1061 feet.

13. Differentiate between a cold front and a warm front as types of frontal lifting.

The leading edge of an advancing air mass is called its front. The leading edge of a cold air mass is a cold front, whereas the leading edge of a warm air mass is a warm front. A cold front is identified on weather maps by a line marked with triangular spikes pointing in the direction of frontal movement along an advancing cP air mass in winter. Warmer, moist air in advance of the cold front is lifted upward abruptly and is subjected to the same adiabatic rates and concepts of stability/instability that pertain to all lifting air parcels. A warm front is denoted on weather maps by a line marked with semicircles facing the direction of frontal movement. The leading edge of an advancing warm air mass is unable to displace cooler, passive air, which is more dense. Instead, the warm air tends to push the cooler, underlying air into a characteristic wedge shape, with the warmer air sliding up over the cooler air.

14. How does a wave cyclone act as a catalyst for conflict between air masses?

Wave cyclones form a dominant type of weather pattern in the middle and higher latitudes of both hemispheres and act as a catalyst for air-mass conflict as they bring contrasting air masses into conflict. A migrating center of low pressure, with converging, ascending air, spiraling inwardly counterclockwise in the Northern Hemisphere (or converging clockwise in the Southern Hemisphere) draws surrounding air masses into conflict in the cyclonic circulation along fronts.

15. What is meant by cyclogenesis? Where does it occur and why? What is the role of upper-tropospheric circulation in the formation of a surface low?

A midlatitude cyclone, or extratropical cyclone, is born along the polar front, particularly in the region of the Icelandic and Aleutian subpolar low-pressure cells in the Northern Hemisphere. Strengthening and development of a wave cyclone is known as cyclogenesis. In addition to the polar front, certain other areas are associated with wave cyclone development and intensification: the eastern slope of the Rockies and other north-south mountain barriers, and the North American and Asian east coasts. As air moves downslope, the vertical axis of the air column extends, shrinking the system horizontally, intensifying wind speed. As the air travels downslope, it is deflected in a cyclonic flow, thus developing new cyclonic systems or intensifying existing ones.

16. Diagram a midlatitude cyclonic storm during its open stage. Label each of the components in your illustration, and add arrows to indicate wind patterns in the system.

See Figures 8-19, 8-21, 8-22, and 8-24.

17. What is your principal source of weather data, information, and forecasts? Where does your source obtain its data? Have you used the Internet and World Wide Web to obtain weather information?

Personal analysis and response.

18. After reading this chapter, please take a moment to assess your "Weather I.Q." How much did you know about the weather and weather forecasts before this chapter as compared to what you are aware of now when you see or read about the next day's weather? Make a brief list of questions or topics about which you want to know more. (Such questions about your learning process represent "critical thinking.")

Personal analysis and response.

19. What constitutes a thunderstorm? What type of cloud is involved? What type of air masses would you expect in an area of thunderstorms in North America?

Tremendous energy is liberated by the condensation of large quantities of water vapor. This process is accompanied by violent updrafts and downdrafts. As a result, giant cumulonimbus clouds can create dramatic weather moments–squall lines of heavy precipitation, lightning, thunder, hail, blustery winds, and tornadoes. Thunderstorms may develop within an air mass, along a front (particularly a cold front), or where mountain slopes cause orographic lifting. Important here are the mT air masses of the Gulf and Atlantic source region.

20. Lightning and thunder are powerful phenomena in nature. Briefly describe how they develop.

Lightning refers to flashes of light caused by enormous electrical discharges–tens to hundreds of millions of volts–which briefly ignite the air to temperatures of 15,000° to 30,000°C (27,000° to 54,000°F). The violent expansion of this abruptly heated air sends shock waves through the atmosphere creating the sonic bangs known as thunder. The greater the distance a lightning stroke travels, the longer the thunder echoes. Lightning is created by a buildup of electrical energy between areas within a cumulonimbus cloud or between the cloud and the ground, with sufficient electrical potential to overcome the resistance of the atmosphere and leap from one surface to the other–it is like a giant spark.

21. Describe the formation process of a mesocyclone. How is this development associated with that of a tornado?

The updrafts associated with a cumulonimbus cloud appear on satellite images as pulsing bubbles of clouds. Because winds in the troposphere blow stronger above Earth's surface than they do at the surface, a body of air pushes forward faster at altitude than at the surface, thus creating a rotation in the air along a horizontal axis that is parallel to the ground. When that rotating air encounters the strong updrafts associated with frontal activity, the axis of rotation is shifted to a vertical alignment, perpendicular to the ground. It is this spinning, cyclonic, rising column of mid-troposphere-level air that forms a mesocyclone. A mesocyclone can range up to 10 km (6 mi) in diameter and rotate over thousands of feet vertically within the parent cloud. As a mesocyclone extends vertically and contracts horizontally, wind speeds accelerate in an inward vortex (much as ice skaters accelerate while spinning by pulling their arms in closer to their bodies). A well-developed mesocyclone most certainly will produce heavy rain, large hail, blustery winds, and lightning; some mature mesocyclones will generate tornado activity.

22. Evaluate the pattern of tornado activity in the United States. Where is "tornado alley?" What generalizations can you make about the distribution and timing of tornadoes?

See Figure 8-30. Of the 50 states, 49 have experienced tornadoes, as have all the Canadian provinces. May is the peak month. A small number of tornadoes are reported in other countries each year, but North America receives the greatest share because its latitudinal position and shape permit conflicting and contrasting air masses to have access to each other. In the United States, 26,458 tornadoes were recorded in the 33 years between 1959 and 1992, with 723 causing 2615 deaths. In addition, these tornadoes resulted in almost 52,000 injuries and damages of over $16 billion (Figures 8-28, 8-29). Interestingly, 1991 set a new record for the most tornadoes in a single year. Of the 1990 tornadoes, a record 14 had winds exceeding 333 kmph (207 mph). Most tornadoes in the United States occur along a four state corridor called "tornado alley", including Texas, Oklahoma, Kansas and Nebraska.

Generally, tornadoes occur in conjunction with a mesocyclone characterized by large cumulonimbus clouds, hail, intense winds and lightening. Timing of tornadoes is still rather unpredictable, yet research using satellites, airplanes and surface measurements are now enabling us to predict the occurrence of tornadoes. With the use of Doppler radar, which can detect the specific flow of moisture in mesocyclones, only 15% of American tornadoes strike without warning. Doppler radar enables forecasters to give tornado warnings 30 minutes to one hour in advance of an oncoming storm.

23. What are the different classifications for tropical cyclones? List the various names used worldwide for hurricanes.

See Table 8-2 and 8-3 and Figure 8-32.

24. What factors contributed to the incredible damage figures produced by Hurricane Andrew? Why have such damage figures increased whereas loss of life has decreased over the past 30 years?

The second greatest dollar loss from any natural disaster in history was caused by Hurricane Andrew as it swept across Florida and on to Louisiana during August 24-27, 1992. Sustained winds were 225 kmph, with gusts to 282+ kmph (140 mph, 175+ mph)–one of the few category five hurricanes this century (Saffir-Simpson scale). Studies recently completed by meteorologist Theodore Fujita estimated that winds in the eyewall reached 320 kmph (200 mph) in small vortices. Property damage exceeded $20 billion.

The tragedy from Andrew is that the storm destroyed or seriously damaged 70,000 homes and left 200,000 people homeless between Miami and the Florida Keys. By April 1993, 60,000 people were still homeless and reconstruction was progressing slowly. The storm is causing continued losses from reduced property assessments. Approximately 8% of the agricultural industry in Florida's Dade County (Miami region) was destroyed outright–exceeding $1 billion in lost sales. About 25% of Louisiana's sugar cane crop was lost. Many plants were killed by wind-driven saltwater that desiccated (dried) leaves that were not already stripped by the winds.

New buildings, apartments, and governmental offices are opening right next to still-visible rubble and bare foundation pads. Unfortunately, careful hazard planning to guide the settlement of these high-risk areas has never been policy. Due to improved technologies that enable us to predict and forecast hurricanes more easily, expansion of urban areas into hazard zones has increased, so that loss of life has been reduced, yet damage is even greater.

Coastal, earthquake, and floodplain hazards, poor human perception of hazards, and the general lack of considered planning and zoning are discussed in Chapters 14, 12, and 16 of *GEOSYSTEMS*.

25. What forecast factors did scientists use to accurately predict the 1995 Atlantic hurricane season?

Listed in Focus Study 8-1.

Overhead Transparencies

As an adopter you are provided with the following figures for overhead projector use.

- Figure 8-4: Principal clouds (illustrated)
- Figure 8-11: Principal air masses
- Figure 8-15: Local heating and convection
- Figure 8-17: Orographic precipitation
- Figure 8-18: Orographic patterns in Washington state
- Figure 8-22: Idealized stages of a midlatitude cyclone (map and art combined)
- Figure 8-27: Mesocyclone and tornado formation
- Figure F.S. 8-1: 1995 Atlantic storm season

9

Water Balance and Water Resources

Overview

This chapter begins by examining the water balance, which is an accounting of the hydrologic cycle for a specific area with emphasis on plants and soil moisture. We discuss the nature of groundwater and look at several examples of this generally abused resource. Groundwater resources are closely tied to surface-water budgets. We also consider the water we withdraw and consume from available resources, in terms of both quantity and quality.

The *Student Study Guide* presents 24 "Learning Objectives" to guide the student in reading the chapter. The *Applied Physical Geography* lab manual has one exercise with six steps that involve aspects of this chapter.

New to the Third Edition

(Note: This section highlights major changes, new features, and additions in the third edition. This does not describe all the rewrite and recast of the text.)

1. A list of key learning concepts begins the chapter.

2. A new introduction discusses the political and environmental significance of water budgets.

3. Discussion of the hydrologic cycle has been incorporated from Chapter 7. Figure 9-2 illustrates the volume of water that moves through the hydrologic cycle, and Figure 9-3 portrays the movement of water through the atmosphere, biosphere and lithosphere.

4. Moved from Chapter 8, Focus Study 1: "Hurricane Camille, 1969 Water Balance Analysis Points to Moisture Benefits," discusses the benefits of moisture abatement provided by hurricanes.

5. News Report #1: "Middle East Water Crisis: Running On Empty," discusses groundwater consumption and availability in the Persian Gulf.

6. The soil-water balance equation is illustrated in Figure 9-4.

7. The soil-water balance equation is compared to a money budget in order to explain inputs, outputs, storage and balance concepts more clearly.

8. New sample water-budget regimes have been added to Figure 9-13, including (a) Berkeley, CA, (b) Saskatoon, SAS, and (c) Seabrook, NJ.

9. Methods used to determine water availability and usage in the United States and Canada are described in News Report #2: "Measuring Up the Water Resources."

10. Figure 9-18 illustrates the interaction between groundwater and streamflow. This includes a clear depiction of effluent and influent stream bases.

11. More elaborate discussion of water usage, including instream, consumptive and nonconsumptive uses. Updated statistics estimating agricultural water reserves are included.

12. News Report #3: "Personal Water Use," describes the amount of water Americans use on a daily basis, indicating urban vs. rural rates of consumption and agricultural vs. industrial usage patterns.

13. Figure 9-20 is a photograph of a desalinization plant in Saudi Arabia.

14. Groundwater pollution is discussed, including sources of contamination, methods of clean-up, and the politics associated with this issue.

15. Figure 9-25 depicts the pollution found in fouled waterways.

16. Table 9-3 illustrates the global availability of water, using a pie-chart format to represent each continent.

17. Description and examples of exotic streams.

18. A new summary and review section ends the chapter.

Key Learning Concepts

1. *Illustrate* the hydrologic cycle with a simple sketch and *label* it with definitions of each water pathway.

2. *Relate* the importance of the water-budget concept to your understanding of the hydrologic cycle, water resources, and soil moisture for a specific location.

3. *Construct* the water-balance equation as a way of accounting for the expenditures of water supply and *define* each of the components in the equation and their specific operation.
4. *Describe* the nature of groundwater and *define* the elements of the groundwater environment.

5. *Define* stream discharge using specific rivers as examples and *explain* the concept of an exotic stream using the Nile and Colorado Rivers.

6. *Identify* critical aspects of freshwater supplies for the future and cite specific issues related to sectors of use, regions and countries, and potential remedies for any shortfalls.

Expanded Outline Discussion

The following headings (boldfaced) match some of the first, second, and third order headings in Chapter 9. Not all text headings are discussed. The narrative under each heading contains information, sources, and anecdotal facts relating to portions of the chapter. Not all text headings are discussed.

To supplement your materials for water resources please get a copy of a first-ever supplement to National Geographic: *Water–The Power, Promise, and Turmoil of North America's Fresh Water*, A National Geographic Special Edition, v. 184, no. 5A, 1993. Page 120 of this supplement presents an entire page of many resources that would be useful in teaching. Sections include: "Supply–Sharing the Wealth of Water," and "California: Desert in Disguise;" "Development–When Human's Harness Nature's Forces," and "James Bay: Where Two Worlds Collide;" "Pollution–Troubled Waters Run Deep," and "The Mississippi River Under Siege;" "Restoration–New Ideas, New Understanding, New Hope." The issue includes a giant map supplement of water resources, surface and ground, for the United States. Call 1-800-638-4077 to order.

THE HYDROLOGIC CYCLE

The precise mechanics of the behavior of water in all its various aspects on Earth is the science of hydrology. Specific subdivisions are sometimes distinguished, such as engineering hydrology or applied hydrology. This discussion on the hydrologic cycle integrates the discussion of the atmospheric component within weather in Chapter 8, and water balance and water resources in Chapter 9. Pointing out each of these related sections should bring relevance to the illustration of the cycle.

As pointed out in the text, note the percentage of water that is surface runoff compared with the percentage that flows subsurface. Then examine these same categories in terms of absolute quantities in Figure 7-4 and Figure 9-2. The dynamic flow of rivers and streams and atmosphere is illustrated in comparison with the more sluggish rates of groundwater movement.

THE SOIL-WATER BUDGET CONCEPT

I include the discussion of soil-water balance in *GEOSYSTEMS* because of its importance to understanding water budgets, its pedagogical value, and its importance to understanding soil moisture regimes in Soil Taxonomy (Chapter 18). Understanding the soil-water budget forms the basis for water resource management decisions, including the use and maintenance of the groundwater resource. The basic water balance approach and bookkeeping (water accounting) procedure assists in understanding the nature of water inputs, outputs, and water storage at a site. In terms of teaching and learning, this chapter should tie together Chapters 7, 8, and 9, setting the stage for the synthesis chapter to follow–Chapter 10: Global Climate Systems.

A full treatment of the subject has been lacking from texts for too long. The Thornthwaite method for estimating potential evapotranspiration (POTET) is still popular because of the ease of obtaining input data and therefore its usefulness for water analyses over large geographic areas. In fact, a water balance was actually calculated for the entire Earth by T.E.A. van Hylckama in 1956 when he took a composite of thousands of sites and regions (*The Water Balance of the Earth*, Laboratory of Climatology, Vol. 9, No. 2, 1956). Let us confine our approach to a smaller unit area. The hydrologic cycle is a general portrait of Earth's water-flow system, whereas the water balance model allows us to examine the hydrologic cycle components in detail at a specific site for any unit of time.

The essential publication for this topic is Thornthwaite, Charles W., and John R. Mather. *The Water Balance*. Publications in Climatology, vol. 1. Centerton, NJ: Drexel Institute of Technology, Laboratory of Climatology, 1955. For additional overview and background, see the "Water Budget Analysis" entry in Oliver, John E., and Rhodes W. Fairbridge, eds. *The Encyclopedia of Climatology*. New York: Van Nostrand Reinhold Co., 1987. This is a good general reference for Chapters 7-10.

The Soil-Water-Balance Equation

A brief definition of each component of the water balance equation appears in Figure 9-4. Then, the discussion of each component is presented in the same sequence for organizational clarity. The students can follow along the equation as they read. Components of the equation are illustrated in Figure 9-4 to provide the students a flow chart of these processes.

The equation provides a convenient method of following the fate of the precipitation input. Students can use the Kingsport, Tennessee, data to work through individual months or the year (Tables 9-1, Figure 9-7, and 9-12).

Students also can perform the bookkeeping (accounting) procedure (PRECIP – POTET), using the various stations in Appendix A to determine the net supply or demand by month. If you assign field capacity values for soil-moisture storage, they can then proceed to work the interaction of this supply and demand against soil moisture to calculate the water balance.

One approach is to obtain the supply and demand data for your area in a dry year and a wet year for contrast and comparison. For more advanced students and labs, you can work from scratch with mean monthly air temperature, latitude, and estimates of soil moisture capacity for various rooting depths. See Thornthwaite, Charles W., and John R. Mather, *Instructions and Tables for Computing Potential Evapotranspiration and the Water Balance*. Publications in Climatology, vol. 3. Centerton, NJ: Drexel Institute of Technology, Laboratory of Climatology, 1957.

A complete list of Publications in Climatology, and of the publications mentioned here in particular, is available from C.W. Thornthwaite Associates, Route 1, Elmer, New Jersey 08318.

Potential Evapotranspiration: Determining POTET

Thornthwaite and Mather's great contribution is the concept of POTET and an easy method for its estimation from generally available data. Thornthwaite prepared a volume of tables to facilitate the calculation procedure, although time-saving computer formats are available today. See Mather, John R., editor, *The Measurement of Potential Evapotranspiration*, vol. 7, No. 1. Seabrook, New Jersey: Johns Hopkins University, 1954.

Another technique for the determination of POTET is through energy balance methods. The Penman Method is used in the agricultural sciences and in engineering, and is a more complex method based on four variables: temperature, wind, net radiation, and humidity. Although results are generally good, the raw data for the input are frequently not available. California (CIMIS Program) and Arizona (ASMET Program) are pressing ahead to increase the availability of Penman input data in their states.

Deficit
(Includes a section on drought)

Thornthwaite and Mather define drought at several different levels of intensity. The more arid areas of the west experience *permanent drought* measured in chronic deficits. During the year, some areas experience *seasonal drought* as patterns of precipitation occur in predictable rainy and dry seasons. *Irregular droughts* occur within the normal variability of precipitation from time to time. And finally, *invisible drought* is defined as an episode when all of POTET is not satisfied even though precipitation is occurring. Invisible drought is caused by the inefficiency of soil moisture removal and use by plants when the soil is less than field capacity. This subtle, almost unseen, drought can be just as damaging to plants and crop yields as visible drought.

NOAA and the National Weather Service publish weekly Palmer Crop Moisture Index reports that give soil moisture conditions for the United States. (See the "Weekly Weather and Crop Bulletin" published jointly by the U.S. Department of Agriculture and the Environmental Data Service of NOAA, and available in single copies or by subscription. The Palmer Crop Moisture Index is overviewed by its author, Wayne C. Palmer in "Keeping Track of Crop Moisture Conditions...The New Crop Moisture Index." *Weatherwise* (August 1968): pp. 156-162.) The Palmer system regards drought as a significant reduction in soil moisture below plant requirements. Palmer established a Drought Severity Index to rate water balance conditions. Severity ratings range from mild (-1), moderate (-2), severe (-3), to extreme (-4). These drought severity indices have measured as far back as A.D. 1700 through tree-ring analysis and demonstrate an interesting periodicity for midlatitude droughts in the American

West and Midwest that coincide with the Hale Sunspot Cycle.

By comparing precipitation and soil moisture with POTET, a much more accurate estimation of drought is possible than with the traditional method of reporting "days without rain." Deficit avoidance provides a useful basis for the management of irrigation applications, so that through careful management not only can obvious drought be deduced for a large area, but small changes in moisture availability to crops also can be estimated. Added irrigation water, if available, augments precipitation, is considered an input to the water balance calculation, and can be handled much as one would a simple bookkeeping procedure.

The drought of 1988 is a recent case in point. Record temperatures for this century took place over much of the United States, whereas precipitation was almost nonexistent except for some relief late in the summer. Most of the United States in 1988 recorded high POTET and low PRECIP and therefore experienced higher than normal soil moisture utilization. Many areas had deficits (unmet POTET) and eventually widespread soil moisture shortages from the Mississippi River to Colorado, and from Lake Tahoe to Texas. Some of the late summer soil moisture recharge that did take place was simply too late to benefit the crops.

One of the climatic effects predicted as a result of global temperature warming over the next century is a reduction of soil moisture in the midlatitudes. NOAA studies have focused on temperature increases and the changing patterns of soil moisture and deficits expected in the next century. Agricultural managers should be concerned about possible losses of soil moisture and increasing occurrences of invisible and irregular droughts if climatic warming continues.

Relative to water balances and drought in Africa, see Glantz, Michael H. "Drought in Africa," *Scientific American* Vol. 256, No. 6 (June 1987): pp. 34-40.

Surplus

Some aspects of urban water balance were discussed at the end of Chapter 4. One study describing the Thornthwaite model and cities was by Robert A. Muller, in "Water Balance Evaluation of Effects of Subdivisions on Water Yield in Middlesex County, New Jersey." *Proceedings of the Association of American Geographers*, Vol. 1, 1969, pp. 121-125. The sealed surfaces of cities produce high runoff values that tend to be underestimated by standard water balance models, whereas these models overestimate actual evapotranspiration. Because of all the sealed surfaces, runoff from urban settings is much higher than it is from undeveloped land. As a result, urban streams can flood readily following even modest rainfall or snowmelt, only to dry quickly when precipitation ceases. More discussion of impact of urbanization on hydrology is found in Chapter 14.

The Thornthwaite method is less satisfactory in areas of permafrost and below freezing temperatures, as analyzed by Dr. Peter J. Kakela in, "Thorthwaite's Climatic Water Balance: Evaluation of Annual Discharge Estimates for Two Subarctic Basins," Canadian Geographer, Vol. 27, No. 2 (1973): pp. 167-179.

Soil Moisture Storage

See Chapter 18 for specific application of the water balance to the Soil Taxonomy classification system.

Sample Water Budgets

Keep in mind that the data are averaged over the month in the text example. If you lived in Kingsport, Tennessee, and a strong thunderstorm hit during a July afternoon, one that rained 50 mm (2 in.), or almost half the monthly average PRECIP, a more accurate approach would be to analyze the water balance for the time span of this single event. Soils waterlogged by heavy precipitation will produce increased runoff and can lead to plant damage. An example of a catastrophic event balancing annual water deficits is discussed in Focus Study 9-1 describing the moisture brought by Hurricane Camille to Mississippi, Virginia, and Delaware.

For a complete sample of water balance and climatic information (PRECIP, POTET, DEFIC, SURPL, ΔSTRGE, ACTET), see C.W. Thornthwaite Associates, *Average Climatic Water Balance Data of the Continents* Vol. XVII, Publications in Climatology, Part V: Europe, Part VI: North America (without U.S.), Part VII: United States. Centerton, New Jersey: Laboratory of Climatology, 1964.

Water Budget and Water Resources

For more mundane water budget applications to which students might relate, see "How Much Water Does Your Lawn Really Need?" *Sunset Magazine* (June 1987): pp. 213-219. This article has an evapotranspiration chart and explanation, and although the focus is generally for the western United States the principles are universal.

As someone living in California, I could have used the California Water Project as an example of the geographical redistribution of water from the north to the south and over the calendar from January to July. Instead, I selected the Snowy Mountain Hydroelectric Scheme in Australia for the illustration of such a vast, multipurpose project. The C.W.P. is massive and even awesome in the water resources it delivers, yet difficulties both political and financial have followed it from the beginning, when voters approved $1.95 billion in bonds in November 1960. To date, the project has exceeded $15 billion in overall cost! Controversy is always at hand about the fair distribution of these captured water resources, with corporate farms, family farms, coop districts, or marketing utilities in the middle. Controversy also emerges from the general taxpayer, who has been asked to assume an enormous financial burden for the apparent benefit of a limited group of special agribusiness interests. Yet California leads the nation and most countries of the world in agricultural output, especially subtropical fruits, nuts, and field crops. The focus study in Chapter 18 introduces the selenium contamination problems associated with irrigation programs in arid and semiarid lands, in particular, California's San Joaquin Valley.

Marc Reisner's book, *Cadillac Desert,; The American West and Its Disappearing Water*, Penguin, 1987, is a must for anyone interested in the history of the California aqueduct system. Reisner also provides much information concerning the water use of various plants, a good connection to Chapters 19 and 20.

GROUNDWATER RESOURCES

The integrated illustration in Figure 9-16 runs along the entire upper portion of 2 pages of text to assist the student to see the overall relationships of groundwater resources and related human impacts. This illustration also appears in your overhead transparency packet. As students read and follow your lecture, they can consult this figure and see most of the basic groundwater structures.

Focus Study 9-2 on the Ogallala aquifer is critical as background to understand all that will be occurring there in the next 30 years. You may want to obtain a copy of the entire USGS Professional Paper 2275 (1984) that is referenced for it is a comprehensive examination of selected water-quality trends and groundwater issues. It is available from any GPO Bookstore or the USGS (see Chapter 1 in this manual).

Water problems in the Persian Gulf states relate to intensive groundwater mining as discussed on pp. 258, 260. See: Priit J. Vesilind, "Water The Middle East's Critical Resource," *National Geographic* , 138, no. 5, May 1993: 38-71. The article presents excellent regional maps and the usual outstanding photographs.

Pollution of the Groundwater

Obtain copies of the four GAO reports mentioned in this chapter's suggested readings for up-to-date background information from a political and legal perspective (see the GAO entry for the address, in Appendix A). See *Groundwater Newsletter*, Water Information Center, 125 E. Bethpage Road, Plainview, NY 11803. Also, see Gordon, Wendy, *A Citizen's Handbook on Groundwater Protection*, Washington, D.C.: Natural Resources Defense Council (1984). (NRDC is listed in Appendix A.) Boyd, Susan, Lynn Fuller, and Reed Wulsin, editors, *Groundwater: A Community Action Guide*, Washington, D.C.: Concern Inc. (1984), available from Concern Inc., 1794 Columbia Road NW, Washington, D.C. 20009.

SURFACE WATER RESOURCES: STREAMS

Chapter 14 discusses stream hydrology and related geomorphic landscape features. This section in Chapter 9 specifically relates to water as a resource. The section includes a discussion of exotic streams–note the inclusion of the Nile River in Table 9-2 despite its rank as 33 in volume. This table might work for a map exercise for students to locate the rivers listed in Table 9-2 and relate them to the global water supply in Table 9-3 and precipitation patterns in Figure 10-3. The area of internal drainage in the Western United States is related to the Basin and Range topography, which is mapped in Figure 15-22.

OUR WATER SUPPLY: Water Supply in the United States, Future Considerations

Using the water balance principles discussed, students can follow the flow paths for the water supply of the 48 states in Figure 9-23. Forty-eight states are used because Alaska and Hawaii's water is not accessible for the remaining 48 states. The three sectors of withdrawn water for the 48 states can be compared with the other countries highlighted in Figure 9-24. The correlation between these water-withdrawal-by-sector bar graphs and primary, secondary, and tertiary economic activity levels within each country can be made. Industrial withdrawal is divided to show usage-specific to steam-electric power stations for cooling and boiling water. A few comments for each sector of withdrawal demand appear below.

Industry. 356 BGD, 32 percent (114 BGD) is withdrawn for industrial process water, and 68 percent (242 BGD) for steam-electric power stations, principally for cooling. Water is a basic raw material in most industrial activity. Water is consumed through actual incorporation into the product, and is used for cooling and the dilution and transport of wastes, and indirectly to produce electricity. Less than one-half of industrial waste that flows into rivers is properly or legally treated. Steam-electric power stations use large amounts of roughly treated water for condenser cooling and smaller amounts of very pure water for boilers and steam generators. A 1000-megawatt plant running at high capacity requires about 300,000 gallons of water per day. Water returned to runoff from these operations is higher in temperature than it was when it was withdrawn, a condition called thermal pollution. Thermal loading of rivers, streams, and coastal ocean water results.

Irrigation. Most of the irrigated land in the United States is west of the 100th meridian of longitude (100°W); about 95 percent is in the arid and semi-arid West. The 50 cm (20 in.) isohyet (a line representing equal values of precipitation) is roughly synonymous with this historic dividing line between the tall (east) and short (west) grass prairies (see reference of this in Figure 18-17).

Irrigation withdrawals have increased by almost 300 percent since 1950 due to, among other factors, public demand for all year availability of fruits and vegetables, supplemental irrigation used for quality and production control, water oversupply used for soil maintenance and the flushing of salts and chemicals associated with agriculture in drier areas, and the great increase of farming in arid lands. Many strategies exist to extend agricultural water resources. Drip irrigation techniques are particularly suited to carefully timed releases of water calculated with water balance techniques. Also, improved metering and conveyance methods, and control of undesirable water-wasting vegetation can help reduce the growth in demand for irrigation water. Recycled water ("gray water") presents possibilities for use. And, of course, water conservation will occur if there are policies that insure we are growing crops that we actually need–ones that are not in chronic oversupply.

Municipalities. 56 BGD is withdrawn for municipal purposes, an amount which will increase to over 100 BGD by A.D. 2020. With more than 23,000 public water systems serving some 60,000 local governing bodies, the difficulty in mere data collection is formidable. And, as each municipality returns portions of its withdrawn water to the runoff stream, it becomes an input into all the water systems downstream. The major use of water in an average home is for personal hygiene, with over 70 percent being used for bathing and toilet operations. In drier areas of the country, summer landscape watering is important. Ways of extending the resource involve quite simple techniques of conservation (cutting use) and efficiency (designing fixtures and landscapes that use less water). With all sectors of withdrawal, motivation to change use patterns is simply implemented through rate pricing strategies that discriminate against waste and excessive use.

A challenge here is to have the students analyze their own daily water use. How many gallons a day are directly withdrawn? If you live in a rural area with your own well, how does your family monitor the water table level in the well? What do you think the difference is between water usage in a home with a septic tank vs. a home on a community sewer system? Is your landscaping appropriate to the climate in which you live, in terms of water demand? How is your water use monitored and billed? The county in which I reside still does not have water meters for residential customers, an incredible situation for a dry climate with chronic moisture deficits.

A fairly extensive survey article by Benedykt Dziegielewski and Duane D. Baumann, is titled "Tapping Alternatives: The Benefits of Managing Urban Water Demands" (*Environment*, 34, no. 9, November 1992: 6-11+). The article is focused on southern California but presents concepts applicable to any urban area. Instead of constantly looking at supply questions, water managers are beginning to analyze *demand management*! This is a unique approach despite the fact that it smacks of great common sense. See Figure 2 in the article (p. 10) for a sector analysis of municipal and industrial water use. This is a good piece to stimulate discussion.

For students to answer: Where does your tap water come from at home? The source? Do you know the name of the utility providing that water? Are there state or provincial laws that require them to provide you with their latest chemical analysis of that tap water? If so, you might want to ask them for a copy. Where does the water originate that you use while you are on campus? How is it regulated?

Glossary Review for Chapter (in alphabetical order)

actual evapotranspiration (ACTET)
aquiclude
aquifer
aquifer recharge area
artesian water
available water
capillary water
cone of depression

confined aquifer
consumptive uses
deficit (DEFIC)
discharge
drawdown
evaporation
evaporation pan
evapotranspiration
exotic stream
field capacity
gravitational water
groundwater
groundwater mining
hydrologic cycle
hygroscopic water
infiltration
interception
internal drainage
lysimeter
overland flow
percolation
permeability
piezometric surface
porosity
potential evapotranspiration (POTET)
precipitation
rain gauge
soil moisture recharge
soil moisture storage (ΔSTRGE)
soil moisture utilization
soil-water budget
surplus (SURPL)
total runoff
transpiration
unconfined aquifer
water detention
water table
wilting point
withdrawal
zone of aeration
zone of saturation

Annotated Chapter Review Questions

1. Explain the simplified model of the complex flows of water on Earth–the hydrologic cycle.

Vast currents of water, water vapor, ice, and energy are flowing about us continuously in an elaborate open global plumbing system. A simplified model of this complex system is useful to our study of the hydrologic cycle (Figure 9-1). The ocean provides a starting point, where more than 97% of all water is located and most evaporation and precipitation occur. If we assume that mean annual global evaporation equals 100 units, we can trace 86 of them to the ocean. The other 14 units come from the land, including water moving from the soil into plant roots and passing through their leaves. Of the ocean's evaporated 86 units, 66 combine with 12 advected from the land to produce the 78 units of precipitation that fall back into the ocean. The remaining 20 units of moisture evaporated from the ocean, plus 2 units of land-derived moisture, produce the 22 units of precipitation that fall over land. Clearly, the bulk of continental precipitation derives from the oceanic portion of the cycle.

2. What are the possible routes that a raindrop may take on its way to and into the soil surface?

Precipitation that reaches Earth's surface follows a variety of pathways. The process of precipitation striking vegetation or other groundcover is called interception. Precipitation that falls directly to the ground, coupled with that which drips onto the ground from vegetation, constitutes throughfall. Intercepted water that drains across plant leaves and down plant stems is termed stem flow and can represent an important moisture route to the surface. Water reaches the subsurface through infiltration, or penetration of the soil surface. It then permeates soil or rock through vertical movement called percolation.

3. Compare precipitation and evaporation volumes from the ocean with those over land. Describe advection flows of moisture and countering surface and subsurface runoff.

More than 97% of Earth's water is in the ocean, and here most evaporation and precipitation occur. 86% of all evaporation can be traced to the ocean. The other 14% comes from the land, including water moving from the soil into plant roots and passing through their leaves by transpiration. Of the ocean's evaporated 86%, 66% combines with 12% advected from the land to produce the 78% of all precipitation that falls back into the ocean. The remaining 20% of moisture evaporated from the ocean, plus 2% of land-derived moisture, produces the 22% of all precipitation that falls over land.

4. How might an understanding of the hydrologic cycle in a particular locale, or a soil-moisture budget of a site, assist you in assessing water resources? Give some specific examples.

A soil-moisture budget can be established for any area of Earth's surface by measuring the precipitation input and its distribution to satisfy the "demands" of plants, evaporation, and soil moisture storage in the area considered. A budget can be constructed for any time frame, from minutes to years. See Figure 9-13 for a specific examples.

5. What does this mean? "the soil-water budget is an assessment of the hydrologic cycle at a specific site."

A water balance can be established for any area of Earth's surface by calculating the total precipitation input and the total of various outputs. The water-balance approach allows an examination of the hydrologic cycle, including estimation of streamflow at a specific site or area, for any period of time. The purpose of the water balance is to describe the various ways in which the water supply is expended. The water balance is a method by which we can account for the hydrologic cycle of a specific area, with emphasis on plants and soil moisture.

6. What are the components that comprise the water-balance equation? Construct the equation and place each term's definition in sequence below the equation.

See figure 9-4.

7. Using the annual water-balance data for Kingsport, Tennessee, work the values through the water-balance "bookkeeping" method. Does the equation balance?

Students can use data from Figure 9-12 and Appendix A to perform the basic PRECIP – POTET bookkeeping (water accounting) procedure to determine net supply or demand. You will need to assign values for soil moisture storage such as the 100 mm (4.0 in.) used in Table 9-1 for Kingsport.

Comparing PRECIP with POTET by month determines whether there is a net supply (+) or net demand (–) for water. We can see that Kingsport experiences a net supply from January through May and from October through December. However, the warm days and months from June through September result in net demands for water.

8. Explain how to derive actual evapotranspiration (ACTET) in the water-balance equation.

The actual amount of evaporation and transpiration that occurs is derived by subtracting DEFIC, or water demand, from POTET. Under ideal conditions, POTET and ACTET are about the same, so that plants do not experience a water shortage. Droughts result from deficit conditions, where ACTET is greater than the available moisture.

9. What is potential evapotranspiration (POTET)? How do we go about measuring this potential rate? What factors did Thornthwaite use to determine this value?

POTET is the amount of moisture that would evaporate and transpire if the moisture were available; the amount lost under optimum moisture conditions–the moisture demand. Both evaporation and transpiration directly respond to climatic conditions of temperature and humidity. For the empirical measurement of POTET, probably the easiest method employs an evaporation pan, or evaporimeter (Figure 9-7). As evaporation occurs, water in measured amounts is automatically replaced in the pan. Screens of various-sized mesh are used to protect against overmeasurements created by wind. A lysimeter is a relatively elaborate device for measuring POTET, for an actual portion of a field is isolated so that the moisture moving through it can be measured.

Thornthwaite found that POTET is best approximated using mean monthly air temperature (measured with a thermometer) and daylength (a function of the measuring station's latitude). Both factors are easily determined, making it relatively simple to calculate POTET and the other water-balance components in the equation indirectly, with fair accuracy for most midlatitude locations.

10. Explain the operation of soil moisture storage, soil moisture utilization, and soil moisture recharge. Include discussion of field capacity, capillary water, and wilting point concepts.

Soil moisture storage (ΔSTRGE) refers to the amount of water that is stored in the soil and is accessible to plant roots, or the effective rooting depth of plants in a specific soil. This water is held in the soil against the pull of gravity and is discussed in more detail in Chapter 18. Soil is said to be at the wilting point when plant roots are unable to extract water; in other words, plants will wilt and eventually die after prolonged moisture deficit stress.

The soil moisture that is generally accessible to plant roots is capillary water, held in the soil by surface tension and cohesive forces between the water and the soil. Almost all capillary water is available water in soil moisture storage and is removable for POTET demands through the action of plant roots and surface evaporation; some capillary water remains adhered to soil particles along with hygroscopic water. When capillary water is full in a particular soil, that soil is said to be at field capacity, an amount determined by actual soil surveys.

When soil moisture is at field capacity, plant roots are able to obtain water with less effort, and water is thus rapidly available to them. As the soil water is reduced by soil moisture utilization, the plants must exert greater effort to extract the same amount of moisture. Whether naturally occurring or artificially applied, water infiltrates soil and replenishes available water content, a process known as soil moisture recharge.

11. In the case of silt loam soil from Figure 9-11, roughly what is the available water capacity? How is this value derived?

See Figure 9-11 and the explanation of soil moisture in the last question. The lower line on the graph plots the wilting point; the upper line plots field capacity. The space between the two lines represents the amount of water available to plants given varying soil textures. Different plant types growing in various types of soil send roots to different depths and therefore are exposed to varying amounts of soil moisture. For example, shallow-rooted crops such as spinach, beans, and carrots send roots down 65 cm (25 in.) in a silt loam, whereas deep-rooted crops such as alfalfa and shrubs exceed a depth of 125 cm (50 in.) in such a soil. A soil blend that maximizes available water is best for supplying plant water needs.

12. In terms of water balance and water management, explain the logic behind the Snowy Mountain Scheme in southeastern Australia.

In the Snowy Mountains, part of the Great Dividing Range in extreme southeastern Australia, precipitation ranges from 100 to 200 cm (40 to 80 in.) a year, whereas interior Australia receives under 50 cm (20 in.), and drops to less than 25 cm (10 in.) further inland. POTET values are high throughout the Australian outback and lower in the higher elevations of the Snowy Mountains. The plan was designed to take surplus water that flowed down the Snowy River eastward to the Tasman Sea and reverse the flow to support newly irrigated farmland in the interior of New South Wales and Victoria. The westward flow of the Murray, Tumut, and Murrumbidgee rivers is augmented, and as a result, new acreage is now in production in what was dry outback, formerly served only by wells drawing on meager groundwater supplies.

13. Are groundwater resources independent of surface supplies or interrelated with them? Explain your answer.

Groundwater is the part of the hydrologic cycle that lies beneath the ground and is therefore tied to surface supplies. Groundwater is the largest potential source of freshwater in the hydrologic cycle–larger than all surface reservoirs, lakes, rivers, and streams combined. Between Earth's surface and a depth of 3 km (10,000 ft) worldwide, some 8,340,000 km^3 (2,000,000 mi^3) of water resides.

14. Make a simple sketch of the subsurface environment, labeling zones of aeration and saturation and the water table in an unconfined aquifer. Next, add a confined aquifer to the sketch.

Utilize Figure 9-16 as the basis for this sketch.

15. At what point does groundwater utilization become groundwater mining? Use the Ogallala aquifer example to explain your answer.

Aquifers frequently are pumped beyond their flow and recharge capacities; groundwater mining refers to this overutilization of groundwater resources. Large tracts of the Midwest, West, lower Mississippi Valley, and Florida experience chronic groundwater overdrafts. In many places the water table or artesian water level has declined more than 12 m (40 ft). Groundwater mining is of special concern today in the Ogallala aquifer, which is the topic of Focus Study 9-2.

16. What is the nature of groundwater pollution? Can contaminated groundwater be cleaned up easily? Explain.

When surface water is polluted, groundwater also becomes contaminated because it is fed and recharged from surface water supplies. Groundwater migrates very slowly compared with surface water. Surface water flows rapidly and flushes pollution downstream, but sluggish groundwater, once contaminated, remains polluted virtually forever. Pollution can enter groundwater from industrial injection wells, septic tank outflows, seepage from hazardous-waste disposal sites, industrial toxic-waste dumps, residues of agricultural pesticides, herbicides, fertilizers, and residential and urban wastes in landfills. Thus, pollution can come either from a point source or from a large general area (a nonpoint source), and it can spread over a great distance, as illustrated in Figure 9-16. Because surface water flows rapidly, it can flush pollution downstream. Yet, if groundwater is polluted, because it is slow moving, once its contaminated, it will remain polluted virtually forever.

17. What differences exist between the freshwater inflow into the Atlantic Ocean and that into the Pacific Ocean? Why?

In volume of runoff, the Western Hemisphere exceeds the Eastern Hemisphere, primarily because the Amazon River carries a far greater volume than any other, and because four of the world's longest river systems are in the Western Hemisphere. The Atlantic Ocean receives about 1.5 times more runoff than the Pacific Ocean.

18. What are the five largest rivers on Earth in terms of discharge? Relate these to the weather patterns in each area, and regional POTET and PRECIP.

See Table 9-2 and Figure 10-3. Relative to water budgets, areas of seasonal and annual surplus produce higher discharges.

19. Contrast a river such as the Amazon River to an exotic stream such as the Nile or Colorado River.

Streams may be perennial (constantly flowing) or intermittent. In either case, the total runoff that moves through them comes from surplus surface-water runoff, subsurface throughflow, and groundwater. Highest runoff amounts are along the equator (see Figure 9-21 and Table 9-1), reflecting continual rainfall along the ITCZ. The Amazon is an example of such a perennial stream. The Amazon River discharges far more water than any other stream in the world.

By contrast, streams which originate in a humid region and flow through an arid region (because of high potential evapotranspiration) may have discharge that decreases with distance. Such streams are called exotic streams. An example of such a stream is the Nile River. This river, drains much of northeastern Africa, but as it courses through the deserts of Sudan and Egypt, it loses water, due to evaporation and extraction for human uses.

20. Describe the principal pathways involved in the water budget of the contiguous 48 states. What is the difference between consumptive use and withdrawal of water resources? Compare these with instream uses.

The 4244 BGD that makes up the U.S. water supply is not evenly distributed across the country or during the calendar year. Figure 9-23 shows that the two principal routes of expenditure of the daily water supply are evapotranspiration from non-irrigated land and surplus. About 71% (3013 BGD) of the daily supply passes through nonirrigated land–farm crops and pastures; forest and browse (young leaves, twigs, and shoots); and noneconomic vegetation–and eventually returns to the atmosphere for another journey through the hydrologic cycle. Only 29%, or 1231 BGD, becomes surplus runoff, available for withdrawal and consumption.

Consumptive uses are those that remove water from the budget at some point, without returning it farther downstream. Withdrawal, or nonconsumptive use, refers to water that is removed from the supply and used for various purposes, then returned to the same supply. Instream uses are those that use streamwater in place; navigation, fishing, hydroelectric power, and ecosystem preservation.

21. Characterize each of the sectors withdrawing water: irrigation, industry, and municipalities. What are the present usage trends in developed and less developed nations?

Figure 9-23 illustrates the three main flows of this withdrawn water in the United States: irrigation (34%), industry (57%), and municipalities (9%). In contrast, Canada uses only 7% of its withdrawn water for irrigation, and 84% for industry. About one-fourth of the annual renewable water worldwide will be actively utilized by the year 2000. By that time, one-half billion people will be depending on the polluted Ganges River alone. Furthermore, of the 200 largest river basins in the world (basins where rivers drain into an ocean, lake, or inland sea), 148 are shared by two nations, and 52 are shared by from three to ten nations. Clearly, water issues are international in scope, yet we continue toward a water crisis without a concept of a world water economy as a frame of reference.

22. Briefly assess the status of world water resources. What challenges are there in meeting future needs of an expanding population and growing economies.

About 30% of the annual renewable water worldwide is presently utilized. The World Bank estimates that $600 billion in capital investment is needed by 2010 to augment current water resources. In addition, aspects of global climate change will worsen water supply conditions for many regions and therefore increase this cost estimate. Available water declines as population increases, and individual demand increases with economic development.

Overhead Transparencies

As an adopter, you are provided with the following figures for overhead projector use.

- Figure 9-1: The hydrologic cycle model
- Mounted together: Figure 9-6- Precipitation in the United States and Canada (top); and, Figure 9-9 potential evapotranspiration demand for U.S. and Canada (bottom)
- Figure 9-12: Sample water budget- Kingsport, Tennessee
- Figure 9-16: Groundwater characteristics (large illustration, split in 2 parts, mounted above and below)

10

Global Climate Systems

Overview

Earth experiences an almost infinite variety of weather. Even the same location may go through periods of changing weather. This variability, when considered along with the average conditions at a place over time, constitutes climate. Climates are so diverse that no two places on Earth's surface experience exactly the same climatic conditions, although general similarities permit grouping and classification. Finally, we take a look at how Earth's temperature system appears to be in a dynamic state of change as concerns about global warming and potential episodes of global cooling are discussed. Chapter 10 serves as a synthesis of content from Chapters 2 through 9–Parts One and Two of the text.

The *Student Study Guide* presents 21 "Learning Objectives" to guide the student in reading the chapter. The *Applied Physical Geography* lab manual has one exercise with six steps that involve aspects of this chapter.

New to the Third Edition

(Note: This section highlights major changes, new features, and additions in the third edition. This does not describe all the rewrite and recast of the text.)

1. A list of key learning concepts begins the chapter.

2. Figure 10-2, a schematic of Earth's climate system, illustrates internal and external processes which influence climatic patterns.

3. A generalized map of Earth climates, Figure 10-5.

4. Climatic regions on a hypothetical continent are illustrated to aid student comprehension of climate patterns in Figure 10-6.

5. Photographs for each climatic region are included with each climograph.

6. Description of the variation in Af climates caused by highland environments.

7. A climograph of Walgett, New South Wales is provided as an example of a BSh climate.

8. News Report #1: "What's in a Boundary?" describes isotherms as transition zones which may be dominated by changing weather conditions.

9. Charrapunji, India is the subject of News Report #2: "Cwa Climatic Region Sets Precipitation Records." This report summarizes the factors behind the record-setting precipitation levels characteristic of this region.

10. Linkages which exist in the Earth-atmosphere-ocean system are clearly displayed in a bullet format.

11. The use of computer models to simulate complex Earth-atmosphere-ocean relationships is discussed.

12. Figure 10-30, a general circulation model scheme (GCM) is included to illustrate how climatologists study climatic change.

13. Review of the 1995 IPCC report, The UN Framework Convention on Climate Change.

14. Description of the affects of climate change upon ice caps, natural resources, and patterns of global health.

15. News Report #3: "Network: Direct to You from the South Pole." This report announces the rejuvenation of the *New South Polar Times*, began in 1901 and rejuvenated in 1994, including excerpts from a 14 month period of Antarctic activity.

16. Recent statistics concerning the disintegration of major ice shelves, and projected increases in sea levels.

17. Figure 10-35, includes a map and satellite image of the Antarctic Ice Shelves.

18. A new summary and review section ends the chapter.

Key Learning Concepts

1. *Define* climate and climatology and *explain* the difference between climate and weather.

2.*Review* the role of temperature, precipitation, air pressure, and air mass patterns used to establish climatic controls.

3, *Review* Köppen's development of an empirical climate classification system and*compare* his with other ways of classifying climate.

4. *Describe* the A, C, D, and E climate classification categories and *generally* locate these regions on a world map.

5. *Explain* the precipitation and moisture efficiency criteria used to determine the B climates and generally *locate* them on a world map.

6. *Outline* future climate patterns from forecasts presented and *explain* the causes and potential consequences.

Expanded Outline Discussion

The following headings (boldfaced) match some of the first, second, and third order headings in Chapter 10. The narrative under each heading contains information, sources, and anecdotal facts relating to portions of the chapter. Not all text headings are discussed.

EARTH'S CLIMATE SYSTEM AND ITS COMPONENTS

As this section is developed, figures and tables from previous chapters can be referenced to formulate the basis of world climate patterns. The effect of temperature and precipitation (abiotic climate controls) is illustrated in Figure 19-10. I suggest having students refer to Table 20-1, for it includes the Köppen classifications as they relate to terrestrial ecosystems and the other physical aspects of the environment. In a broader sense this is also the function of the synthesis in Chapter 20–Terrestrial Biomes.

For an interesting comparison of similar elements that evolved to produce extremely different climates, see Robert M. Haberle, "The Climate of Mars," *Scientific American* (May 1986): pp. 54-62.

Climatic information can also be found on the Internet. *The Climatic Data Catalogue* is accessible at:

http://rainbow.ldgo.columbia.edu.

This site includes global atmospheric circulation statistics, Navy bathometry, and surface climatologies. Another good site is the Climate Diagnosis Center located at:

http://noaacdc.colorado.edu/cdu/cdchome.html. This site has weather statistics spanning centuries. Lastly, the Climatic Research Unit at the University of East Anglia, England has information about current weather, climatic research and links to other climate research facilities around the world. Contact them at:

http://www.cru.uea.ac.uk.

CLASSIFICATION OF CLIMATIC REGIONS

A necessary resource is *The Encyclopedia of Climatology* edited by John E. Oliver and Rhodes W. Fairbridge (New York: Van Nostrand Reinhold Co., 115 Fifth Avenue, New York, NY 10003, 1987). The section on climatic classification covers the basics of classification, empirical systems, the Köppen system, the Thornthwaite system, and genetic systems. This volume also contains extensive sections on climatic data, with a list of helpful references. Appropriate chapters in Glenn T. Trewartha's several textbooks contain valuable material on climates and classification modification that he established (4 titles with McGraw-Hill). Glenn T. Trewartha's *The Earth's Problem Climates* (Madison: The University of Wisconsin, 1961) examines climates worldwide, particularly those that do not conveniently fit expected global patterns. Also refer to the chapter on climatic classification in Howard J. Critchfield's *General Climatology*, Englewood Cliffs: Prentice-Hall, 1966. He describes both the Köppen and Thornthwaite systems. Also consult the Hidore and Oliver text mentioned earlier.

Figure 2-13 demonstrates the input of 400 to 450 W/m^2 per day at the top of the atmosphere above the equator that produces the nearly constant temperatures that exist along the equator. Seasonal variation of input increases with latitude, so that in the midlatitudes, winter to summer energy inputs vary from 150 to 475 W/m^2 a day, whereas the polar regions range from twilight and darkness for six months to dawn and constant low-angle sunlight for six months and a low overall annual input.

The Köppen Classification System

Please refer to Arthur A. Wilcox's "Köppen After Fifty Years." *Annals of the Association of American Geographers* 58, No. 1 (March 1968): 12-28, for it provides interesting back-ground on Köppen and the development of his classification. Although it is an empirical classification, aspects of his early genetic considerations involving plants are discussed. He selected temperature criteria in an attempt to match vegetation patterns; his classification remains based on climatic data.

The Köppen system is best viewed for what it is: a valuable tool for general understanding, best limited to small scale hemispheric and world maps showing general climatic relationships and patterns. Most criticism of his system stems from asking the classification model to do what it was not designed to do, that is, produce specific climatic descriptions for local areas.

For arguments that the Thornthwaite system should be allowed to supersede the Köppen system, see Douglas B. Carter, "Farewell to the Köppen Classification of Climate" from Abstracts of the papers presented at the 63rd annual meeting of the Association of American Geographers, St. Louis, Missouri, 11-14 April 1967 in *Annals of the Association of American Geographers* Vol. 57, No. 4 (December 1967). Following is some background on the Thornthwaite system.

The development of a simple method for the determination of potential evapotranspiration led Thornthwaite to his climate classification system. Thornthwaite was critical of Köppen's choice of criteria for his climatic boundaries, and especially, the boundaries between the humid and dry climates. Temperature efficiency and precipitation effectiveness were concepts contributed by Thornthwaite. His 1948 classification introduced a moisture index concept as a basis for classification. Thornthwaite's classification is marred only by its complexity and lack of widespread use. Otherwise, the system in several ways is more accurate than is the Köppen system in its depiction of the humid-dry boundaries, especially those in North America.

The key to Thornthwaite's approach is that temperature and precipitation alone are not the most active factors in the distribution of vegetation; rather, Thornthwaite regarded POTET and its relation to precipitation and plant moisture needs as the critical factor. Whereas Köppen used average annual temperature and precipitation for the determination of a moisture index, Thornthwaite used POTET and established a moisture index based on calculations of water balance moisture surpluses and moisture deficits as discussed in Chapter 9. The moisture index can range from +100 as a measure of the degree PRECIP exceeds POTET, to a low of −100, where no PRECIP is received. When the moisture index is at zero, it is at the midpoint along the boundary line between the humid and dry climates. Thornthwaite established five climate classifications: arid, semiarid, subhumid, humid, and perhumid. A specific area to compare Thornthwaite with the Köppen world map is the transition region between the dry and humid climates from the western prairies to the Rocky Mountains. Köppen extended the humid climates, both C and D, west of the 100°W meridian providing poor definition of the region. In reality, the region goes through a transition from tall grass to short grass to semiarid steppe to arid zones (see Chapter 20, p. 641). On the Thornthwaite map, this transition area is clearly divided into moist subhumid, dry subhumid, and semiarid classifications, thereby providing greater clarification.

Classification Criteria Köppen's Climatic Designations

You may have used or even developed your own Köppen classification key. Using the climatic data in Appendix A, students can determine the Köppen classification for several stations. Figure 10-5 and the world map in Figure 10-7 can be used for assistance and location. I hope you find the revised and enlarged climate map helpful. Figure 10-5 is designed to highlight each climate type with a distribution map along with the specific defining criteria. Figure 10-6 illustrates climatic patterns on a hypothetical climate, to add students comprehension of climatic patterns. The *Student Study Guide* contains a couple of stations for the student to classify. The *Applied Physical Geography* lab manual presents a new Köppen climate key and related exercise.

Note that throughout *GEOSYSTEMS* the names of each climate type are spelled out in italics and presented along with the Köppen climatic classification symbol wherever they are mentioned.

Global Climate Patterns

Use Figures 10-5 and 10-7 together with each of the following brief discussions of each climate type. And, as stated, these tie-in with Table 20-1 (specific column in table) and the world biome map (in Figure 20-4). You may want to put a tab or post-it note on these pages for easy reference. I hope your classroom is equipped with a wall map of climates that you can reference throughout the semester; if not, then students may want to place a tab on those pages containing frequently used maps and tables.

Marine West Coast Climates (Cfb, Cfc)

The effect of elevation on climate type is demonstrated in Marlinton, West Virginia (38° N, 655 m, 2150 ft). Marlinton features a Dfb climate type similar to more northern stations. The climate of this area of the Appalachians is similar to the cooler marine west-coast regions than it is to the surrounding region. To the east of this city, the climates are Cfa, to the west are Dfa climates, and to the south Cfb occurs. The pattern of surplus in relation to lower potential evapotranspiration demands is evident in Marlinton; PRECIP (119.5 cm, 47 in.) is almost double the annual average POTET (61.4 cm, 24.2 in.).

Mediterranean Dry Summer Climates (Csa, Csb)

The productivity within this climatic type is exemplified by California, which is the number one producer of 42 different commodities in the United States, and for at least 20 of these, California production is 99 percent of the total for the entire nation. The need for irrigation was perceived early, with the construction of canals in the Central Valley of California in the 1850s. Those Mediterranean climates that are bounded by mountains have an additional advantage in that the mountains capture winter snowpack, which melts gradually throughout the spring, releasing needed waters in the dry months of summer. The construction of reservoirs to delay runoff and capture winter precipitation is common in these regions. The need for enormous expenditures of capital and technology is obvious.

Microthermal D Climates: Humid Continental Mild Summer Climates (Dfb, Dwb); and Subarctic Climates (Dfc, Dwc, Dwd).

See new sections on periglacial landscapes in Focus Study 1, Chapter 17 and with related biomes in Chapter 20.

Polar Climates

The polar marine (EM) climates are described in Shear, James A., "The Polar Marine Climate," *Annals of the Association of American Geographers* Vol. 54, No. 3 (September 1964). As with all classifications, they are portrayed as a compromise between detail and simplicity. Shear states that, "It is a climate so cold that snow may fall on any day of the year, although warm enough for the mean temperature of no month to be below freezing."

Dry Arid and Semiarid Climates

Chapter 15 and the desert biome section in Chapter 20 describe these unique arid places in greater detail. Graphs to assist in the determination of the B-climate classifications appear as part of Figures 10-5 and 10-28. Each station graphed differs according to the seasonal distribution of precipitation.

GLOBAL CLIMATE CHANGE Developing Climate Models

For much valuable information and background on leading scientists and their work, see the National Science Foundation's *MOSAIC* Vol. 19 No. 3/4 Fall/Winter 1988, with the entire issue being devoted to "Global Change-A *MOSAIC* Double Issue." Twenty years have passed since meteorologist Edward N. Lorenz delivered his paper at a AAAS meeting titled "Does the Flap of a Butterfly's Wings in Brazil Set Off a

Tornado in Texas?" The articles in *MOSAIC* address various aspects of global systems with articles titled, "Laying the Foundation," "Grappling with Coupled Systems," "Dirtying the Infrared Window," "The Sum of Its Parts," "One Model To Fit All," "Wheels within Wheels," "Forging the Tools," "Energizing the Climate Cycles," " A Global Chemical Flux," and "Learning the Language of Climate Change." This should be available from NSF, Washington, D.C., 20550; Ph-202-357-9498 (single copies are $2.50). The critical concept for students to grasp is the profound interconnectedness of all elements of the physical environment, especially when viewed through the spatial perspective of physical geography.

For a general and colorful overview, with many images, photographs, and systematic models, obtain a copy of *Earth System Science-A Preview* and the more complete *Earth System Science-A Closer View* as part of Earth System Science-A Program for Global Change, a report to the Earth System Sciences Committee, NASA Advisory Council, NASA, Washing-ton, D.C., available from Office of Interdisciplinary Earth Studies, Univer-sity Corporation for Atmospheric Re-search, P.O. Box 3000, Boulder, Colorado 80307.

The 1992 IPCC Update presents many GCM forecast maps in color and discussion of the results. See: Houghton, J.T., Jenkins, G.J. and J.J. Ephraums, editors. *Climate Change 1992–The Supplementary Report to the IPCC Scientific Assessment*. World Meteorological Organization/United Nations Environment Programme. New York: Cambridge University Press, 1992. (Paper-ISBN: 0-521-43829-2; phone 1-800-431-1580).

A new text by John J. Hidore and John E. Oliver features a chapter on "Future Climates" that includes an in-depth discussion of analog models, the nature of GCMs, and model projections. A table compares and summarizes the various results of the GCMs assuming a doubling of atmospheric CO_2. Their book is *Climatology–An Atmospheric Science*, New York: Macmillan College Publishing Company, 1993.

Climate is ever-changing as inputs and outputs in the system vary. Studies of longterm climatic change treat climate as a system, with negative feedback information about carbon dioxide increases or increased pollution providing impetus for humans to begin questioning their impact on this process-response system. Society can observe these climatic signals and should contemplate these apparent changes, and possibly use this system information as a basis for policy change.

FUTURE CLIMATE PATTERNS

Scientifically, firm evidence of a rising sea level could constitute a kind of "smoking gun" relative to climatic warming. During the research for this text, several U.S. government scientists told me candidly that either they or their colleagues had lost research funds or had encountered bureaucratic difficulties in their attempts to study sea-level rise. We may need to look to the European Community for leadership in such studies. I cannot make a generalization based on the few statements I encountered; I only relate them to perhaps alert you to watch for certain items in your reading and research.

For an interesting overview and policy analysis from one state that has exceeded the federal effort in planning, obtain a copy of *Global Climate Change: Potential Impacts and Policy Recommendations* (Draft–1991), Nancy J. Deller, Deputy Director, Chuck Mizutani, Acting Manager, and Kari Smith, Project Manager. Sacramento: California Energy Commission (Energy Technology Development Division, Technology Evaluation Office, 1516 9th Street, Sacramento, California 95814-5512).

See Topping, John C., Jr., ed. "Coping with Climate Change." *Proceedings of the Second North American Conference*

On Preparing for Climate Change. Washington: Climate Institute, Suite 402, 316 Pennsylvania Avenue, SE, Washington, D.C. 20003 (202-547-0101), 1989. This volume includes sections on policy overview, scientific overview, understanding atmospheric and oceans processes, human health implications of climate change and stratospheric ozone depletion, agricultural implications, ecological implications, water resource implications, among other sections in this 700-page report. As mentioned in an earlier part of this manual, the Climate Institute publishes *Climate Alert* ten times a year. The newsletter provides valuable background on current events, listings of all conferences and meetings worldwide, and a summary of what is being done in many other countries.

As cited in the text, there is a consensus (reached in 1990 and reaffirmed in 1992) among scientists participating in the Intergovernmental Panel on Climate Change (IPCC) that human-induced global warming is occurring. Although we cannot dismiss the degree of uncertainty as to severity and specific timing of potential continued warming, for there are scientists that are in disagreement with a global warming scenario. The IPCC conclusions deserve our attention and the attention of the general public, policy makers, and the transnational corporate community. IPCC Working Groups I (Science) and II (Impacts) concluded that:

> We are certain that emissions resulting from human activities are substantially increasing the atmospheric concentrations of the greenhouse gases: carbon dioxide, methane, chlorofluocarbons (CFCs), and nitrous oxide. These increases will enhance the greenhouse effect, resulting on average in an additional warming of the Earth's surface...The long-lived gases would require immediate reductions in emissions from human activities of over 60 percent to stabilize their concentrations at today's levels.*

* *Climate Change—The IPCC Response Strategies*, Working Group III, World Meteorological Organization/United Nations Environment Programme. Covelo, CA: Island Press, 1991, p. xxv.

Intergovernmental Panel on Climate Change reports are available from the sources listed below.

Scientists looking back over the past 1000 years find the rate of temperature change over the last 100 years to be greatly accelerated. A recent report demonstrated that atmospheric greenhouse gas concentrations and aerosol loading are responsible for the warming of the past 100 years and that they are not some aspect of a natural interaction of the atmosphere-ocean system (see: R.J. Stouffer, S. Manabe, and K. Ya. Vinnikov, "Model Assessment of the Role of Natural Variability in Recent Global Warming," *Nature*, 367, no. 6464, February 17, 1994: 634-636.

Stephen H. Schneider, in his book *Global Warming: Are We Entering the Greenhouse Century* (New York: Random House, 1989) states, "We are insulting the atmospheric environment faster than we are comprehending the effects of those insults."

There is a tremendous amount of material available, too numerous to mention here. Many of the organizations to contact are listed in Appendix A of *GEOSYSTEMS*. Also, available is a general directory by Janet Wright, *Directory of Global Climate Change Organizations*, National Agricultural Library, 10301 Baltimore Blvd., Beltsville, MD 20705; 301-504-5755. A few additional sources among these many include publications from:

Worldwatch Institute, 1776 Massachusetts Ave., N.W., Washington, D.C., 20036; 202-452-1999: e.g., Christopher Flavin *Slowing Global Warming: A Worldwide Strategy*.

World Resources Institute, 1709 New York Ave., N.W., Washington, D.C.; 800-822-0504: e.g., Mark C. Trexler, *Minding the Carbon Store: Weighing U.S. Forestry Strategies to Slow Global Warming*; Mark C. Trexler, *Keeping it Green: Tropical Forestry and the*

Mitigation of Global Warming; Jessica Tuchman Mathews, ed., *Greenhouse Warming: Negotiating a Global Regime*; Francesca Lyman and James MacKenzie, *The Greenhouse Trap: What We're doing to the Atmosphere and How We Can Slow Global Warming.*

Island Press, 1718 Connecticut Ave., N.W., Suite 300, Washington, DC 20009; 202-232-7933: e.g., Lynne T. Edgerton, *The Rising Tide: Global Warming and World Sea Levels*; Dean E. Abrahamson, ed., *The Challenge of Global Warming*;

Yale University Press, 302 Temple Street, 92 A Yale Station, New Haven, CT 06520; 203-432-0960: John Firor, *The Changing Atmosphere: A Global Challenge*; Robert L. Peters, *Global Warming and Biological Diversity.*

Global Warming

Global warming, however uncertain in severity of impact and timing, is certainly a spatial problem and deserves analysis within the context of physical geography. Examine the aspects of our lives that are influenced by this one topic of possible human-induced temperature change: food production and climate patterns, energy resources and transnational corporations, national economies, sea-level change, housing types and location, geopolitics and the interaction of nations, and unknown psychological ramifications.

As cited in the text, there is a consensus (reached in 1990) among scientists participating in the Intergovernmental Panel on Climate Change (IPCC) that human-induced global warming is occurring. Without control of emissions, IPCC predicted that average temperatures will increase by at least 1C° (1.8F°) over present averages by the year A.D. 2025. This position was affirmed in the IPCC 1992 update. Uncertainties involve specifics as to severity and timing of temperature increases in the next century. There are still scientists unconvinced as to the connection between greenhouse gases and warming, or if a warming is occurring at all.

There is uncertainty as to the precise degree and timing of change, with this secondary uncertainty being politically shifted to the question of whether there is any radiatively forced warming. The argument is then stated that no action be taken unless scientific certainty is at a 100% level. The Bush Administration refused to sign the November 1990 international agreement by 130 nations on reductions in CO_2 emissions–yet we produce 27% of the excessive CO_2 each year with only 5% of the world's population. Meanwhile, between 1987 and 1991, carbon dioxide increased at an average annual rate of 0.5% (1.71 ppm), with a maximum increase in 1987 of a full 1% (measured at the Climate Monitoring and Diagnostic Laboratory of NOAA, Boulder, CO).

Carbon Dioxide and Global Warming

I assume that between all of us we have quite a file on this subject. Almost 35 years have passed since Roger Revelle, then-director of the Scripps Institution of Oceanography, and his colleague Hans E. Suess presented the hypothesis that the oceans would not be able to absorb all the excessive CO_2 that industrial society was dumping into the atmosphere. The measurement of CO_2 at the Mauna Loa Observatory began that year. Today we see a 12% increase in the overall CO_2 concentration in the lower atmosphere since this monitoring began ("Profile Dr. Greenhouse–Roger Revelle started the world thinking about global warming," *Scientific American* Vol. 263, 6, December 1990: pp. 33-36).

One question is whether the temperature changes that are occurring are part of a natural pattern or one cause by human activity, namely increased greenhouse gases in the atmosphere. Among many sources check the recent study: R. J. Stouffer, S. Manabe, and K. Ya. Vinnikov, "Model Assessment of the Role of Natural Variability in Recent Global Warming," *Nature*, 367, February 17, 1994: 634-36. They state: "Assuming that the model is realistic, these results suggest that the observed trend is not a natural feature of the interaction between the atmosphere and oceans." This places the focus on human-induced forcing of temperatures. In addition, the rate of change is faster than at any time in the past 1000 years. It is the sustained rate that is remarkable.

Here is a partial chronological list of some items of interest from across the span of years (not otherwise listed with selected readings at the end of Chapters 5 and 10). This brief list is not in any way meant to be comprehensive.

Broecker, Wallace S., "Climatic Change: Are We on the Brink of a Pronounced Global Warming," *Science* Vol. 189 (8 August 1975): 460-463.

Siegenthaler, U. and H. Oescher, "Predicting Future Atmospheric Carbon Dioxide Levels," *Science* Vol. 199 (27 January 1978): 388-395.

Woodwell, George M., "The Carbon Dioxide Question," *Scientific American* Vol. 238, No. 1 (January 1978).

Kellogg, William W., "Is Mankind Warming the Earth," *Bulletin of the Atomic Scientists* (February 1978): 10-19.

Hansen, J., et al., "Climate Impact of Increasing Atmospheric Carbon Dioxide," *Science* Vol. 213 (28 August 1981): 957-966.

Olson, Steve, "Computing Climate Global Forecast" *Science* 82 (May 1982): 52-60. An interesting look at a very early version of the GISS (NASA) general circulation model.

Revelle, Roger, "Carbon Dioxide and World Climate" *Scientific American* Vol. 247 No. 2 (August 1982): 35-43.

Manabe, S. and R.T. Wetherald, "Reduction in Summer Soil Wetness Induced by an Increase in Atmospheric Carbon Dioxide," *Science* Vol. 232 (2 May 1986): 626-629.

Mintzer, Irving M. *A Matter of Degrees: The Potential for Controlling the Greenhouse Effect.* Research Report No. 5. Washington: World Resources Institute, 1987. (Phone: 1-202-638-6300.)

Nordhaus, W. D. "An Optimal Transition Path for Controlling Greenhouse Gases." *Science* 258, no. 5086 (20 November 1992): 1315-19.

Patrusky, Ben, "Dirtying the Infrared Window," *MOSAIC* Vol. 19 No. 3/4 (Fall/ Winter 1988): 24-37.

Peltier, W.R. and A.M. Tushingham, "Global Sea Level Rise and the Greenhouse Effect: Might they be Connected?" *Science* Vol. 244 (19 May 1989): 806-810.)

Peters, R. L. and T. E. Lovejoy. eds. *Global Warming and Biological Diversity.* New Haven: Yale University Press, 1992. (ISBN: 0-300-05056-9; phone 1-203-432-0940.

Schneider, Stephen H., "The Greenhouse Effect: Science and Policy," *Science* Vol. 243 (10 February 1989): 771-781.

Schneider, Stephen H., (reviewer) "Three Reports of the Intergovernmental Panel on Climate Change," *ENVIRONMENT* Vol. 33 No. 1 (January/February 1991): 25-30.

Warrick, R. A. E. M. Barrow, and T. M. L. Wigley, eds. *Climate and Sea Level Change–Observations, Projections, and Implications.* Cambridge, England: Cambridge University Press, 1993.

White, Robert M., "The Great Climate Debate," *Scientific American* Vol. 263, No. 1 (July 1990): 36-43.

Yan, Xiao-Hai, Chung-Ru Ho, Quanan Zheng, and Vic Klemas. "Temperature and Size Variabilities of the Western Pacific Warm Pool." *Science* 258, no. 5088 (4 December 1992): 1643-45.

Solutions

An example of an attempt to reduce the combustion of fossil fuels to lower the output of radiatively active gases, called the *Small Appliances Efficiency Act*, struggled for passage in the United States in the mid-1980s. The bill was passed and vetoed three times, finally emerging as law. An estimated impact of implementation of this act would be the elimination of the need for 22 new 1000 MWe (1 million-watt capacity, electric rating) power plants. Think of the reduction in coal combustion. Of interest is that the act was favored by business groups and appliance manufacturers, as well as environmental groups. Who do you think lobbied to defeat passage and stall enactment?

Many resources are available, such as, *Soft Energy Paths: Toward A Durable Peace* and *Non-Nuclear Futures: The Case for an Ethical Energy Strategy* by Amory B. Lovins (the latter co-authored with John Price); both books are available from Friends of the Earth International (Ballinger Publishing Company, Cam-bridge, MA). The interesting thing about Amory Lovins, and about his Rocky Mountain Institute (see Appendix B), is that he is now hired as a consultant by "both sides" of the energy debate–soft path and hard path.

Estimates for 1990, from Charles Komanoff of the Council on Economic Priorities and the Worldwatch Institute, place the costs of conservation and efficiency at 4¢ per kilowatt hour or less, cogeneration and power sharing at 5¢ per kwh, biomass conversion at 7¢ per kwh, and wind power at 12¢ per kwh, all cheaper than new coal production and combustion which costs 15¢ per kwh, with new nuclear electrical generation coming in at even higher prices. Coal per unit consumed produces more carbon dioxide than any other fuel. This data poses difficult questions for society and for the industrialized countries as they lead the developing world. See the discussion in Chapters 3 and 5 in the text and this resource manual.

Also, check the focus studies in Chapter 4 on solar energy applications and Chapter 6 on wind energy resources and this resource manual under the same chapters.

FUTURE TEMPERATURE TRENDS

Some basic issues: First, this subject is heavily politicized, especially by those corporations who think that they have the most to lose by any changes from traditional, centralized, energy-production patterns. Unfortunately, most traditional energy production is based on fuels that are linked to the production of radiatively active gases, especially carbon dioxide. In the previous chapter of this resource guide, two lists of centralized and decentralized energy source characteristics were presented. The centralized list is the one being promoted for the world by the transnational corporations (see the *National Energy Strategy* issued by the White House, February 1991). In an editorial AAAS stated that this strategy document of the former administration "was not impressive" and in fact it abandoned the two-year hearing record and adopted a program devoid of any decentralized strategies.

Second, we are asked to treat radiatively active gases, such as CO_2, as if they were somehow an arrestee with full civil rights to a trial and a presumption of innocence until guilt is proven; or, a presumption of open-ended risk assessment. With such serious issues, we might want to ask why our representative political system places most risk on the side of *error* rather than on the side of *caution*. The burden of proof is placed on the public (the potential victim) to prove that something is disruptive to a sustainable environment.

An objective observer might wonder if modern society's unconscious proposition seems simply to be, remodel the ecological texture of North America and the planet and then proceed with an experiment in adaptation! Before we embark on such a one-time only experiment perhaps the perpetrators need to prove a higher level of sustainability for their actions. Admittedly, this requires a longterm view, such as the one you sense when you look into a child's eyes.

The Climate Institute publishes *Climate Alert* ten times a year. The newsletter provides valuable background on current events, listings of all conferences and meetings, and a summary of what is being done in many other countries. The newsletter is available from the Climate Institute, Suite 402, 316 Pennsylvania Avenue, S.E., Washington, D.C. 20003.

Intergovernmental Panel on Climate Change reports (these are softcover ISBN numbers; all are available in hardback as well):

1. Houghton, J.T., Jenkins, G.J. and J.J. Ephraums, editors. *Climate Change–The IPCC Scientific Assessment*, Working Group I, World Meteorological Organization/United Nations Environment Programme. Port Chester, N.Y.: Cambridge University Press, 1991. (Paper-ISBN: 0-521-40720-6; phone 1-800-431-1580).
2. Intergovernmental Panel on Climate Change. *Climate Change–The IPCC Response Strategies*, Working Group III. World Meteorological Organization/United Nations Environment Programme. Covelo, CA: Island Press, 1991. (Paper-ISBN: 1-55963-102-3; phone 1-800-828-1302).
3. Tegart, W.J. McG., Sheldon, G.W. and D.C. Griffiths. *Climate Change–The IPCC Impacts Assessment*, Working Group II. World Meteorological Organization/United Nations Environment Programme. Portland, OR: International Specialized Book Service/Australian Government, 1991. (Paper-ISBN: 0-644-13497-6; phone 1-503-287-3093.)
4. Houghton, J.T., Jenkins, G.J. and J.J. Ephraums, editors. *Climate Change 1992–The Supplementary Report to the IPCC Scientific Assessment.* World Meteorological Organization/United Nations Environment Programme. New York: Cambridge University Press, 1992. (Paper-ISBN: 0-521-43829-2; phone 1-800-431-1580). This report contains many color GCM maps.

Global Cooling

The Scientific Committee on Problems of the Environment (SCOPE) has been examining the potential effects of nuclear war–environmental consequences of nuclear war (ENUWAR)–in conjunction with the International Council of Scientific Unions. Several of the scenarios developed by them are being used to understand the possible impact of the Kuwaiti oil-well fires of 1991. See Warner, Sir Frederick, "The Environmental Consequences of the Gulf War," *ENVIRONMENT*, Vol. 33, No. 5 (June 1991): 7-9, 25-26. And, Renner, Michael G., "Military Victory, Ecological Defeat," *WORLD WATCH*, Vol. 4, No. 4 (July-August 1991): pp. 27-33.

Following the 1991 Persian Gulf War, much discussion took place as to the effect of all the oil-well fires on climate. The Mauna Loa Observatory, beginning in May 1991, recorded soot from Kuwaiti and Iraqi oil fires and from U.S. bombing in the war. Scientific analysis of the soot is being hampered because of the political nature of the aftermath of the war. The oil-fire effects appear to be only regional in extent at this time, although they are locally significant. No global cooling seems to be occurring in response to these increases in atmospheric turbidity.

The nuclear winter hypothesis is also mentioned in Chapter 21. Given the worldwide expenditures on armaments, and on strategic and tactical nuclear weapons and defenses, one can imagine the political impact of the 1983 October conference and the research article in *Science* (December 1983) describing a potential "nuclear winter" as a possible consequence of modern technological warfare. The same research team published an updated research article in

Science (January 1990). (See the suggested readings section in the text.) This hypothesis has triggered much research and debate. We all can hope that the potential atmospheric experiment associated with the detonation of a small amount of nuclear weapons in the atmosphere will never take place. Historically, war is supposed to have winners and losers, but such an event as this would produce only losers, as well as an unknown outcome to a global atmospheric experiment.

Glossary Review for Chapter10 (in alphabetical order)

classification
climate
climatic regions
climatology
climographs
empirical classification
general circulation model (GCM)
genetic classification
Köppen-Geiger climate classification
nuclear winter hypothesis
paleoclimatology

Annotated Chapter Review Questions

1. Define climate and compare it with weather. What is climatology?

Earth experiences an almost infinite variety of weather. Even the same location may go through periods of changing weather. This variability, when considered along with the average conditions at a place over time, constitutes climate. In a traditional framework, early climatologists faced the challenge of identifying patterns as a basis for establishing climatic classifications. Currently at the forefront of scientific effort by climatologists are developing models that can simulate the vast interactions and causal relationships of the atmosphere and hydrosphere. Climatology, the study of climate, involves analysis of the patterns in time and space created by various physical factors in the environment.

2. Explain how a climatic region synthesizes climate statistics.

Weather observations, gathered simultaneously from different points within a region, are plotted on maps and are compared to identify climatic regions. The weather components which combine to produce climatic regions include insolation, temperature, humidity, seasonal precipitation, atmospheric pressure and winds, air masses, types of weather disturbances, and cloud coverage. Similar climatic regions experience many of the same weather components.

3. How does the El Niño phenomenon produce the largest interannual variability in climate? What are some of the changes and effects that occur worldwide?

Normally, as shown in Figure 6-24, the region off the west coast of South America is dominated by the northward-flowing Peru Current. These cold waters move toward the equator and join the westward movement of the south equatorial current. The Peru current is part of the overall counter-clockwise circulation that normally guides the winds and surface ocean currents around the subtropical high-pressure cell dominating the eastern subtropical Pacific (visible on the world pressure maps in Figure 6-13). Occasionally, and for as-yet unexplained reasons, pressure patterns alter and shift from their usual locations, thus affecting surface ocean currents and weather on both sides of the Pacific. Unusually high pressure develops in the western Pacific and lower pressure in the eastern Pacific. This regional change is an indication of large-scale ocean-atmosphere interactions. Trade winds normally moving from east to west weaken and can be replaced by an eastward (west-to-east) flow. Sea-surface temperatures off South America then rise above normal, sometimes becoming more than 8C° (14F°) warmer, replacing the normally cold, upwelling, nutrient-rich water along Peru's coastline.

4. How do radiation receipts, temperature, air pressure inputs, and precipitation patterns interrelate to produce climate types? Give an example from a humid environment and one from an arid environment.

Uneven insolation over Earth's surface, varying with latitude, is the energy input for the climate system (Chapter 2 and Figure 2-13 and 2-14.) Daylength and temperature patterns vary diurnally and seasonally. Average temperatures and daylength are the basic factors that help us approximate POTET (potential evapotranspiration). The principal controls of temperature are latitude, altitude, land-water heating differences, and the amount and duration of cloud cover. The pattern of world temperatures and annual temperature ranges are discussed in Chapter 5 and portrayed in Figures 5-11, 5-13, and 5-14. The moisture input to climate is precipitation in the forms of rain, sleet, snow, and hail. Figure 9-6 portrays the distribution of mean annual precipitation in North America; Figure 10-3 shows the worldwide distribution of precipitation and identifies several patterns. Average air pressure for January and July is portrayed in Figure 6-13.

5. Evaluate the relationships among a climatic region, ecosystem, and biome.

One type of climatic analysis involves discerning areas of similar weather statistics and grouping these into climatic regions that contain characteristic weather patterns. Climate classifications are an effort to formalize these patterns and determine their related implications to humans. A community is formed by interacting populations of plants and animals in an area. An ecosystem involves the interplay between a community of plants and animals and its abiotic physical environment. A biome is a large, stable terrestrial ecosystem.

6. What are the differences between a genetic and an empirical classification system?

Classification is the process of ordering or grouping data or phenomena in related classes. A classification based on causative factors–for example, the genesis of climate based on the interaction of air masses–is called a genetic classification. An empirical classification is based on statistics or other data used to determine general categories. Climate classifications based on temperature and precipitation data are empirical classifications.

7. Describe Köppen's approach to climatic classification. What are the factors used in his system?

The Köppen classification system, widely used for its ease of comprehension, was designed by Wladimir Köppen (1846-1940), a German climatologist and botanist. First published in stages, his classification began with an article on heat zones in 1884. By 1900, he was considering plant communities in his selection of some temperature criteria, using a world vegetation map prepared by French plant physiologist A. de Candolle in 1855. Letter symbols then were added to designate climate types. Later he reduced the role played by plants in setting boundaries and moved his system strictly toward climatological empiricism. The first wall map showing world climates, co-authored with his student Rudolph Geiger, was introduced in 1928 and soon was widely used.

8. List and discuss each of the principal climate designations. In which one of these general types do you live? Which classification is the only type

associated with the distribution and amount of precipitation?

See Figure 10-5 for the climatic classification system and Figure 10-7 for the portrayal of these climates on a world map.

9. What is a climograph, and how is it used to display climatic information?

Climographs for cities that exemplify particular climates are presented in the text. These climographs show monthly temperature and precipitation, location coordinates, annual temperature range, total annual precipitation, annual hours of sunshine (as an indication of cloudiness), the local population, and a location map.

10. Which of the major climate types occupies the most land and ocean area on Earth?

In terms of total land and ocean area, tropical A climates are the most extensive, occupying about 36% of Earth's surface. The A climate classification extends along all equatorial latitudes, straddling the tropics from about 20° N to 20° S and stretching as far north as the tip of Florida and south-central Mexico, central India, and southeast Asia.

11. Characterize the tropical A climates in terms of temperature, moisture, and location.

The key temperature criterion for an A climate is that the coolest month must be warmer than 18°C (64.4°F), making these climates truly winterless. The consistent daylength and almost perpendicular Sun angle throughout the year generates this warmth. Subdivisions of the A climates are based upon the distribution of precipitation during the year. Thus, in addition to consistent warmth, an Af climate is constantly moist, with no month recording less than 6 cm (2.4 in.) of precipitation. Indeed, most stations in Af climates receive in excess of 250 cm (100 in.) of rainfall a year. Not surprisingly, the water balances in these regions exhibit enormous water surpluses, creating the world's largest stream discharges in the Amazon and Congo (Zaire) Rivers.

12. Using Africa's tropical climates as an example, characterize the climates produced by the seasonal shifting of the ITCZ with the high Sun.

Those areas that are covered by the ITCZ during all 12 months of the year constitute the Af climates. As the ITCZ shifts with the high Sun, new areas are affected, whereas opposite latitudes of lower Sun angles are abandoned by the rains of the ITCZ. Generally, regions that are covered from 6 to 12 months (essentially along coastal areas) are Am; those receiving coverage by the ITCZ less than 6 months of the year are principally the Aw climatic regions.

13. Mesothermal C climates occupy the second-largest portion of Earth's entire surface. Describe their temperature, moisture, and precipitation characteristics.

The word mesothermal suggests warm and temperate conditions, with the coldest month averaging below 18°C (64.4°F) but with all months averaging above 0°C (32°F). The C climates, and nearby portions of the D climates, are regions of great weather variability, for these are the latitudes of greatest air-mass conflict. The C climatic region marks the beginning of true seasonality, with contrasts in temperature as evidenced by vegetation, soils, and human lifestyle adaptations. Subdivisions of the C classification are based on precipitation variability as given in Figure 10-5.

14. Explain the distribution of the Cfa *humid continental* and Csa *Mediterranean dry-summer* climates at similar latitudes and the difference in precipitation patterns between the two types. Describe the difference in

vegetation associated with these two climate types.

Cfa climates are located in the eastern and east-central portions of the continents and are influenced during the summer by the maritime tropical air masses generated over warm coastal waters off eastern coasts. The warm, moist, unstable air forms convectional showers over land. In fall, winter, and spring, maritime tropical and continental polar air masses interact, generating frontal activity and frequent midlatitude cyclonic storms.

Across the planet during summer months, shifting cells of subtropical high pressure block moisture-bearing winds from adjacent regions. As an example, in summer the continental tropical air mass over the Sahara in Africa shifts northward over the Mediterranean region and blocks maritime air masses and cyclonic systems. This shifting of stable, warm-to-hot, dry air over an area in summer and away from these regions in the winter creates a pronounced dry-summer and wet-winter pattern. The designation *s* specifies that at least 70% of annual precipitation occurs during the winter months. Cool offshore currents (the California Current, Canary Current, Peru Current, Benguela Current, and West Australian Current) produce stability in overlying maritime tropical air masses along west coasts, poleward of subtropical high pressure. The world climate map shows Csa and Csb climates along the western continental margins.

15. Which climates are characteristic of the Asian monsoon region?

C (mesothermal) w (winter dry) a (hot summer, warmest month above 22°C) climates are related to the winter-dry, seasonal pulse of the monsoons and extend poleward from the Aw climates. Köppen identified the wettest Cwa summer month as receiving 10 times more precipitation than the driest winter month. Cherrapunji, India, mentioned in Chapter 8 as receiving the most precipitation in a single year, is an extreme example of this classification. In that location the contrast between the dry and wet monsoons is most severe, ranging from dry winds in the winter to torrential rains and floods in the summer. Downstream from the Assam Hills, such heavy rains produced floods in Bangladesh in 1988 and 1991, among other years. A representative station of this humid subtropical wet summer and dry winter regime is Ch'engtu, China. Figure 10-19 demonstrates the strong correlation between precipitation and the high Sun of summer.

16. How can a marine west coast climate type (Cfb) occur in the Appalachian region of the eastern United States? Explain.

An interesting anomaly relative to the marine west coast climate occurs in the eastern United States. Increased elevation in portions of the Appalachian highlands moderates summer temperatures in the Cfa *humid continental* classification, producing a Cfb *marine west coast* designation. The climograph for Bluefield, West Virginia (Figure 10-17) shows that its temperature and precipitation patterns are a marine west coast type, despite its location in the east. Vegetation similarities between the Appalachians and the Pacific Northwest are quite noticeable, enticing many emigrants who relocate from the East to settle in this climatically familiar environment in the west.

17. What role do offshore currents play in the distribution of the *marine west coast Csb* climate designation? What type of fog is formed in these regions?

The dominant air mass in Cfb and Cfc *marine west coast* climates is the maritime polar, which is cool, moist, and unstable. Weather systems forming along the polar front access these regions throughout the year, making weather quite unpredictable. Coastal fog, totaling a month or two of days per year, is a part of the moderating marine influence. Frosts are possible and tend to shorten the growing season. These are advection fogs as discussed in Chapter 8.

18. Discuss the climatic designation for the coldest places on Earth outside the poles. What do each of the letters in the Köppen classification indicate?

The Dwc and Dwd *microthermal subarctic* climates occur only within Russia. Köppen selected the tertiary letter *d* for the intense cold of Siberia and north-central and eastern Asia; it designates a coldest month with an average temperature lower than –38°C (–36.4°F). A typical Dwd station is Verkhoyansk, Siberia (Figure 10-22; Figure 5-12). For four months of the year average temperatures fall below –34°C (–29.2°F). Verkhoyansk frequently reaches minimum winter temperatures that are lower than –68°C (–90°F). However, as pointed out in Chapter 5 (Figure 5-12), higher summer temperatures in the same area produce the world's greatest annual temperature range from winter to summer, a remarkable 63C° (113.4F°).

19. In general terms, what are the differences among the four desert classifications? How are moisture and temperature distributions used to differentiate these subtypes?

See the temperature criteria and moisture distribution inset graphs in Figure 10-5. The major subdivisions are the BW *deserts*, where PRECIP is less than one-half of POTET, and the BS *steppes*, where PRECIP is more than one-half of POTET. In an effort to better approximate the dry climates, Köppen developed simple formulas to determine the usefulness of rainfall based on the season in which it falls–whether it falls principally in the summer with a dry winter, or in the winter with a dry summer, or whether the rainfall is evenly distributed. Winter rains are the most effective because they fall at a time of lower POTET.

20. Relative to the distribution of dry climates, describe at least three locations where they occur across the globe, and the reasons for their presence in these locations.

The B *dry, desert and steppe* climates occupy more than 35% of Earth's land area, clearly the most extensive climate over land. The world climate map reveals the pattern of Earth's dry climates, which cover broad regions between 15° and 30° N and S. In these areas the subtropical high-pressure cells predominate, with subsiding, stable air and low relative humidity. Under generally cloudless skies, these subtropical deserts extend to western continental margins, where cool, stabilizing ocean currents operate offshore and summer advection fog forms.

21. Explain climate forecast scenarios. What is implied in *Scenario A* as compared to *Scenario B*?

The text discusses various scenarios of the expected temperature changes up to the year 2050 in three scenarios. *Scenario A* assumes that activities continue with no real change in the behavior of society, which will produce a doubling of CO_2 by 2030 (assuming that no large volcanic eruptions inject into the atmosphere material that reflects or absorbs insolation). *Scenario B* represents a modification in human activities, leading to limited emissions frozen at current levels; this merely delays the doubling of CO_2 to the middle of the century. An unrealistic, yet most desirable, *Scenario C* involves drastic reductions in emissions and would produce significant delays in climatic warming.

22. Describe the potential climatic effects of global warming on polar and high-latitude regions. What are the implications of these climatic changes for persons living at lower latitudes?

Perhaps the most pervasive climatic effect of increased warming would be the rapid escalation of ice melt. The additional water, especially from continental ice masses that are grounded, would raise sea level worldwide. Scientists are currently studying the ice sheets of Greenland and Antarctica for possible changes in the operation of the hydrologic cycle, including snowlines and the rate at which icebergs break off (calve) into the sea. The key area being watched is the West Antarctic ice sheet, where the Ross Ice Shelf holds back vast grounded ice masses.

A loss of polar ice mass, augmented by melting of alpine and mountain glaciers, will affect sea-level rise. A quick survey of world coastlines shows that even a moderate rise could bring change of unparalleled proportions. At stake are the river deltas, lowland coastal farming valleys, and low-lying mainland areas, all contending with high water, high tides, and higher storm surges. There will be both internal and international migration of affected populations, spread over decades, away from coastal flooding if sea levels continue to rise.

23. How is climatic change affecting agricultural and food production? Natural environments? Forests? The possible spread of disease?

Modern single-crop agriculture is more delicate and susceptible to temperature change, water demand and irrigation needs, and soil chemistry than is traditional multicrop agriculture. Specifically, the southern and central grain-producing areas of North America are forecast to experience hot and dry weather by the middle of the next century as a result of higher temperatures. An increased probability of extreme heat waves is forecast for these U.S. grain regions. Also, available soil moisture is projected to be at least 10% less throughout the midlatitudes over the next 30 years than present levels. Scientists are considering changing to late-maturing, heat-resistant crop varieties and adjusting fertilizer applications and irrigation.

Biosphere models predict that a global average of 30% of the present forest cover will undergo major species redistribution, the greatest change occurring in high latitudes. Many plant species are already "on the move" to more favorable locations. Land dwellers must also adapt to changing forage. Warming is already stressing some embryos as they reach their thermal limit. Particularly affected are amphibians, whose embryos develop in shallow water. The warming of large bodies of water may benefit some species, and harm others.

Recent studies suggest that climate change may affect health on a global basis. Populations previously unaffected by malaria, schistosomiasis, sleeping sickness, and yellow fever may be at greater risk in subtropical and midlatitude areas.

24. What are the possible global consequences of nuclear war?

Many ecological, biological and climatic impacts would be associated with the detonation of a small number of nuclear warheads within the biosphere. Resulting urban firestorms would produce a great pall of insolation-obscuring smoke. Surface heating would be drastically reduced. Temperatures could drop 25°C (45°F) and perhaps more in the midlatitudes.

Overhead Transparencies

As an adopter you are provided with the following figures for overhead projector use.

- ------------: A blank climograph (in the style of 10-9)
- Figure 10-2: A schematic of Earth's climate system
- Figure 10-3: Worldwide average annual precipitation

- Figure 10-6: Arrangement of Earth's climates of a hypothetical continent
- Figure 10-7:. World climates according to Köppen-Geiger classification system
- Figure 10-33: Global temperature trends

PART THREE
Earth's Changing Landscapes

Overview–Part Three

Earth is a dynamic planet whose surface is actively shaped by physical agents of change. Part Three is organized around two broad systems of these agents–the internal (endogenic) and external (exogenic). The endogenic system (Chapters 8 and 9) encompasses internal processes that produce flows of heat and material from deep below the crust, powered by radioactive decay. This is the solid realm of Earth. "The Ocean Floor" chapter-opening illustration that begins Chapter 9 is used as a bridge between these two endogenic chapters. The exogenic system (Chapters 10–14) includes external processes that set air, water, and ice into motion, powered by solar energy. This is the fluid realm of Earth's environment. These media are sculpting agents that carve, shape, and reduce the landscape. The content is organized along the flow of energy and material or in a manner consistent with the flow of events.

To assist in preparing lecture materials for Part 3, I recommend that you obtain a copy of *Geomorphology from Space–A Global Overview of Regional Landforms*, edited by Nicholas M. Short and Robert W. Blair, Jr. Washington, D.C.: Scientific and Technical Information Branch, National Aeronautics and Space Administration, 1966. This is available from the Superintendent of Documents, U.S. GPO. The 700-page volume has chapters on regional landform analysis associated with tectonic, volcanic, fluvial, deltaic, coastal, karst, lacustrine, eolian, glacial, and planetary landform processes. Also included are sections on geomorphological mapping and a futures look at global geomorphology. This book contains thousands of images and photographs, many in full color. Each chapter is accompanied by an informative text, references, and detailed source information for the images used. It is a great resource for teaching, lecture preparation, and classroom media development. Also there are two other basic reference works: the essential glossary–Bates, Robert L. and Julia A. Jackson eds., *Glossary of Geology*, 3rd ed., Alexandria, VA: American Geological Institute (AGI), 1987; and a topical source book: Smith, David G. ed., *The Cambridge Encyclopedia of Earth Sciences*, New York (London): Cambridge University Press, 1981.

11

The Dynamic Planet

Overview–Chapter 11

The twentieth century is a time of great discovery about Earth's internal structure and dynamic crust, yet much remains undiscovered. This is a time of revolution in our understanding of how the present arrangement of continents and oceans evolved. One task of physical geography is to explain the spatial implications of all this new knowledge and its effect on Earth's surface and society.

The *Student Study Guide* presents 21 "Learning Objectives" to guide the student in reading the chapter. The *Applied Physical Geography* lab manual has one exercise with seven steps that involve aspects of this chapter.

New to the Third Edition

(Note: This section highlights major changes, new features, and additions in the third edition. This does not describe all the rewrite and recast of the text.)

1. A list of key learning concepts begins the chapter.

2. Periods of geologic time are referred to more often throughout the chapter.

3. News Report #1: "Radioactivity: Earth's Time Clock," describes radiometric dating techniques which validate plate tectonic theory.

4. More elaborate description of Figure 11-4, illustrating the size and scale of Earth's interior.

5. New information concerning convection currents and how they occur.

6. Elements of the crust's composition and systems which affect the geologic cycle have been put into a bullet format for greater clarity.

7. News Report #2: "Drilling the Crust to Record Depths," describes the International Ocean Drilling Program (ODP) and other attempts which have been made to penetrate the Moho discontinuity and the mantle.

8. New photos of mineral types have been included in Table 1 and Table 2, as part of a more elaborate description of mineral and rock types.

9. New photos of rock types are located in Figure 11-9 (intrusive igneous), 11-10 (sedimentary), and 11-13 (metamorphic).

10. Mono Lake is renown for its unique tufa deposits, created by the reaction between carbonates and calcium-rich hot springs in the lake. News Report #3: "The Rescue of Mono Lake," describes the fate of this unique rock formation as changing politics affect the lake's water level. News Report #3, Figure 1, depicts this landscape.

11. News Report #4: "Concerns About Coal Reserves," describes estimated coal reserves, focusing on the rate of consumption and rate of extraction, which may deplete this resource in the next 100 years.

12. News Report #5: "Canyon Rocks Harder Than Steel," describes the ancient mountain roots, metamorphic rocks harder than steel, which are located in the Grand Canyon.

13. New examples of subduction trenches are used in discussion of plate convergence.

14. Figure 11-9, a global anomaly map which illustrates sea surface heights, taken from satellite by Scripps Institute of Oceanography and NOAA. A section of this map has been added to Figure 11-22 to complement the illustration of hot spot activity creating the Hawaiian Islands.

15. News Report #6: "Yellowstone on the Move," identifies hot spot activity across Idaho, Oregon, and Washington as the North American plate moves westward.

16. A new summary and review section ends the chapter.

Key Learning Concepts

1. *Distinguish* between the endogenic and exogenic systems, the driving force for each, and *explain* the pace at which these systems operate.

2. *Diagram* Earth's interior in cross section and *describe* each distinct layer.

3. *Illustrate* the geologic cycle and *relate* the rock cycle and rock types to endogenic and exogenic processes.

4. *Describe* Panagea and its breakup and *relate* several physical proofs that crustal drifting is continuing today.

5. *Portray* the pattern of Earth's major plates and *relate* this to the occurrence of earthquakes, volcanic activity, and hot spots.

Expanded Outline Discussion

The following headings (boldfaced) match some of the first, second, and third order headings in Chapter 11. The narrative under each heading contains information, sources, and anecdotal facts relating to portions of the chapter. Not all text headings are discussed.

THE PACE OF CHANGE

You can tie in Table 2-1, that describes the evolution of Earth's atmosphere, early life, first photosynthesis, and more recent events as ecosystems emerged. I added the percentages and details of Earth's life history to the geologic time scale in Figure 11-1 to communicate the idea of the relative scale of Earth history since the Cambrian period and the long evolutionary development of life forms.

For interesting historical background see Lawrence Badash's "The Age-of-the-Earth Debate," *Scientific American* (August 1989): pp. 90-96. Also see a good succinct review in John Thackray's *The Age of the Earth*, London: Her Majesty's Stationery Office for the Institute of Geological Sciences, 1980 (ISBN 0-11-884077-0), available in the U.S. from Cambridge University Press, New York. Although it is filled with European terminology, a good treatment of chronostratic and chronometric time scales and related dating methods is in Harland, W. Brian, Armstrong, Richard L., et al., *A Geologic Time Scale*, New York (London): Cambridge University Press (1990).

The *Student Study Guide* presents the geologic time scale for the student to fill in and work with the labels and dates.

EARTH'S STRUCTURE AND INTERNAL ENERGY: Earth in Cross Section

Figures 11-3 and 11-4 portray Earth's interior; the exploded wedge relates to the portion of the interior noted. To give the student some idea of spatial perspective, a similar wedge is compared to the distance from Halifax to San Francisco. The thinness of the crust and the enormous volume of the core and mantle should be apparent. Present research indicates that the transition zones between the regions within Earth are not smooth but instead are rough and even undulating with some sharp features. Figure 11-3 is available in the overhead transparency packet.

Earth's Magnetism

This section on Earth's magnetism and magnetic-field generation in the outer core sets the discussion of magnetic reversals and the patterns deduced by Matthews and Vine in the sea floor as shown in Figure 11-15. This ties Earth's interior to surface phenomena. Also see Kenneth A. Hoffman, "Ancient Magnetic

Reversals: Clues to the Geodynamo," *Scientific American*, May 1988: pp. 76-83. Figure 11-16 portrays the relative age of the oceanic crust placed below Figure 11-15.

Recent studies indicate that Earth's magnetic field tends to phase gradually to near zero then when it returns the field quickly reaches full strength and in opposite polarity. This uneven pattern has repeated over millions of years.

Earth's magnetic field presently is losing strength at the rate of approximately 7% per 100 years. The field was about 40% stronger 2000 years ago according to the latest published research.

Earth's Mantle Lithosphere, and Crust

The depth of convection currents in the mantle is still being investigated. Indirect evidence suggests that there are influences as deep as the Gutenberg discontinuity at the outer core-mantle boundary. Some of the undulating peaks and valleys at that transitionary boundary are perhaps triggering some mantle motion. Principal regions of heating and diapir (upward moving hot plumes) formation are still probably within 300 km (185 mi) of the surface. See: Jason Phipps Morgan and Peter M. Shearer, "Seismic constraints on mantle flow and topography of the 660-km discontinuity: evidence for whole-mantle convection," *Nature*, 365, October 1993: 506-11; among many published studies in *Nature* and *Science*.

Various metaphors can be used here: The continental plates are barges or rafts floating on the denser asthenosphere. I will sometimes ask if any students have soft-cover textbooks that I can use for a demonstration. Then, standing next to a world physiographic map or "The Ocean Floor" mural map, I will use one book as the Nazca plate and the other as the South American plate. As the plates migrate and collide (book spine to spine), I allow the sea-floor book to subduct beneath the continental book. The uplifted edge of the one book becomes the Andes crest.

The original Moho project, designed to drill through oceanic crust, was abandoned in the 1960s. The Russians have been working on the deepest penetration of the crust since 1970–the Kola well under the control of the Russian Interdepartmental Council for the Study of Earth's Interior and Superdeep Drilling. The Russians have started several deep wells. Depths are now approaching 12,000 m (39,400 ft, 7.5 mi). The deepest well in the United States is over 9000 m (29,529 ft). See: Kozlovsky, Ye. A. "The World's Deepest Well," *Scientific American* (December 1984): pp. 98-104, and the new paragraph in the text on page 318.

GEOLOGIC CYCLE

Figure 11-6 presents the geologic cycle concept as an illustration and in a schematic–featuring the tectonic cycle, rock cycle, and hydrologic cycle. Using the illustration, you can point out the elements in these 2 chapters covering endogenic systems and the 5 chapters of the exogenic system. The dynamic endogenic fluid flows that affect Earth's surface are discussed in Lawrence M. Cathles III, "Scales and Effects of Fluid Flow in the Upper Crust," *Science* Vol. 248, No. 4953 (20 April 1990): pp. 323-329.

Figure 11-6 is in the overhead packet of transparencies and allows you to follow materials from the sea-floor spreading center to subduction, reprocessing, and eventual crustal intrusion in this generalization.

Rock Cycle

Some background on economic aspects. Many different types of rock are useful and of economic importance to society. Clays of certain grades are used for pipe-making and pottery, some very fine clays are used to coat paper, such as the paper on which the textbook is printed. Pure sands, high in quartz content, are processed in glass making. Sands and gravels are an important aggregate in cement and for building construction. Lime derived from limestone is used in the making of cement and in agriculture. Phosphates from marine shales and limestone are important in the making of fertilizer. Gypsum, an evaporite derived from deposits related to sea water, is used in plaster. Marble, granite, and limestone are used as construction material, with pure white marble preferred for sculpture and decorative building facades. And, of course, salt (NaCl) is of importance to civilization throughout history. The rock cycle is intricately woven into society in many ways.

An **ore** is a body of rock which contains minerals sought by society. If the concentration of the desired mineral is high enough, mining becomes economically feasible. A major concentration process in nature for iron, lead, zinc, mercury, copper, and other minerals involves hydrothermal solutions. Sometimes associated with an intrusive igneous pluton, and sometimes with contact metamorphism, high temperature moisture solutions dissolve minerals and carry them through cracks, joints, and fractures. These valuable elements precipitate out in vein formations. In many places, individual veins of ore are traceable back to the parent igneous pluton. Sometimes these ore deposits occur in association with other vein-filling minerals such as quartz and calcite.

In other areas, *hydrothermal* solutions disseminate the mineral precipitate throughout an ore body, producing a low concentration of the desired mineral in a large mass of ore. Large-scale mining methods are used to extract such disseminated minerals. The Bingham Open Pit Copper Mine west of Salt Lake City, Utah is a prime example. The copper ore was so low-grade that economics dictated location of the concentrator, smelter, and refinery near the mine, thus reducing transportation costs. To get 6.4 kg of copper, almost 900 kg of ore had to be processed, which required the removal of 2.04 metric tons of overburden (14 lbs. of copper, 2000 lbs. of ore, 4500 lbs of overburden). Low copper prices worldwide kept the operation, along with other western copper mines closed during much of the 1980s.

Relative to the formation of mineral deposits on the ocean floor, current scientific thinking points directly to the actions of plate tectonics. For the first time, scientists aboard deep-sea submersibles saw mineral deposits actually forming on the ocean floor. Hot solutions of minerals spew from ocean floor vents that are associated with the mid-ocean ridge system. Minerals precipitate out when the hot solution comes into contact with the near-freezing temperatures of the deep ocean. Plate tectonics carries these accumulated deposits toward eventual collision and subduction beneath the continents. In areas where some of the "cargo" on the plate is not subducted but is instead raised and pasted onto the continental mass, the mineral-rich content is readily visible. On the island of Cypress in the Mediterranean Sea, deposits of copper were mined for the past 4000 years by various civilizations. Cypress is composed of old sea floor pressed upward by the African-Eurasian collision.

The deposits that do subduct melt with the diving plate and work their way toward the surface, cooling, and dissolving according to the specific nature of each element or mineral involved. Hot water trapped in the rock does the rest, dispersing the minerals in veins or disseminating the mineral ore bodies as we discussed above.

Other formation processes were at work in Sudbury, Ontario, Canada. Nickle, iron, and copper came up with a mafic (high in magnesium and iron), intrusive body of magma that cooled and began crystallization. These elements, along with other minerals, settled out in the magma chamber in specific layers, forming a very rich resource body and the basis for an active mining district.

Igneous Processes

Figure 11-8 is presented in an easy-to-use format for the student and is included in your overhead transparency packet. Aspects of silica content, resistance to weathering, increased potassium and sodium or increased calcium, iron, magnesium, melting temperatures, and coloration each is plotted along a continuum. Figure 12-24 compares dacite and basalt and can be pointed out as students see the position of these two as listed on this table.

See: Millard F. Coffin and Olav Eldholm, "Large Igneous Provinces," *Scientific American*, October 1993: 42-9. The illustration on page 48 of this article demonstrates the various effects of periods of active volcanism on the atmosphere, ocean, and crust. The figure on page 46 shows different configurations of rising plumes in the mantle, asthenosphere, and upper mantle. A world map is presented showing large igneous provinces.

Sedimentary Processes Metamorphic Processes

Photos of actual landforms and some actual samples to pass around the room could augment this discussion. New inset photographs appear in the second edition to better identify these rock types (Figure 11-10 and Table 11-4).

The example of evaporites forming in Death Valley, shown in the documentary pair of photos in Figure 11-11, was a dramatic experience. Another pair of photos taken over these same two days appears in Figure 15-16. During the 1982-83 El Niño event the southwest and California experienced record precipitation. On a day in 1983 Death Valley received 2.57 cm, or about 55% of the normal expected amount for an entire year. I waited overnight at a road block because the road into the valley was closed with playa flooding over the pavement. My wait was rewarded with incredible scenes of fluvial action in the desert.

The lake you see in Figure 11-11(a) is approximately 3 km wide and 8 km long and only several centimeters deep. I walked out about a kilometer. The natural sorting process that began at the top of the alluvial fans had left large rocks and grains behind, leaving only a very fine clay–as fine as face powder–to squish between my toes. The scene was strange because the blocked roads had let few into the valley making the aloneness, quiet, and stillness almost overwhelming. I returned one month later and matched the pictures using a slide viewer in one hand and a camera in the other to produce the picture in (b). The vast reflective water surface was replaced by a bed of borated-salt precipitate approximately 2 cm thick. The salt surface cracked underfoot exposing those very fine clays. As a result of the two trips I have matching pairs of about 50 scenes throughout Death Valley demonstrating that water is the major erosional force in the desert, however infrequent it occurs!

PLATE TECTONICS: A Brief History

The unlocking of the mysteries of Earth's dynamic crust and interior is a major scientific revolution in this century. I have found it effective to present the development of continental drift in an historic context, layering the progression of the debate and discoveries and proofs as they occurred in time. Plate tectonics is the correct inclusive terminology, which encompasses the original continental drift proposal, the sea-floor spreading mechanism, subduction, hot spots, and transform faults. If your school has the *Planet Earth* video series that appeared on PBS, your students could benefit from the numerous segments that portray this development. For instance, Drs. Vine and Matthews are interviewed as they describe their discovery of magnetic reversal patterns on the sea floor.

A further interesting account of an earlier continental drift proposal is in article by James O. Berkland, "Elisée Reclus–Neglected geologic pioneer and first (?) continental drift advocate," *GEOLOGY* Vol. 7 (April 1979): pp. 189-192.

A 16th-Century geographer suggested the basic idea of continental drift in 1596. Abraham Ortellus suggested the separation of matching coastlines in the 3rd edition of his work *Thesaurus Geographicus* (Leaf Nnn verso, Plantin, Antwerp, 1596). This slightly predates the observations of Sir Francis Bacon noted in the text. The preparation of more accurate world maps led to these early speculations because the maps disclosed a matching symmetry to landmass coastlines. Ortellus published his world map in 1598. For a survey article of this important geographer see: James Romm, "A new forerunner for continental drift," *Nature*, 367, no. 6462, February 1994: 407-08.

There are many multi-media resources available on the subject of plate tectonics. A superb interactive CD-ROM is *Interactive Plate Tectonics,* which includes videos, animation, photos, maps, and descriptive text to expose students to detailed case studies. Another CD-ROM is useful for classroom display, *The Earth*, includes photos and illustrations which may provide greater visuals to your lectures. Both of these CD-ROMs are available from Crystal Productions @ (1-800-255-8629).

Sea-Floor Spreading and Production of New Crust

The relative age of the oceanic crust is portrayed on "The Bedrock Geology of the World" map in Figure 11-16 and is available from various sources, including GSA. It is a large wall map, almost 2 m wide, with detailed legend, bibliography, and accompanying documents. The related processes are shown in Figure 11-14 that illustrates sea-floor spreading, upwelling, subduction, and plate movements. The pattern of magnetic reversals on the sea floor south of Iceland are detailed in Figure 11-15 and ties in with the patterns shown in Figure 11-16. For further detail and excellent illustrations see Kenneth C. Macdonald and Paul J. Fox, "The Mid-Ocean Ridge," *Scientific American*, June 1990: 72-9.

Subduction

The relationship of these subduction zones and inland locations of explosive volcanic activity and composite volcanic cones can be mentioned here to tie in elements of Chapter 12, and specifically, Figure 12-23. These oceanic trenches formed by subduction processes are the deepest single features on Earth's surface.

The Formation and Breakup of Pangaea; Pre-Pangaea; Pangaea; Pangaea Breaks Up; Modern Continents Take Shape; The Continents Today

Figure 11-17 takes the student through 5 progressive illustrations from 465 million years ago to the present. In Figure 11-17(e), exploded diagrams relate the mechanism of convergent, divergent, and transform plate boundaries to the world map. In Robert S. Dietz and John C. Holden's "The Breakup of Pangaea," *Scientific American* (October 1970), a map is presented that suggests the location of the continents 50 million years hence. Although not included in this edition of *GEOSYSTEMS* it makes a good discussion item to add to this lecture sequence. A global plate map is featured in Figure 11-18 with arrows to indicate rates of movement, and a world map correlating these plate boundaries to earthquake, volcano, and hot spot occurrences is in Figure 11-21. In lecture, tie these two maps, mounted together in the overhead packet, to the sequential progression in the multiple-map composite in Figure 11-17.

The idea of a "supercontinent cycle," that is, the formation and breakup of a supercontinent such as Pangaea over some irregular time sequence, is discussed in J. Brendon Murphy and R.

Damian Nance, "Mountain Belts and the Supercontinent Cycle," *Scientific American*, April 1992: 84-91.

In the *Applied Physical Geography* lab manual the students are asked to cut out continental shapes from 225 m.y.a. and assemble them properly on an elliptical projection of the world.

Plate Boundaries

Plate motions are complex. Imagine that each plate segment has its own spreading axis or axis-of-plate rotation. Added to this is Earth's differential rotation speed with latitude–higher velocities toward the equator. Each transform fault is aligned along parallels, with spreading centers roughly perpendicular to the direction of plate motion. On the Chapter 12 opening ocean-floor map these patterns can be seen.

Transform Faults

The horizontal movement (right- or left-lateral) occurs only along the portion of the plate boundary *between the spreading centers*. Beyond the spreading center, the two sides of the plate are joined along long fracture zones and move in the same direction.

The development of the San Andreas fault system in California is shown in Figure 12-17, ties in with the illustration in Figure 11-20, and can be pointed out on the ocean-floor illustration that opens Chapter 12.

Earthquake and Volcanic Activity

To augment the maps in Figures 11-18 and 11-21, the U.S. Geological Survey has a large, inexpensive wall map with art illustration that contains this information in greater detail: "The Dynamic Planet–World Map of Volcanoes, Earthquakes, and Plate Tectonics," by Tom Simkin, Robert Tilling, James Taggart, William James, and Henry Spall. Washington, DC: USGS, 9 July 1989. The map commemorates the 28th International Geological Congress and is available at USGS or USGPO.

Hot Spots

This section can also tie in to Chapter 12 and the section including Hawaii. The chemistry of these effusive eruptions is related to the asthenosphere as described, and illustrated in the geologic cycle figure 11-6, and the cross-section in 11-14.

Glossary Review for Chapter (in alphabetical order)

asthenosphere
basalt
batholith
catastrophism
coal
continental drift
core
crust
endogenic system
exogenic system
geologic cycle
geologic time scale
granite
hot spots
igneous rock
isostasy
lava
limestone
lithification
magma
magnetic reversal
mantle
metamorphic rock
mid-ocean ridge
mineral
Mohorovicic discontinuity
Pangaea
plate tectonics
pluton
rock
rock cycle
sea-floor spreading
sedimentary rock

Pangaea
plate tectonics
pluton
rock
rock cycle
sea-floor spreading
sedimentary rock
seismic waves
stratigraphy
subduction zone
tectonic processes
transform faults
uniformitarianism

Annotated Chapter Review Questions

1. To what extent is Earth's crust actively building at this time in its history?

The U.S. Geological Survey reports that, in an average year, continental margins and seafloors expand by 1.9 km^3 (0.46 mi^3). But, at the same time, 1.1 km^3 (0.26 mi^3) are consumed, resulting in a net addition of 0.8 km^3 (0.2 mi^3) to Earth's crust. The results are irregular patterns of surface fractures, the occurrence of earthquakes and volcanic activity, and the formation of mountain ranges.

2. Define the endogenic and the exogenic systems. Describe the driving forces that energize these systems.

The endogenic system (Chapters 11 and 12) encompasses internal processes that produce flows of heat and material from deep below the crust, powered by radioactive decay. This is the solid realm of Earth. The exogenic system (Chapters 13-17) includes external processes that set air, water, and ice into motion, powered by solar energy. This is the fluid realm of Earth's environment. These media are sculpting agents that carve, shape, and reduce the landscape–all under the pervasive influence of gravity.

3. How is geologic time scale organized? What is the basis for the time scale in relative and absolute terms? What era, period, and epoch are we living in today?

The geologic time scale (Figure 11-1) reflects currently accepted names and the relative and absolute time intervals that encompass Earth's history (eons, eras, periods, and epochs). The sequence in this scale is based upon the relative positions of rock strata above or below one another. An important general principle is that of superposition, which states that rock and sediment always are arranged with the youngest beds "superposed" near the top of a rock formation and the oldest at the base–if they have not been disturbed. The absolute ages on the scale, determined by scientific methods such as dating by radioactive isotopes, are also used to refine the time-scale sequence. The figure presents important events in Earth's life history along with the geologic time scale.

4. Contrast uniformitarianism and catastrophism as models for Earth's development.

Uniformitarianism assumes that the same physical processes active in the environment today have been operating throughout geologic time. The phrase "the present is the key to the past" is an expression coined to describe this principle. In contrast, the philosophy of catastrophism attempts to fit the vastness of Earth's age and the complexity of its rocks into a shortened time span. Because there is little physical evidence to support this idea, catastrophism is more appropriately considered a belief rather than a serious scientific hypothesis.

5. Make a simple sketch of Earth's interior, label each layer, and list the physical characteristics, temperature, composition, and

range of size of each on your drawing.

See details in Figures 11-3 and 11-4 as a basis for this sketch.

6. What is the present thinking on how Earth generates its magnetic field? Is this field constant, or does it change? Explain the implications of your answer.

The fluid outer core generates at least 90% of Earth's magnetic field and the magnetosphere that surrounds and protects Earth from the solar wind. A present hypothesis by scientists from Cambridge University details spiraling circulation patterns in the outer core region that are influenced by Earth's rotation; this circulation generates electric currents, which in turn induce the magnetic field. An intriguing feature of Earth's magnetic field is that it sometimes fades to zero and then returns to full strength with north and south magnetic poles reversed! In the process, the field does not blink on and off but instead oscillates slowly to nothing and then slowly regains its strength. (New evidence suggests the field fades slowly to zero then when it returns it tends to do so abruptly.) This magnetic reversal has taken place nine times during the past 4 million years and hundreds of times over Earth's history. The average period of a magnetic reversal is 500,000 years, with occurrences as short as several thousand years possible.

7. Describe the asthenosphere. Why is it also known as the plastic layer? What are the consequences of its convection currents?

The extreme upper mantle, just below the crust, is known as the asthenosphere, or plastic layer. It contains pockets of increased heat from radioactive decay and is susceptible to convective currents in these hotter (and therefore less dense) materials. The depths affected by these convection currents are the subject of much scientific speculation. Because of this dynamic condition, the asthenosphere is the least-rigid region of the mantle, with densities averaging 3.3 g/cm^3. This section of the mantle is known as the plastic layer due to its dynamic activity. About 10% of the asthenosphere is molten in asymmetrical patterns and hot spots. Think of Earth's outer crust (densities of 2.7 g/cm^3 for continental crust and 3.0 g/cm^3 for oceanic crust) as floating on the denser layers beneath, much as a boat floats on water. With a greater load (e.g., ice, sediment, mountains), the crust tends to ride lower in the asthenosphere. Convection currents in the asthenosphere disturbs the overlying crust and creates tectonic activity. In return, the movement of the crust-collision, divergence, etc. may influence currents in the mantle.

8. What is a discontinuity? Describe the principal discontinuities within Earth.

A discontinuity is a place where a change in physical properties occurs between two regions deep in Earth's interior. A transition zone of several hundred kilometers marks the top of the outer core and the beginning of the mantle. Scientists at the California Institute of Technology analyzed the behavior of more than 25,000 earthquakes and determined that this transition area is bumpy and uneven, with ragged peak-and-valley-like formations. Some of the motions in the mantle may be created by this rough texture at what is called the Gutenberg discontinuity. The boundary between the crust and the rest of the lithospheric upper mantle is another discontinuity called the Mohorovicic discontinuity, or Moho for short, named for the Yugoslavian seismologist who determined that seismic waves change at this depth, owing to sharp contrasts of materials and densities.

9. Define isostasy and isostatic rebound, and explain the crustal equilibrium concept.

The principle of buoyancy (that something less dense, like wood, floats in denser things like water) and the principle of balance were further developed in the 1800s into the important principle of isostasy to explain certain movements of Earth's crust. The entire crust is in a constant state of compensating adjustment, or isostasy, slowly rising and sinking in response to its own weight, and pushed and dragged about by currents in the asthenosphere (Figure 11-5).

10. Diagram the upper mantle and crust. Label the density of the layers in gm/cm^3. What two types of crust were described in the text in terms of rock composition. See Figure 11-5, 11-3(b) as the basis for this diagram. The two types of crust discussed in the text were oceanic crust, composed of basalt, a rock high in silica and magnesium (earning its name as *simatic* crust), and continental crust, composed mostly of granite, a rock high in silica and aluminum (earning its name as *sialtic* crust).

11. Illustrate the geologic cycle and define each component: rock cycle, tectonic cycle, and hydrologic cycle.

See Figure 11-6 as the basis of this illustration.

12. What is a mineral? A mineral family? Name the most common minerals on Earth. What is a rock?

A mineral is an element or combination of elements that forms an inorganic natural compound. A mineral can be described with a specific symbol or formula and possesses specific qualities. Silicon (Si) readily combines with other elements to produce the silicate mineral family, which includes quartz, feldspar, amphibole, and clay minerals, among others. Another important mineral family is the carbonate group, which features carbon in combination with oxygen and other elements such as calcium, magnesium, and potassium. Of the nearly 3000 minerals, only 20 are common, with just 10 of those making up 90% of the minerals in the crust. A rock is an assemblage of minerals bound together (such as granite, containing silica, aluminum, potassium, calcium, and sodium) or sometimes a mass of a single mineral, such as rock salt.

13. Describe igneous process. What is the difference between intrusive and extrusive types of igneous rocks?

Rocks that solidify and crystallize from a molten state are called igneous rocks. Most rocks in the crust are igneous. They form from magma, which is molten rock beneath the surface (hence the name igneous, which means *fire-formed* in Latin). Magma is fluid, highly gaseous, and under tremendous pressure. It is either intruded into preexisting crustal rocks, known as country rock, or extruded onto the surface as lava. The cooling history of the rock–how fast it cooled, and how steadily the temperature dropped–determines its texture and degree of crystallization. These range from coarse-grained (slower cooling, with more time for larger crystals to form) to fine-grained or glassy (faster cooling). Table 11-2 illustrates various igneous forms.

14. Characterize felsic and mafic minerals. Give examples of both coarse- and fine-grained textures.

See Table 11-2. Felsic igneous rocks are derived both in composition and name from feldspar and silica (SiO_2). Felsic minerals are generally high in silica, aluminum, potassium, and sodium, with low melting points. Rocks formed from felsic minerals generally are lighter in color and density than mafic mineral rocks. Mafic igneous rocks are derived both in composition and name from magnesium and ferric (Latin for *iron*). Mafic minerals are low in silica and high in magnesium and iron, with high melting points. Rocks formed from mafic minerals are darker in color and of greater density than felsic mineral rocks.

15. Briefly describe sedimentary processes and lithification. Describe the sources and particle sizes that comprise sedimentary rocks.

Most sedimentary rocks are derived from preexisting rocks, or from organic materials, such as bone and shell that form limestone, mud that becomes compacted into shale, and ancient plant remains that become compacted into coal. The exogenic processes of weathering and erosion generate the material sediments needed to form these rocks. Bits and pieces of former rocks–principally quartz, feldspar, and clay minerals–are eroded and then mechanically transported (by water, ice, wind, and gravity) to other sites where they are deposited. In addition, some minerals are dissolved into solution and form sedimentary deposits by precipitating from those solutions; this is an important process in the oceanic environment. The cementation, compaction, and hardening of sediments into sedimentary rocks is called lithification. Various cements fuse rock particles together; lime ($CaCO_3$, or calcium carbonate) is the most common, followed by iron oxides (Fe_2O_3), and silica (SiO_2). Particles also can be united by drying (dehydration), heating, or chemical reactions. The two primary sources of sedimentary rocks–the mechanically transported bits and pieces of former rock and the dissolved minerals in solution–are known as clastic sediments and chemical sediments, respectively. See Table 11-3 for the range of clast sizes and the form these materials take as lithified rock.

16. What is metamorphism and how are metamorphic rocks produced? Name some original parent rocks and their metamorphic equivalents.

Any rock, either igneous or sedimentary, may be transformed into a metamorphic rock by going through profound physical and/or chemical changes under increased pressure and temperature. (The name metamorphic comes from the Greek, meaning to *change form*.) Metamorphic rocks generally are more compact than the original rock and therefore are harder and more resistant to weathering and erosion. See Table 11-4.

17. Briefly review the history of continental drift, sea-floor spreading, and the all-inclusive plate tectonics theory. What was Alfred Wegener's role?

In 1912, German geophysicist and meteorologist Alfred Wegener publicly presented in a lecture his idea that Earth's landmasses migrate. His book, *Origin of the Continents and Oceans*, appeared in 1915. Wegener today is regarded as the father of the concept called continental drift. Wegener postulated that all landmasses were united in one supercontinent approximately 225 million years ago, during the Triassic period, Figure 11-17(b). The fact that spreading ridges and subduction zones are areas of earthquake and volcanic activity provides further evidence for plate tectonics, which by 1968 had become the all-encompassing term for these crustal processes.

18. Define upwelling and describe related features on the ocean floor. Define subduction and explain the process.

The worldwide submarine mountain ranges, called the mid-ocean ridges, were the direct result of upwelling flows of magma from hot areas in the upper mantle and asthenosphere. When mantle convection brings magma up to the crust, the crust is fractured and new seafloor is formed, building the ridges and spreading laterally. When continental crust and oceanic crust collide, the heavier ocean floor will dive beneath the lighter continent, thus forming a descending subduction zone (Figure 11-14). The world's oceanic trenches coincide with these subduction zones and are the deepest features on Earth's surface.

19. What was Pangaea? What happened to it during the past 225 million years?

See the sequence of illustrations in Figure 11-17, a - e. The supercontinent of Pangaea and its subsequent breakup into today's continents represents only the last 225 million years of Earth's 4.6 billion years, or only the most recent 1/23 of Earth's existence.

20. Characterize the three types of plate boundaries and the actions associated with each type.

The boundaries where plates meet are clearly dynamic places. Divergent boundaries are characteristic of seafloor spreading centers, where upwelling material from the mantle forms new seafloor, and crustal plates are spread apart. Convergent boundaries are characteristic of collision zones, where areas of continental and/or oceanic crust collide. These are zones of compression. Transform boundaries occur where plates slide laterally past one another at right angles to a sea-floor spreading center, neither diverging nor converging, and usually with no volcanic eruptions.

21. What is the relation between plate boundaries and volcanic and earthquake activity?

Plate boundaries are the primary location of Earth's earthquake and volcanic activity, and the correlation of these phenomena is an important aspect of plate tectonics because they are produced by plate/asthenosphere interactions at these boundaries. Earthquakes and volcanic activity are discussed in more detail in the next chapter, but their general relationship to the tectonic plates is important to point out here. Earthquake zones and volcanic sites are identified on the world plate map in Figure 11-21.

22. What is the nature of motion along a transform fault? Name a famous example of such a fault.

Transform boundaries occur where plates slide laterally past one another at right angles to a sea-floor spreading center. These plates are not diverging or converging, and there is usually no volcanic activity associated with transform boundaries. A famous example of a transform fault is the San Andreas fault, see Figures 12-15 and 12-16.

Overhead Transparencies

As an adopter you are provided with the following figures for overhead projector use.

- Figure 11-1: Geologic Time Scale
- Figure 11-3: Earth in cross-section
- Figure 11-5: Isostatic adjustment of the crust
- Figure 11-6: The geologic cycle rock cycle, tectonic cycle, hydrologic cycle
- Figure 11-8: Igneous rock types
- Mounted together: Figure 11-18 Earth's 14 lithospheric plates and their motions(top); and Figure 11 21-earthquake and volcanic activity locations (bottom)

12

Tectonics and Volcanism

Overview

The chapter opening illustration, "The Ocean Floor," is a bridge between Chapter 11 and this chapter. The plate map in Figure 11-18 and the map of earthquake and volcanic occurrence in Figure 11-21, as well the world structural map in Figure 12-15, all correlate within this opening illustration.

Tectonic activity has repeatedly deformed, recycled, and reshaped Earth's crust during its 4.6-billion-year existence. The principal tectonic and volcanic zones lie along plate boundaries. The arrangement of continents and oceans, the origin of mountain ranges, and the locations of earthquake and volcanic activity are all the result of dynamic endogenic processes. In this chapter we build the continental crust as a basis of world structural regions in Figure 12-15. Then the often dramatic earthquakes and volcanoes are covered.

The *Student Study Guide* presents 20 "Learning Objectives" to guide the student in reading the chapter.

New to the Third Edition

(Note: This section highlights major changes, new features, and additions in the third edition. This does not describe all the rewrite and recast of the text.)

1. A list of key learning concepts begins the chapter.

2. An updated introduction, including recent tectonic activity, demonstrates the significance of studying geography.

3. News Report #1: "James Michener On Terranes and Tectonics in *Alaska*," describes the 1988 book which addresses accreted terranes.

4. Are the number of earthquakes increasing in frequency or has our technology to identify quakes simply given this appearance? This topic is addressed in News Report #2: "Are Earthquakes on the Increase?"

5. News Report #3: " A Tragedy in Kobe, Japan - The Hyogo-ken Nanbu Earthquake," describes the earthquake which devastated Kobe, Japan in January, 1995. Photos of the Kobe disaster are included.

6. New statistics on recent earthquakes and volcanic explosions have been included in this chapter.

7. Figure 12-20 plots seismic activity in the United States between 1899 and 1990.

8. The Moment Magnitude Scale, in use since 1993, is explained.

9. Earthquake damage usually fades with distance from the epicenter. News Report #4: "Damage Strikes Far From Epicenters," describes two exceptions to this rule, areas of unstable bedrock in Mexico City and San Francisco.

10. News Report #5: "Seismic Gaps, Nervous Animals, Dilitancy, and Radon Gas," describes the science of forecasting earthquakes.

11. A new map illustrating the distribution of extensive igneous provinces is found in Figure 12-28.

12. Focus Study 1 includes an updated discussion of volcanic eruption and recent recovery in the Mount St. Helens region. See Focus Study, Figure 2.

13. More elaborate discussion of the Cordilleran system, the Alpine system, the Euro-Himalayan system, and continental shields.

14. New illustrations to clarify Figures 12-12, horst and graben landscapes, and 12-13, a cross-section of the Tetons.

15. The global effects of Mount Pinatubo's eruption are identified; altering pollution levels, weather patterns, and the amount of radiation received at the surface.

16. News Report #6: "Is the Long Valley Caldera Next," describes the history of the Long Valley Caldera. In this popular recreation area, the combination of volcanic gas production and earthquake activity may suggest a possible eruption within the next century.

17. More elaborate description of plateau basalts and downwarping.

18. A new summary and review section ends the chapter.

Key Learning Concepts

1. *Describe* first, second, and third orders of relief and *relate* examples of each from Earth's major topographic regions.

2. *Describe* the several origins of continental crust and *define* displaced terranes.

3. *Explain* compressional processes and folding; *describe* four principal types of faults and their characteristic landforms.

4. *Relate* the three types of plate collisions associated with orogenesis and *identify* specific examples of such.

5. *Explain* the nature of earthquakes, their measurement, and the nature of faulting.

6. *Distinguish* between an effusive and an explosive volcanic eruption and *describe* related landforms using specific examples.

Expanded Outline Discussion

The following headings (boldfaced) match some of the first, second, and third order headings in Chapter 12. The narrative under each heading contains information, sources, and anecdotal facts relating to portions of the chapter. Not all text headings are discussed.

THE OCEAN FLOOR

Use the chapter opening illustration as a focal point for integrating all the elements of plate tectonics and the present state of Earth's surface and ocean basins. Each section can be referenced back to this illustration, e.g., San Andreas fault, types of plate convergence, types of volcanic processes and features.

Several figures in Chapters 11 and 12 can be referenced at this point in the text and correlated to this opening illustration: Figure 11-14, sea-floor spreading, etc.; Figure 11-15, relative age of the oceanic crust; Figures 11-18, 11-20, 11-21, and 11-22, various features related to plate tectonics; Figure 12-15, world structural regions; and Figure 12-22, illustrating the mechanisms of volcanic activity, along with insets 12-22 and 12-23.

An ocean floor illustration is available from the National Geographic Society (17th and M Streets N.W., Washington, D.C. 20036; 1-800-638-4077). A world physical map without water or ice titled "World Ocean Floor" (#02683, Mercator projection) costs under $10 and is almost 1.5 m (5 ft) wide; although not the Marie Tharp version, it serves well as a reference wall map. NGS has also prepared a new rendition in their *Atlas of the World*, 6th edition.

EARTH'S SURFACE RELIEF FEATURES: Crustal Orders of Relief

A good idea when discussing Earth's topographic regions (Figure 12-2) is the consideration of your local region's location. An introduction to world regional geomorphology is in Bridges, Edwin M., *World Geomorphology*, New York (London): Cambridge University Press (1990). Following a brief 30-page introduction to Earth's interior structure, time scale, orders of relief, continental drift and plate tectonics, the next 230 pages present a regional treatment of the geomorphology of each plate.

CRUSTAL DEFORMATION PROCESSES: Continental Crust; Terranes

This section follows the construction of the crust, beginning with the shields, adding the terranes, then the deformation processes of folding and faulting, orogenesis, and ending with the world structural regions map in Figure 12-15. An excellent color map of the terranes surrounding the entire Pacific region appears in David G. Howell's "Terranes," *Scientific American* (November 1985): pp. 116-125.

An entire issue of *EPISODES*-International Geoscience Newsmagazine, Vol. 10, No. 4 (December 1987): 238-295 (from International Union of Geological Sciences, Room 177, 601 Booth Street, Ottawa, Ontario, Canada K1A 0E8), is devoted to the formation and structure of North America, and the United States in particular. The issue includes excellent maps and discussions on Alaska's geological framework, the Hawaiian Islands, geological hazards research programs, minerals and fossil fuels, and an overview of geological research and educational institutions.

For more on the growth of continents, terranes, and orogenic collage formation see A.M.C. Sengör, B.A. Natal'in, and V.S. Burtman, "Evolution of the Altaid tectonic collage and Paleozoic crustal growth in Eurasia," *Nature*, 364, July 22, 1993: 299-307. The article features color maps of Eurasian topography, tectonics, and identification of individual terranes added in a vast collage along the periphery of the continental heart.

Deformation Processes: Folding; Faulting

Earth's crust bends and breaks along zones of *strain*, which are *stress*-induced zones in deformed rock. The strain is caused by *compression* (shortening or folding), *tension* (stretching or faulting), and *shearing* (tangential or transform stress created when two pieces slide past each other), as shown in Figure 12-6.

Orogenesis

Examination of a world map reveals two large mountain chains. The relatively young mountains along the western margins of the North and South American plates stretch from the tip of Tierra del Fuego to the massive peaks of Alaska, comprising the Cordilleran system. The mountains of southern Asia, China, and northern India continue in a belt through the upper Middle East to Europe and the European Alps, constituting the Eurasian-Himalayan system (see Figure 12-15). A good overview of mountain building and the types of mountains is in Peter Molnar's "The Structure of Mountains," *Scientific American* (July 1986): pp. 70-79.

Appalachians

Again, having the American Geological Institute (AGI) *Glossary of Geology*, 3rd ed., is useful. The mountain-building event, the Alleghany orogeny, that deformed rocks of the Valley and Ridge province and the Allegheny Plateau are spelled with an "a" and an "e," respectively. H.P. Woodward (1957, 1958) introduced the term's spelling (see Figure 12-14 for illustration).

EARTHQUAKES

Earthquakes and volcanism naturally attract student interest. I opened this section using the setting of baseball's 1989 World Series to relate seismology to an experience that many people shared that evening. It is possible to overstress the "gee whiz" aspects of damaging physical events but it seems appropriate in an effort to grab the attention of our movie-going student audience, which is so accustomed to visual action. For a brief yet thorough overview with many illustrations and excellent photographs see Susanna van Rose, *Earthquakes*, London: Her Majesty's Stationery Office for the Institute of Geological Sciences, 1983 (ISBN 0-11-884066-5), available in the U.S. from Cambridge University Press, New York.

Students at California State University, Northridge, in the northern portion of the San Fernando Valley of southern California, certainly do not need reference to an interrupted World Series game to get their attention. On January 17, 1994, their campus was the approximate epicenter for the most devastating earthquake in U.S. history in terms of property damage–the 6.6 Northridge earthquake and thousands of aftershocks. Rebuilding the extensively damaged campus is estimated to cost $300 million! Spring classes still began but started late in February in several hundred portable trailers. More on this a bit later.

Earthquake Essentials
The Nature of Faulting

To teach the essentials it works well to use specific earthquakes as illustrations of specific seismic phenomena: 1906 San Francisco for transform faults, 1985 Mexico and 1989 Loma Prieta for the distant damage magnified by unstable materials many miles from the epicenter, 1964 Alaskan and others for subduction zones and filling of seismic gaps, 1971 San Fernando for overthrust motions, or the October 1991 earthquake in India that killed nearly 2000 (6.1R) as part of the continuing collision of India and Asia. An additional reference: Scholz, Christopher H., *The Mechanics of Earthquakes and Faulting*, New York (London): Cambridge University Press, 1990. Among many topics, asperities and the asperity model are presented (pp. 45-53, 216-19). The last chapter of this work is entitled "Earthquake Prediction and Hazard Analysis."

I recommend that you obtain a copy of a USGS book which contains great color maps and schematics, photographs, and a series of articles on the San Andreas fault: *The San Andreas Fault System, California*, USGS Robert E. Wallace, ed., Professional Paper 1515, Washington: USGS, 1990, 283 pp.

The San Francisco Earthquakes

In keeping with the perception and political theme of Figure 12-21, one of the initial efforts following the 1906 quake by city officials was to put a proper "spin" on the earthquake itself. Public relation efforts centered on blaming fire as the main destructive agent–a preventable danger–rather than the unknown danger of future earthquakes that could destroy the growth and development of San Francisco in the century ahead. These efforts were documented by an archivist from the San Francisco Library.

See the serial publication *Earthquakes and Volcanoes* Vol. 21, No. 6 (1989), USGS, Department of the Interior. Not only does the issue summarize all aspects of the Loma Prieta earthquake of 17 October 1989, but it also includes 9 pages of invaluable information listings, general readings, personal preparedness suggestions, regional planning overviews, seismic zoning and land-use listings, safety planning information, sources of damage estimates, and a complete listing of map references. Subscriptions to this journal are $6.50 per year.

You also might want to obtain a copy of "The Loma Prieta Earthquake of October 17, 1989" from USGS, November 1989 (Revised 1990), 16 pages, prepared by Peter Ward and Robert Page, and including the same extensive references that are in the related issue of *Earthquakes and Volcanoes*. This publication includes an excellent map of earthquake occurrences in the United States as well as a dimensional fault-plane solution illustration that is the basis of Figure 12-18 in the text. Several pages are devoted to an analysis of how the quake filled a seismic gap, as discussed in the next section of this chapter. A lot of the material in this publication is suitable for use in overhead transparencies.

An interative videodisc entitled, "The Great Quake of 1989," is available from Crystal Productions (1-800-255-8629). This disc investigates the causes of earthquakes, includes footage of the Loma Prieta earthquake, the 1906 San Francisco earthquake, and offers tips on earthquake preparedness.

One last item, from the California Division of Mines and Geology (1416 Ninth Street, Room 1341, Sacramento, CA 95814, 916-445-1825) their serial publication *California Geology*, is free of charge in single copies. The Loma Prieta earthquake is detailed in one issue, Vol 43, No. 1 (January 1990) and has excellent illustrations and seismological analysis of the fault plane solution, depth-distance graphs of shocks, and accelerogram records.

Southern California Earthquakes of 1992–and the Future

This section focuses on the sequence of earthquakes taking place along the San Andreas system in interior southern California. Figure 12-17 illustrates in a general way the formation of the San Andreas fault system of California as a series of NW - SE trending transform faults (right-lateral, strike-slip motion). A blow-up inset map details the epicenters of the 1992 quakes.

The Northridge earthquake in January 1994 occurred along a deeply buried thrust fault north of the Santa Monica fault (see inset map in text). The Los Angeles basin is being subjected to intense compression as portions of the Pacific Plate press northward against the east-west trending San Gabriel Mountains. The most damaging earthquakes to strike this region may indeed be on these thrust faults beneath the surface rather than north and east of Los Angeles along the San Andreas. Note on page 358, the text states, "Beneath the Los Angeles basin, such overthrust faults produce a high risk of earthquakes," and p. 370, "Scientists want to determine if these events are building up to the proverbial *big one* for southern California."

The role of space geodesy (GPS-global positioning system) coupled with geological and geophysical observations can provide a powerful tool to predict rates and types of fault movement–this played a role in the Northridge quake. A paper sent to *Nature* in April 1993 and accepted for publication in October 1993 identified the mechanism and mapped the suspected fault. The research paper is: Andrea Donnellan, Bradford H. Hager, and Robert W. King, "Discrepancy between geological and geodetic deformation rates in the Ventura basin, " *Nature*, 366, November 25, 1993: 299-301, 333-36. The authors state: "Our modeling suggests that the faults bounding the basin are locked at the surface, but are slipping at depths below 2-5 km." In

Robert S. Yates' summary of the research several color cross-section solutions of the thrust fault mechanisms are presented. Be sure to get a copy of this interesting article and look at our modern Earth system sciences in operation.

The following illustration (also Figure 12-19) is a preliminary and generalized sketch of a candidate fault-plane mechanism for the January 17, 1994, Northridge earthquake.

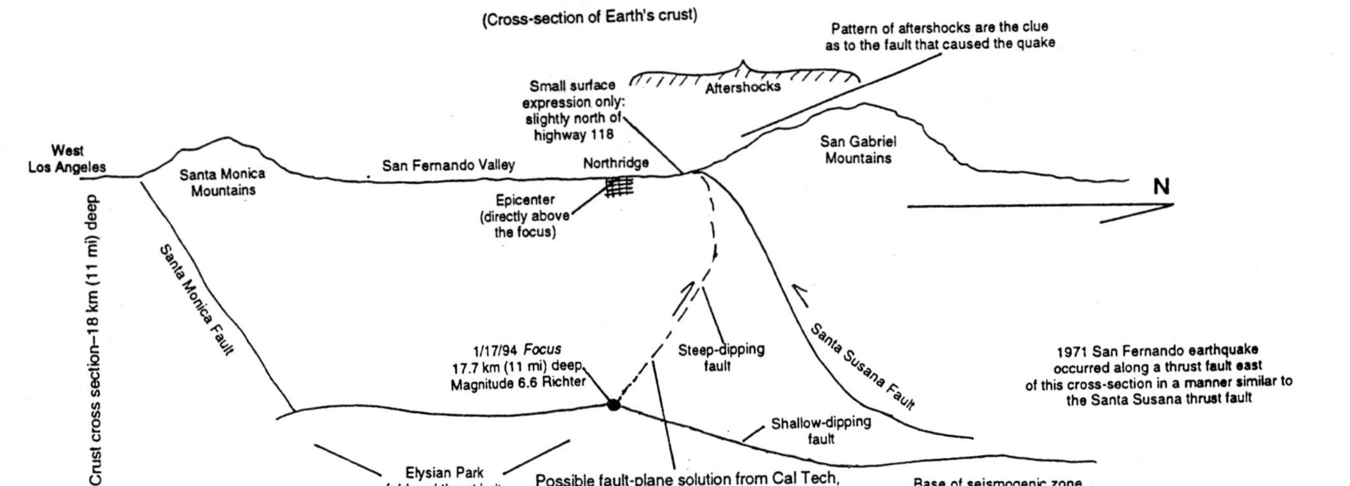

Paleoseismology and Earthquake Forecasting

An analysis of the 8.1 Mexico earthquake that hit 19 September 1985 was written by J.G. Anderson, P. Bodin, *et al.*, "Strong Ground Motion from the Michoacan, Mexico, Earthquake," *Science* Vol. 233 (5 September 1986): pp. 1043-1049. This expected quake occurred in a well-established seismic gap as did the 1964 Alaskan and the 1989 Loma Prieta quakes.

Up-to-date information on the most recent earthquake activity is available from a variety of Internet sources. Contact the USGS for the most recent quakes in Northern California at: quake@andreas.wr.usgs.gov, quake information for Southern California is available at: quake@scec.caltech.edu. Worldwide quake information can be found at:

quake@geophys.washington.edu.

Weekly and historical seismic reports are available from: Seismology Resources under the USGS Seimiology and Tectonophysics Information heading at: earth.nwu.edu.

Earthquake Preparedness and Planning

We should not use terms such as earthquake prevention or even use of the words "earthquake-proof" when referring to earthquakes–these terms are not used in geoscience nor in civil engineering disciplines; rather they are more frequently used in the political arena. Detailed discussion of hazards and hazard perceptions appear in the writings of Ian Burton (University of Toronto) and Robert W. Kates (Clark University).

Figure 12-21, "Socio-economic impacts and adjustment to earthquakes," can be a powerful teaching tool relative to hazard perception and the political-economic arena. As the text states:

> **A valid and applicable generalization seems to be that humans are unable or unwilling to perceive hazards in a familiar environment. In other words, we tend to feel secure in our homes and communities, even if they are sitting on a quiet fault zone. Such an axiom of human behavior certainly helps explain why large populations continue to live and work in earthquake-prone settings in developed countries. Similar questions also can be raised about populations in areas vulnerable to other disasters in areas vulnerable to floods, droughts, coastal storm surge, or hurricanes (p. 370).**

Given this axiom, try substituting other natural and human hazards in the "earthquake prediction" box in Figure 12-21. You can place floodplain zoning, global warming prediction, stratospheric ozone depletion, prevention of toxic waste dumping by industry, air bags in cars, cigarette smoking–whatever–in the box and read the litany of socioeconomic impacts as assessed by related special interests. On a global scale, try placing "world peace and disarmament," in the box and the students can see the economic impact posed. This illustrates the oft-cited "that strategy will hurt jobs and the economy" or, "that just can't be valid–it would be too damaging to the economy." I have included this chart in the text because of this versatility and the political and economic reality it portrays. The irony is that proper planning, zoning, and proactive tactics would actually save lives and money in the long run.

Las Vegas shook and quaked on command, not from the "action" in the city, but from human-made earthquakes induced by underground testing of nuclear warheads at the Nevada Test Site north of Las Vegas. The first indication that humans could create earthquake-like shock waves came from these tests in the 1950s through the 1980s.

Denver, Colorado had not experienced earthquakes in 90 years, at least nothing greater than a 3.0 Richter. Early in 1962, quakes mysteriously began to occur and continued for four years, some strong enough to cause damage. The spatial distribution of the quake epicenters were focused north of Denver. In late 1965, the announcement was made that the recent quake activity was created by the U.S. Army. The Army was pumping chemical wastes into a 3658 m (12,000 ft) deep well. These wastes were causing fluid pressure in basement rock that increased sliding along fracture zones– fluids under pressure were "lubricating" shear zones. The movements were felt in Denver as earthquakes. Pumping records for the wells confirmed the correlation; the Army reluctantly halted the pumping. A confirming experiment was conducted at an abandoned oil field near Rangely, Colorado between 1969 and 1973. Water was pumped in and subsequently out of wells in the field, which was specially instrumented for the tests. The USGS reported a strong correlation between fluid injection, producing fluid pressure, and earthquakes. When the water was withdrawn, the earthquakes ceased.

The knowledge of the role of fluid pressure and earthquakes is helping us to understand the safety of large dams and their possible seismic effects. In August 1975, the largest earthfill dam in North America was rocked by a 5.7 Richter earthquake that created little damage but much interest. Evidently, fluid pressure related to the rapid filling of the reservoir the previous spring had caused the release of accumulated strain along a susceptible fault zone. The Oroville Dam in east central California, 236 m (775 ft) high with a capacity of 4300 million m^3, was filled by early 1969. Although conclusions are still unresolved, Dr. Bruce Bolt of the University of California later wrote:

> **Undoubtedly, it [the nearby reservoir] sent a pressure pulse through the water in the rocks of the crust. Perhaps, as the pressure pulse spread out by percolation through the crustal rocks nearby, it eventually reached a weak place along an already existing fault zone....it may have been sufficient to open microcracks just enough to allow fault slip (*Earthquakes: A Primer*, W.H. Freeman, 1978, pp. 129-130).**

This event added an important parameter to help determine the behavior of a dam and aspects of dam safety. Estimates are that 5 percent of the world's large dams are candidates for seismic effects, although fewer than a dozen cases have been thoroughly studied. On the other hand, other researchers have found evidence that filling a reservoir creates a "loading" situation that actually reduces earthquake frequency in the area of the reservoir body.

It is interesting to compare all this with such older publications as Frank Press's "Earthquake Prediction" *Scientific American* Vol. 232, No. 5 (May 1975): pp. 14-23, or with the cover story of *TIME* Magazine, 1 September 1975, pp. 37-42, entitled "Forecast: Earthquake." The hazard perception wheel is reinvented over and over, and we are destined to repeat history until education turns the tide on perception and consciousness! More recently, Robert L. Wesson and Robert E. Wallace, in "Predicting the Next Great Earthquake in California," *Scientific American* Vol. 252, No. 2 (February 1985): pp. 35-43, discuss asperities and asperity breaks and possible prediction methods.

At this time the cost of the Northridge quake will exceed by at least $10 billion the record $20 billion price tag caused by Hurricane Andrew!

VOLCANISM

The 1991 eruption of Mount Pinatubo in the Philippines has exceeded the 1982 El Chichón eruption in the production of aerosols and stratospheric sulfuric acid mist. This eruption exceeded others this century, including Mount Katmai, Alaska, that extruded 12 km^3 of pyroclastics. NASA is measuring the volcano's impact with the TOMS (Total Ozone Mapping Spectrometer) on the *Nimbus-7* satellite, that can detect the AOT and the interference in the ozone signal caused by sulfur dioxide from the eruption as it spreads worldwide. *GEOSYSTEMS* presents remotely sensed images of this eruption in Figure 6-1. This scale of an eruption potentially can confuse warming signals of the greenhouse effect, although October 1991 was one of the warmest Octobers in instrumental history. Mount Pinatubo's eruption began in June 1991 and is providing further evidence of the dynamic circulation of the atmosphere. See an initial analysis by Richard A. Kerr, "Huge Eruption May Cool the Globe," in *Science* Vol. 252, No. 5014 (28 June 1991): p. 1780.

Many USGS materials are available in the form of reports, maps, images, posters, and slides for your presentation. For example, Robert Tilling, Christina Heliker, and Thomas Wright's, USGS, *Eruption of Hawaiian Volcanoes: Past, Present, and Future*, 1987; the U.S. Forest Service, USGS, and State of Washington's, *Mount St. Helens and Vicinity*, 1:100,000-scale map series (large topo map); and, of course, the 2 volumes, Decker, Robert W., Thomas L. Wright, and Peter H. Stauffer, *Volcanism in Hawaii*, 2 vols. U.S. Geological Survey Professional Paper 1350, 1987; as well as Robert and Barbara Decker's book, *Mountains of Fire–The Nature of Volcanoes*, New York (London): Cambridge University Press (1991). The Deckers' book has an insert of very dramatic photographs of volcanic processes and activities. A good, large wall map-mural is called "The Dynamic Planet-World Map of Volcanoes, Earthquakes, and Plate Tectonics" USGS and Tom Simkin, Robert Tilling, et al., 9 July 1989. Despite its size, color, and complexity, it costs under $5. Relative to the Hawaiian volcanoes, a direct phone call to Hawaii Volcanoes National Park can yield slide sets and maps by mail (808-967-7311). You can obtain current eruption information directly for Hawaiian volcanoes by calling 808-967-7977.

There are a number of interactive videodiscs available from Crystal Productions (1-800-255-8629) on the subject of volcanism. *Volcanoes* is a two part videodisc which examines volcanoes is Hawaii, Iceland, and Mexico, and includes discussion on how volcanic eruptions affect people and the environment. *Inside Hawaiian Volcanoes,* is a videodisc from the Smithsonian. This disc includes great photos of volcanic eruptions, and also illustrates the volcanic activity under the surface to explain the causes of such activity.

Locations of Volcanic Activity

The boundaries of the plates are places of dynamic action in the form of earthquakes, volcanic eruptions, deformation of the crust, and mountain building. Hot spots of activity take place at plate boundaries as well as far inland. The Hawaiian Islands are produced by such a hot spot. Figure 11-22 illustrates these relationships.

Types of Volcanic Activity (Effusive Eruptions, Explosive Eruptions)

I simplify the classification of volcanoes into effusive eruptions (shield volcanoes, fissure eruptions, plateau (flood) basalts, lava flows) and explosive eruptions (composite volcanoes, conical mountains, dramatic outbursts of pyroclastics and incandescent gases). The variety of forms among volcanoes makes them hard to classify; most fall in transition between one type and another. Even during a single eruption, a volcano may behave in several different ways. The two primary factors in determining an eruption type are the magma's chemistry, which is related to its originating processes, and its viscosity. Viscosity refers to a magma's resistance to flow, or degree of fluidity; it ranges from low viscosity (very fluid) to high viscosity (thick and flowing slowly).

Materials from the mantle rise at the spreading centers, flow outward as ocean floor, and collide with the continents, only to plunge back into the mantle, remelt, and work back up to form more continent as intrusive igneous rock. The upwelling material along the mid-ocean ridges and hot spots, originating in the asthenosphere, is less than 50% silica and is rich in iron and magnesium. The resulting magma is quite fluid and forms effusive eruption features. However, a subducting oceanic plate works its way under a continental plate, taking with it sediment and trapped water, plus various elements that are melted from the crust and drawn into the subducting mixture. As a result, the molten magma that migrates upward from a subducted plate contains 50-75% silica, is high in aluminum, and has a viscous (thick) texture. Bodies of viscous magma may reach the surface in explosive volcanic eruptions, or they may stop short and become subsurface intrusive bodies in the crust, cooling slowly to form crystalline plutons. You can draw these contrasts with the maps in the book and poster-maps suggested. See: Robert S. White and Dan P. McKenzie, "Volcanism at Rifts," *Scientific American*, July 1989: 62-71.

See Robert I. Tilling and John J. Dvorak, "Anatomy of a basaltic volcano," *Nature*, 363, May 13, 1993: 125-33, for an analysis of the Kilauea volcano in Hawaii. Magma rises from a depth of 80 km to the surface and erupts through cracks and rifts created by the pressures generated. The article has several maps and illustrations of these processes.

Glossary Review for Chapter 12
(in alphabetical order)

anticline
caldera
cinder cone
circum-Pacific belt
composite volcano
continental platforms
continental shield
crater
earthquake
effusive eruption
elastic-rebound theory
explosive eruption
faulting
folding
geothermal energy
graben
horst
lava
moment magnitude scale
normal fault
ocean basins
orders of relief
orogenesis
plateau basalts
pyroclastics
relief
reverse fault
Richter scale
ring of fire
seismograph
shield volcano
strike-slip fault
syncline
terranes
thrust fault
topography
volcano
Wrangellia terranes

Annotated Chapter Review Questions

1. How does the ocean floor map (chapter - opening illustration) bear the imprint of the principles of plate tectonics? Briefly analyze.

The illustration that opens this chapter is a striking representation of Earth with its blanket of water removed. The scarred ocean floor is clearly visible, its sea-floor spreading centers marked by over 64,000 km (40,000 mi) of oceanic ridges, its subduction zones indicated by deep oceanic trenches, and its transform faults stretching at angles between portions of oceanic ridges. The correlation with various figures in Chapters 11 and 12 is clearly visible.

2. What is meant by an "order of relief"? Give an example from each order.

Geographers group the landscape's topography into three orders of relief. These orders classify landscapes by scale, from vast ocean basins and continents down to local hills and valleys. The first order of relief consists of continental platforms and oceanic basins. Examples of first order features would be the Pacific Ocean basin and the African continent. Intermediate landforms are considered to be second orders of relief, such as continental masses, mountain masses, plains and lowlands. A few examples are the Alps, Canadian and American Rockies, west Siberian lowland, and the Tibetan Plateau. Great rock cores, or shields, are second order features, see Figure 12-3. In ocean basins, second order features include rises, slopes, abyssal plains, mid-ocean ridges, and submarine trenches. A few examples of these would be the Sohm Abyssal Plain, the Mid-Atlantic Ridge, and the Peru-Chile Trench. Third order features are the most detailed forms of relief, consisting of individual mountain, cliffs, valleys and other landforms of smaller size, such as Mount Diablo in California, and Ayres Rock in Australia (Figure 15-20).

3. Explain the difference between relief and topography.

Relief refers to vertical elevation differences in the landscape, examples include the low relief of Nebraska and high relief in the Himalayas. Topography is the term used to describe Earth's overall relief, its changing surface form, effectively portrayed on topographic maps.

4. What is a craton? Relate this structure to continental shields and platforms, and describe these regions in North America.

All continents have a nucleus of old crystalline rock on which the continent grows. Cratons are the cores, or heartland regions, of the continental crust. They generally are low in elevation and old (Precambrian, more than 570 million years in age). Those regions where various cratons and ancient mountains are exposed at the surface are called continental shields. Figure 12-3 shows the principal areas of shield exposure.

4. What are migrating terranes and how do they add to the formation of continental masses?

Each of Earth's major plates is actually a collage of many crustal pieces acquired from a variety of sources. Accretion, or accumulation, has occurred as crustal fragments of ocean floor, curved chains (or arcs) of volcanic islands, and other pieces of continental crust have been swept aboard the edges of continental shields. These migrating crustal pieces, which have become attached to the plates, are called terranes. At least 25% of the growth of western North America can be attributed to the accretion of terranes since the early Jurassic period (190 million years ago). A good example is the Wrangell Mountains, which lie just east of Prince William Sound and the city of Valdez, Alaska. The Wrangellia terranes–a former volcanic island arc and associated marine sediments from near the equator–migrated approximately 10,000 km (6200 mi) to form the Wrangell Mountains and three other distinct areas along the western margin of the continent (Figure 12-5).

6. Briefly describe the journey and destination of the Wrangellia Terranes.

The Wrangellia Mountains, which lie just east of Prince William Sound and the city of Valdez, Alaska, are made up of a former volcanic arc and associated marine sediments from near the equator. These terranes migrated approximately 10,000 km (6200 mi) to form distinct formations along the western margin of North America.

7. Diagram a simple folded landscape in cross section, and identify the features created by the folded strata.

See Figure 12-7.

8. Define the four basic types of faults. How do these relate to earthquakes and seismic activity?

See Figure 12-10 and related text section. When rock strata are strained beyond their ability to remain a solid unit, they fracture, and one side is displaced relative to the other side in a process known as faulting. Thus, fault zones are areas of crustal movement. At the moment the fault line shifts, a sharp release of energy occurs, called an earthquake or quake.

9. How did the Basin and Range Province evolve in the western United States? What other examples exist of this type of landscape?

The Basin and Range Province, in the interior western United States, experienced tensional forces caused by uplifting and thinning of the crust, which cracked the surface to form aligned pairs of normal faults and a distinctive landscape (please refer to Figures 15-22 and 15-23). The term horst is applied to upward-faulted blocks; graben refers to downward-faulted blocks. Examples of horst and graben landscapes include the Great Rift Valley of East Africa (associated with crustal spreading), which extends northward to the Red Sea, and the Rhine graben through which the Rhine River flows in Europe.

10. Define orogenesis. What is meant by the birth of mountain chains?

Orogenesis literally means the *birth of mountains* (*oros* comes from the Greek for *mountain*). An orogeny is a mountain-building episode that thickens continental crust. It can occur through large-scale deformation and uplift of the crust in episodes of continental plate collision such as the formation of the Himalayan mountains from the collision of India and Asia. It also may include the capture of migrating terranes and cementation of them to the continental margins, and the intrusion of granitic magmas to form plutons. These granite masses often are exposed following uplift and removal of overlying materials. Uplift is the final act of the orogenic cycle. Earth's major chains of folded and faulted mountains, called orogens, bear a remarkable correlation to the plate tectonics model.

11. Name some significant orogenies.

Major orogens include the Rocky Mountains, produced during the Laramide orogeny (40-80 million years ago), the Sierra Nevada in the Sierra Nevadan orogeny (35 million years ago, with older batholithic intrusions dating back 130-160 million years), the Appalachians and the Valley and Ridge Province formed by the Alleghany orogeny (250-300 million years ago, preceded by at least two earlier orogenies), and the Alps of Europe in the Alpine orogeny (20-120 million years ago and continuing to the present, with many earlier episodes). See the geologic time scale in Figure 11-1, where these are listed.

12. Identify on a map Earth's two large mountain chains. What processes contributed to their development?

The mountain chain which includes the Andes, the Sierra of Central America, the Rockies, and other western mountains was created from the collision between oceanic plates and continental plates. The collision of the Nazca plate with the South American plate creating the Andes, is a good example of subduction causing folded sedimentary formations, with intrusions of magma forming granitic plutons characteristic of explosive volcanism. Another mountain chain is the Himalayas, created by the collision of the India plate with the Eurasian plate 45 million years ago. This is an example of continental plate-continental plate collision.

13. How are plate boundaries related to episodes of mountain building? Do different types of

plate boundaries produce differing orogenic episodes and differing landscapes?

Figure 12-12 illustrates the plate-collision pattern associated with each type of orogenesis and points out an actual location on Earth where each mechanism is operational. Shown in (a) is the oceanic plate-continental plate collision type of orogenesis. This occurred along the Pacific coast of the Americas and has formed the Andes, the Sierra of Central America, the Rockies, and other western mountains. Shown in (b) is the oceanic plate-oceanic plate collision, where two portions of oceanic crust collide. This has formed the chains of island arcs and volcanoes that continue from the southwestern Pacific to the western Pacific, the Philippines, the Kurils, and portions of the Aleutians. Shown in (c) is the continental plate-continental plate collision, which occurs when two large continental masses collide. Here the orogenesis is quite mechanical; large masses of continental crust are subjected to intense folding, overthrusting, faulting, and uplifting. As mentioned earlier, the collision of India with the Eurasian landmass produced the Himalayan Mountains.

14. Relate tectonic processes to the formation of the Appalachians and the Alleghany orogeny.

The old, eroded fold-and-thrust belt of the eastern United States contrasts with the younger mountains of the western portions of North America. As noted, the Alleghany orogeny followed at least two earlier orogenic cycles of uplift and the accretion of several captured terranes. (In Europe this is called the Hercynian orogeny.) The original material for the Appalachians and Valley and Ridge Province resulted from the collisions that produced Pangaea.

15. Describe the differences in human response between the two earthquakes that occurred in China in 1975 and 1976.

In the Liaoning Province of northeastern China, ominous indications of possible tectonic activity began in 1970. Foreboding symptoms included land uplift and tilting, increased numbers of minor tremors, and changes in the region's magnetic field–all of this after almost 120 years of quiet. These precursors of coming tectonic events continued for almost five years before Chinese scientists took the bold step of forecasting an earthquake. Finally, on 4 February 1975 at 2:00 P.M., some 3 million people were evacuated in what turned out to be a timely manner; the quake struck at 7:36 P.M., within the predicted time frame. Ninety percent of the buildings in the city of Haicheng were destroyed, but thousands of lives were saved, and success was proclaimed–an earthquake had been forecasted and preparatory action taken for the first time in history.

Only 17 months later, at Tangshan in the northeastern province of Hebei (Hopei), an earthquake occurred on 28 July 1976 without any warning precursors from which a forecast could be prepared. Consequently, this quake destroyed 95% of the buildings and 80% of the industrial structures, severely damaged more than half the bridges and highways, and killed a quarter of a million people! The jolt was strong enough to throw people against the ceilings of their homes. An old, undetected fault had ruptured and offset 1.5 m (5 ft) along an 8 km (5 mi) stretch through Tangshan, devastating large areas just 145 km (90 mi) southeast of Beijing, China's capital city.

16. Differentiate between the Mercalli and moment magnitude and amplitude scales. How are these used to describe an earthquake? Why has the Richter scale been updated and modified?

Earthquake intensity is rated on the arbitrary Mercalli scale, a Roman numeral scale from I to XII representing "barely felt" to "catastrophic total destruction." It was designed in 1902 and modified in 1931 to be more applicable to conditions in North America. Intensity scales are useful in classifying and describing terrain, construction, and local damage conditions following an earthquake.

Earthquake magnitude is estimated according to a system originally designed by Charles Richter in 1935. In this method, the amplitude of a seismic wave is recorded on a seismograph located at least 100 km (62 mi) from the epicenter of the quake. That measure is then charted on the Richter scale, which is open-ended and logarithmic; that is, each whole number on the scale represents a 10-fold increase in the measured wave amplitude. Translated into energy, each whole number demonstrates a 31.5-fold increase in the amount of energy released.

Today, the Richter scale has been improved and made more quantitative. The need for revision was because at higher magnitudes, the scale did not properly measure or differentiate between quakes of high intensity. The moment magnitude scale, in use since 1993, is considered more accurate than Richter's scale for large earthquakes. Moment magnitude considers the amount of fault slippage produced by the earthquake, the size of the surface or subsurface area that ruptured, and the nature of the materials that faulted, including how resistant they were to failure.

17. What is the relationship between an epicenter and the focus of an earthquake?

The subsurface area along a fault plane, where the motion of seismic waves is initiated, is called the focus, or hypocenter (Figure 12-18). The area at the surface directly above this subsurface location is the epicenter. Shock waves produced by an earthquake radiate outward from both the focus and epicenter. An aftershock may occur after the main shock, sharing the same general area of the epicenter. A foreshock is also possible preceding the main shock.

18. What local conditions in Mexico City and San Francisco severely magnified the energy felt in their respective earthquakes?

If the ground is unstable, distant effects can be magnified, as they were in Mexico in 1985 and in San Francisco in 1989. Mexico City, currently the world's largest, is positioned on the soft, moist sediments of an ancient lake bed. On 19 September 1985, during rush-hour traffic, two major earthquakes (8.1 and 7.6 on the Richter scale) struck 400 km (250 mi) southwest of Mexico City. The epicenter was on the seafloor off Mexico and Central America, and the old lake bed beneath Mexico City magnified the shock waves by more than 500%. Likewise, in the Marina District of San Francisco and at the Cypress Freeway structure in Oakland, areas of landfill magnified earthquake energy, resulting in catastrophic damage.

19. How do the elastic-rebound theory and asperities help explain the nature of faulting?

The elastic-rebound process is described by the elastic-rebound theory. Generally, two sides along a fault appear to be locked by friction, resisting any movement despite the powerful motions of adjoining pieces of crust. This stress continues to build strain along the fault surfaces, storing elastic energy like a wound-up spring. When movement finally does occur as the strain build-up exceeds the frictional lock, energy is released abruptly, returning both sides of the fault to a condition of less strain. Think of the fault plane as a surface with irregularities that act as sticking points, preventing movement, similar to two pieces of wood held together by drops of glue rather than an even coating of glue. Research scientists at the USGS and the University of California have identified these small areas of high strain as asperities. They are the points that break and release the sides of the fault.

20. Describe the San Andreas fault and its relationship to ancient sea-floor spreading movements along transform faults.

The San Andreas is a good example of the evolution of a spreading center overridden by an advancing continental plate (Figure 12-17). In the figure, the East Pacific Rise developed as a spreading center with associated transform faults (a), while the North American plate was progressing westward after the breakup of Pangaea. Forces then shifted the transform faults toward a northwest-southeast alignment along a weaving axis (b). Finally, the western margin of North America overrode those shifting transform faults (c). In relative terms, the motion along the fault is right lateral, whereas in absolute terms, the North American plate is still moving westward.

21. How does the seismic gap concept relate to expected earthquake occurrences? Have any gaps correlated with earthquake events in the recent past? Explain.

One approach to earthquake prediction is to examine the history of each plate boundary and determine the frequency of earthquakes in the past, a study called paleoseismology. Paleoseismologists construct maps that provide an estimate of expected earthquake activity. An area that is quiet and overdue for an earthquake is termed a seismic gap, that is, an area that forms a gap in the earthquake occurrence record and therefore a place that possesses accumulated strain. The area along the Aleutian Trench subduction zone had three such gaps until the great 1964 Alaskan earthquake (8.6 on the Richter scale) filled one of them.

22. What do you see as the biggest barrier to effective earthquake prediction?

See Figure 12-21 that shows the socio-economic impacts and adjustments to earthquake prediction.

23. What is a volcano? In general terms, describe some related features.

A volcano forms at the end of a central vent or pipe that rises from the asthenosphere through the crust into the volcanic mountain, usually forming a crater, or circular surface depression at the summit. Magma rises and collects in a magma chamber deep below the volcano until conditions are right for an eruption. Other features related to volcanic activity are; calderas, large basin-shaped depressions formed when summit material on a volcanic mountain collapses inward after eruption or loss of magma, cinder cones, small cone-shaped hills with a truncated top formed from cinders that accumulate during moderately explosive eruptions, and shield volcanoes, which are created by effusive volcanism, similar in shape to a shield of armor lying face up on the ground.

24. Where do you expect to find volcanic activity in the world? Why?

See Figure 12-22 and insets 12-23 to 12-29. Also, review the plate tectonic map and volcanic activity in Figure 11-21. The location of volcanic mountains on Earth is a function of plate tectonics and hot spot activity. Volcanic activity occurs in three areas: along subduction boundaries at continental plate-oceanic plate or oceanic plate-oceanic plate convergence; along sea-floor spreading centers on the ocean floor and areas of rifting on continental plates; and at hot spots, where individual plumes of magma rise through the crust.

25. Compare effusive and explosive eruptions. Why are they different? What distinct landforms are produced by each type? Give examples of each.

Effusive eruptions are the relatively gentle eruptions that produce enormous volumes of lava on the seafloor and in places like Hawaii. Direct eruptions from the asthenosphere produce a low-viscosity magma that is very fluid and yields a dark, basaltic rock (less than 50% silica and rich in iron and magnesium). Gases readily escape from this magma because of its texture. A typical mountain landform built from effusive eruptions is gently sloped, gradually rising from the surrounding landscape to a summit crater, similar in outline to a shield of armor lying face up on the ground, and is therefore called a shield volcano.

Volcanic activity along subduction zones produces the well-known explosive volcanoes. Magma produced by the melting of subducted oceanic plate and other materials is thicker (more viscous) than magma from effusive volcanoes; it is 50-75% silica and high in aluminum. Consequently, it tends to block the magma conduit inside the volcano, allowing pressure to build and leading to an explosive eruption. The term composite volcano, or composite cone, is used to describe explosively formed mountains. Composite volcanoes tend to have steep sides and are more conical than shield volcanoes, and therefore they are also known as composite cones.

26. Describe several recent volcanic eruptions.

We are reminded of Earth's internal energy by the recent violent eruptions of Mount Pinatubo and Mayon volcano (Philippines, 1991 and 1993 respectively), Mount Unzen (Japan, 1991), Mount Hudson (Chile, 1992), Mount Spurr and Mount Redoubt (Alaska, 1992), and Galeras volcano (Colombia, 1993), Rabual Caldera (Papua, New Guinea, 1994), and Mount Etna (Italy, 1995). And, consult Table 12-3, as well as the Volcanism section in Chapter 12 for lists of other examples.

Overhead Transparencies

As an adopter you are provided with the following figures for overhead projector use.

- Figure 12-6: Three kinds of stress, strain and resulting surface expressions
- Figure 12-10: Types of faults
- Figure 12-12: Three types of plate convergence
- Figure 12-17: San Andreas fault formation; southern California map inset
- Figure 12-22: Tectonic settings of volcanic activity

13

Weathering, Karst Landscapes, and Mass Movement

Overview

Chapter 13 begins the treatment of the exogenic system. Chapters 11 and 12 covered the endogenic system. The distinction is shown in Figure 11-6, which can be referenced at this point. As the landscape is formed, a variety of exogenic processes simultaneously operate to wear it down. The endogenic system builds and creates initial landscapes, while the exogenic system works towards low relief, little change, and the stability of sequential landscapes.

Students can be asked if they have noticed highways in mountainous and cold climates that appear rough and broken. Roads that experience freezing weather seem to pop-up in chunks each winter. Or, maybe they have seen older marble structures, such as tombstones, etched and dissolved by rainwater. Similar physical and chemical weathering processes are important to the overall reduction of the landscape and the release of essential minerals from bedrock. A simple examination of soil gives evidence of weathered mineral grains from many diverse sources. In addition, mass movement of surface material rearranges landforms, providing often-dramatic reminders of the power of nature.

The *Student Study Guide* presents 20 "Learning Objectives" to guide the student.

New to the Third Edition

(Note: This section highlights major changes, new features, and additions in the third edition. This does not describe all the rewrite and recast of the text.)

1. A list of key learning concepts begins the chapter.

2. Figure 13-1 is a photograph of Delicate Arch from Arches National Park, Utah, a dramatic example of differential weathering.

3. Figure 13-4, a photograph of a Wyoming hillslope.

4. Plant roots growing in the joints of rocks are a good examples of organic physical weathering, Figure 13-6.

5. New photos of physical and chemical weathering are found in Figures 13-10 and 13-12.

6. The description of chemical weathering processes have been reworded to ease comprehension. Chemical equations have been added to support discussion of these processes.

7. News Report #1: "Alabama Hills-A Popular Movie Location." The unique appearance of spheroidal weathering in the Alabama Hills creates a great backdrop for movies.

8. Speleology, the scientific study of caves, is dominated by amateurs. News

Report #2: "Amateurs Make Cave Discoveries," illustrates the type of conditions "spelunkers" face.

9. The factors which create karst landscapes have been put into a bullet format for greater clarity.

10. New examples of mass movement have been included to help students classify types of mass movement more easily.

11. Focus Study 1, "Vaiont Reservoir Landslide Disaster," describes one of the worst dam disasters in history. This study gives students background information concerning the areas slope stability and climate, in order to recognize signs which may have predicted the disaster.

12. News Report #3: "Landslides Beneath the Sea," describes mass movement along the Hawaiian Island which creates large depositional features and may generate tsunamis.

13. The use of Global Positioning Systems (GPS) as a method to measure slight land shifts is discussed. This technology may allow greater prediction of mass movement.

14. News Report #4: "Debris Avalanche Destroys Colombian Town," addresses the magnitude of damage that mass movements can inflict.

15. Malibu, California has faced six federally-declared disasters in the last four years. Should we allow homes to be rebuilt there? News Report #5: "Malibu 90263," addresses the issue of zoning in hazardous areas.

16. Figures 13-24 and 13-26 are new photographs of mass movement.

17. The process of scarification is discussed in greater depth, including a new photograph of scarification on Figure 13-28.

18. A new summary and review section ends the chapter.

Key Learning Concepts

1. *Define* the science of geomorphology.

2. *Illustrate* the forces at work on materials residing on a slope.

3. *Define* weathering and *explain* the importance of the parent rock and joints and fractures in rock.

4. *Describe* frost action, crystallization, hydration, pressure-release jointing, and the role of freezing water as physical weathering processes.

5. *Describe* the susceptibility of different minerals to the chemical weathering processes called hydration.

6. *Review* the processes and features associated with karst topography.

7. *Portray* the various types of mass movements and *identify* examples of each in relation to moisture content and speed of movement.

Expanded Outline Discussion

The following headings (boldfaced) match some of the first, second, and third order headings in Chapter 13. The narrative under each heading contains information, sources, and anecdotal facts relating to portions of the chapter. Not all text headings are discussed.

LANDMASS DENUDATION

The rocks of Earth's crust are broken down and dissolved in the surface environment by physical and chemical weathering. Contact with the atmosphere, hydrosphere, and biosphere initiates remarkable changes in rock, produces basic sediments for erosion and transport, and generates the basic ingredients that are important to soil formation.

The Grand Canyon of Arizona is an example of form and process: The erosional processes that reduce a landscape are balanced against the resistance of the materials that make up the landscape. Weathering and erosional forces naturally oscillate, especially in the desert, with high rainfall variability coming in episodic thunderstorms. Each rainfall event at the Grand Canyon operates on available slopes and cliffs. The river receives materials and discharges its flow of water and sediment load. The variation in rock resistance is responsible for the pattern of cliffs and slopes visible in the rock surfaces: more resistant rock cliffs, less resistant rock slopes.

The tremendous uplift of the Colorado Plateau over the past 10 million years rejuvenated the profile of the Colorado River and facilitated the downwasting of the surrounding landscape. The river drops 670 m (2200 ft) in elevation as it makes its way through the Grand Canyon region. Today, the river is almost 1.6 km (1 mi) below the rim of the canyon; the cliffs and slopes have retreated to the extent that the canyon is over 16 km (10 mi) wide. What is revealed in the exposed strata is an incredible history of that entire region, covering a time span of over 2 billion years of Earth's existence. The sea has come and gone at least seven times, each episode clearly marked by layers of limestone and beach sands hardened (lithified) to sandstone.

Endogenic periods of uplift did not take place in a smooth sequence; quiet periods occurred and are recorded in rock formations by plateau-like terraces called esplanades in the canyon–the Tonto Plateau. The roots of old mountains are represented by metamorphic rocks in the inner gorge. Imagine, as you look at a photograph of the canyon, the constant and on-going adjustments that are occurring in each portion of the landscape. Think of the geomorphic thresholds that are breached following thunderstorms and flash floods throughout the canyon. Yet what we witness is a beautiful, quiet, and seemingly motionless landscape that is constantly changing over a grand time scale.

Early Hypotheses, Dynamic Equilibrium Approach

An interesting exercise is to seek additional background from related disciplines to see how they handle landmass denudation concepts and the various theories that have been put forward. We want to make sure that a derivative science such as physical geography correctly reflects the content of allied disciplines. Relative to landmass denudation you might want to review portions of several works: Easterbrook, Donald J., *Surface Processes and Landforms*, New York: Macmillan College Publishing Company, 1993, pp. 125-31, 164-79; Chorley, Richard J., Stanley A. Schumm, and David E. Sugden, *Geomorphology*, New York: Methuen, 1985, pp. 22-28; Rice, R. J., *Fundamentals of Geomorphology*, 2d ed. White Plains, NY: Longman, 1988, pp. 387-391; or Press, Frank, and Raymond Siever, *Earth*, 4th ed. New York: W. H. Freeman, 1986, pp. 145-148.

Interactions between the structural elements of the land and denudation processes are complex, and represent a constant struggle between internal and external processes. An important question to ask is whether or not this dynamic interplay is progressive, evolving and building landforms in an orderly manner through stages? Do landscapes initially form and subsequently age in graceful stages until they are flat? Or does the interplay of forces fluctuate back and forth across a never achieved steady-state equilibrium? The debates in geomorphology are fueled by the fact that landscapes evolve on a much longer time span than does a human life, or a research grant, or thesis project, or the edition of a textbook. Modern geomorphology has moved away from simple descriptive classifications.

A complete model has yet to replace the cyclic model of Davis, although landscapes are generally viewed as operating in a state of dynamic equilibrium. In Chapter 14, additional discussion of William Morris Davis' cyclic model is presented and includes the following statement:

> **However, as suggested by Schumm and Lichty, two modern geomorphologists, the validity of cyclic or functional landscape models may depend on the time frame. Over the long span of geologic time, cyclic models of evolutionary development might explain, for example, aspects of the disappearance of entire mountain ranges through denudation. But these generalizations would not be applicable to *steady time*, the adjustments ongoing in a portion of a drainage basin. In between these two time frames lies the realm of *graded time*, or the conditions of dynamic equilibrium.**

As dynamic equilibrium models have replaced the Davisian cyclic model, we must not lose sight of the teaching value that is derived from the pursuit of a theory: the searching propelled and the learning stimulated. Failure occurs when a theory is too rigidly accepted as reality, which by definition is not the purpose of a theory.

Dynamic Equilibrium Approach to Landforms; Slopes

Slopes seek an angle of equilibrium among the forces described in the text (Figure 13-3, p. 389). Strahler (1950) actually calculated slope equilibrium angles and found great uniformity for the specific area he studied. In other words, slopes exhibit maximum inclines that balance conflicting forces. A geomorphic threshold (change point) is reached if one of the conditions in the balance is altered. When this happens all the forces on the slope compensate by adjusting to a new dynamic equilibrium. A slope is an open system responding to variable inputs and producing variable outputs.

In a generic sense and under most climatic regimes, slopes have the same four identifiable traits: a waxing slope, a free-face scarp, a debris slope, and a waning slope or pediment. Variation in the resistance of rock strata or climatic conditions may alter the conformations in this ideal model. Weathering and mass movement principally work on the upper portions of the slope, while running water predominates on the lower slopes. A new Figure 13-3(b) illustrates these principal elements of slope form. The *waxing slope* is a forming slope of convex shape which is consumed by the enlarging *free face scarp*, and retreats as the landscape is reduced. The *debris slope* receives rock fragments and materials from above, with continuously moving water in humid climates carrying away material as it arrives. In arid climates, graded debris slopes persist where moving water is infrequent. And finally, the *waning slope*, or *pediment*, has a graded form in a river valley, or is a wide, flat intermittently dry lake bed in arid areas.

In Figure 13-2, the dam that failed was a coffer construction dam at the abandoned site of the Auburn Dam project on the upper American River east of Sacramento, California. The 80 m+ coffer dam was about 7 years beyond its design limits when record rainfall in February 1986 topped the dam and quickly eroded the earthfill structure to bedrock, transporting the material to the next reservoir downstream. This surge placed strain on the downstream dam, dikes, and levees. Incidentally, in the newspapers the next day officials put their version–a "correct spin"–on events by saying that the coffer dam had "broken successfully"! The construction dam was designed with a 10m soft plug along one side of its upper layers that was meant to erode to relieve any floodwaters. However, the draining flood waters tore the dam away and in an unplanned manner eroded the entire structure to bedrock.

WEATHERING PROCESSES

We start the exogenic system with weathering because weathering processes are essential for the production of materials for erosion, transportation, and deposition. Weathered materials are key to soil formation (Chapter 18). Weathering processes release minerals for resource exploitation and production. Weathering is the response of materials now at the surface environment. These materials were fixed (at equilibrium with conditions in the lithosphere) in a variety of stable forms at or near Earth's surface, where atmospheric, hydrospheric, and biospheric processes loosened them and made them available for erosion and transport by wind, water, and ice. The degradation and aggradation of landforms is initially dependent on weathering processes.

Figure 13-7 relates climatic elements to weathering processes and shows the relationship among temperature, rainfall, and the various types of related weathering.

Discuss the importance of jointing in rock to physical and chemical weathering. These fractures and separations increase the surface area of rock exposed to weathering processes.

Physical Weathering Processes

Several of the physical weathering examples are particularly effective with granite, especially granular disintegration and pressure-release jointing. The formation of domes on Earth's exposed batholiths is the result of this latter process, involving sheeting and exfoliation.

The role of crystals in physical weathering processes principally involves the behavior of salt crystals and ice crystals, with ice crystals being the more effective. You might want to refer to those portions in Chapter 7 to refresh the students on the behavior of water as it cools and freezes, expanding up to 9 percent of its volume (Chapter 7), creating frost action.

Hydration can be enhanced when it is followed by temperature change and rehydration over short periods of time. A hydration–dehydration cycle can effectively disaggregate rock.

An important point: thermal stress is not presented as a physical weathering process. Laboratory simulation has shown that daily and seasonal temperature differences on exposed granite do not produce significant expansion and contraction in the rock. Thermal stress does not appear to be a significant weathering force, although research is continuing. However, students might think that this seems intuitively accurate. After all, the action of ice formation and frost wedging cracks rock, so why not heat?

Chemical Weathering Processes

Chemical processes occur at the molecular level and are key to weathering, as opposed to large-scale physical-gravitational-geomorphological processes–the realm of the geochemist over the geomorphologist. The effectiveness of chemical weathering is directly related to bond strengths in the minerals and the arrangement of these bonds, and represents a more complex set of interrelated processes than physical weathering.

The present increase in acid precipitation is enhancing chemical weathering rates. Limestone, marble, plaster, and cement are dissolved and etched by the low-pH values now characteristic of rain and snow in many parts of the world. The facades of cathedrals, translocated ruins such as Cleopatra's needle in Central Park, and tombstones are literally wearing away. Chemical weathering studies are conducted on such things as the depth of the carved lettering on the tombstones at Arlington National Cemetery since it is easy to establish a time line.

Related to chemical weathering and deposition, the travertine deposit (a form of calcium carbonate) at Mammoth Hot Springs in Yellowstone National Park is an example of surface hydrothermals. In various places in the world, hydrothermal activity produces tufa, a sedimentary rock formed by chemical precipitates. At Mono Lake, an ancient lake in the interior Basin and Range of the western U.S. near the Nevada-California border, alkaline lake water high in carbonates and sulfates, with an overall salinity today of 95% (the ocean is 35 %), interacts with calcium-rich hot springs to produce precipitate and form depositional features called tufa towers as pictured in Chapter 11, News Report #3, Figure 1. These processes are discussed in Chapter 11 with the rock cycle.

Hydrolysis

Compared to hydration, a physical process in which water is simply absorbed, the hydrolysis process involves active participation of water in chemical reactions to produce different compounds and minerals. Water dissociates into H^+ and OH^- ions which then react with mineral ions. The pH of water determines how many of these ions are available for hydrolysis. (See discussion of pH with acid rain in Chapter 3 and with soils in Figure 18-7).

Karst Topography and Landscapes

The section on carbonation and solution sets up the following new first-order heading "Karst Topography and Landscapes." Limestone is so prevalent on Earth that inclusion of an expanded section on that subject seems appropriate and reflects widespread interest among geographers. The name is derived from a region of Yugoslavia which is comprises about 25 percent calcareous landscape and exceeds 4000 m in thickness. Other such landscapes include the 700 m thickness in Jamaica and 200 m in Yorkshire, England.

In the earlier part of this century when evolutionary models dominated geomorphology, limestone landscapes were used as a prime example of the concept that landscapes evolve from youth to maturity to old age. Today, support for such idealized and simplified evolutionary models is lacking. Instead, each limestone region is a unique by-product of its own physical conditions: availability and fluctuation of groundwater, climate change, and moisture regimes, as well as the variety of joint and fracture patterns and other structural controls. It is anachronistic to persist in evolutionary scenarios when the related literature supports a more landscape-specific, dynamic explanation.

Caves and Caverns

The single largest explored cave system in the world is in the Swiss Alps. Mammoth Cave, Kentucky is second, with over 70 km (45 mi) of explored caves. Carlsbad Caverns in New Mexico is the place with the largest single room: the Big Room is 500 m (1500 ft) long and 90 m (285 ft) high.

The picture in the text was taken near the Big Room (Figure 13-18a). It is an enjoyable visit, because they have transmitters embedded in the trail with individual receivers that you carry on a lanyard around your neck. In this way, people can pace themselves individually, listening to explanations and taking time for the underworld environment to seep into consciousness. As with most of the Planet Earth series, the book on calcareous caves is a good overview: see Jackson, Donald Dale, and the editors of Time-Life Books. *Underground Worlds.* Planet Earth Series. Alexandria, VA: Time-Life Books, 1982.

MASS MOVEMENT PROCESSES

I have used several dramatic episodes to illustrate specific principles such as the lahar produced by the eruption of Nevado del Ruiz in central Colombia, South America. Perhaps localized events in your area have occurred that could serve this purpose. The text makes the distinction between *mass movement* and *mass-wasting* terminologies. The term mass movement is sometimes used interchangeably with mass wasting, although the latter also is applied to gravitational movement of nonunified material. According to the *AGI Glossary*, mass movement specifically refers to unit movement of a portion of land. To combine the concepts, we can say that mass movement of material works to waste slopes and provide raw material for erosion, transportation, and deposition.

A particularly useful chapter, "Mass Movement" Chapter 10, appears in Chorley, Richard J., Stanley A. Schumm, and David E. Sugden. *Geomorphology*. New York: Methuen, 1985. Two texts from Chapman and Hall, New York and London, have sections on mass movement hazards: David Alexander, *Natural Disasters*, 1993, "Landslides" pp. 242-275; and, G.J.H. McCall, D.J.C. Laming, and S.C. Scott eds., *Geohazards–Natural and man-made*, 1992, "Landslides Hazards" pp. 115-158.

Classes of Mass Movements

Fall, slide, flow, and creep mass movements are categorized in a new process-system illustration in Figure 13-20. The vertical axis grades from faster to slower movement, the horizontal axes grade from drier to wetter moisture content. The inset illustration demonstrates each type. Note the two debris avalanche photographs provided by George Plafker, USGS. The Peruvian example is particularly tragic because of its reoccurrence within an eight-year period. The contrast between avalanche materials and the white surface of the Cascade Glacier produces a stark drama in Alaska (Figure 13-23).

Human-Induced Mass Movements

Strip mining for coal produces a mass wasting of the landscape. Overlying layers of soil and rock must be removed to expose the coal-bearing strata. As mentioned earlier, great controversy surrounds the removal of this overburden and whether it should be replaced or the topography restored. Even if the topography is restored, the disturbed soils usually prove difficult to manage. Vast mining excavations are underway in western coal fields. Tens of thousands of acres in Kentucky, Indiana, Ohio, West Virginia, and Pennsylvania are now scarred from years of surface strip mining. That very mining activity provided jobs and a local economy at the time the mining took place, although the areas remain economically depressed, spoiled, and neglected in the mining aftermath.

The various levels of human health misery and environmental problems associated with coal mining in the Appalachians is well-established. Over 100,000 jobs disappeared in those coal mines between 1981 and 1991, with increasing political problems of falsified safety records, medical tests, and failed inspection procedures–all disputed by the mine owners and the federal Mine Safety and Health Administration and the Office of Surface Mining. See Ted Gup's "The Curse of Coal–Vanishing jobs, a ruined economy, broken lives and broken bodies," *TIME* Magazine, 4 November 1991, pp. 54-64, which focuses on the Appalachians, and Logan County, West Virginia in particular.

The basic modes of transportation to facilitate scarification at Bingham Canyon open-pit copper mine were electric railroads that ran on moveable tracks. The Berkeley Open Pit, at Butte, Montana, used large 150-to-250 ton trucks to haul (mass move) the material from the pit. Unstable markets and international competition, coupled with very low grade ores, led to closure of these two open pit mines. Rising copper prices will no doubt bring these sites into operation again, and scarification mass movements will begin anew.

Glossary Review for Chapter 13 (in alphabetical order)

angle of repose
bedrock
carbonation
debris avalanche
denudation
differential weathering
dynamic equilibrium model
exfoliation dome
frost action
geomorphic threshold
geomorphology
hydration
hydrolysis
jointing
karst topography
landslide
mass movement
mass wasting
mudflows
oxidation
parent material
physical weathering
regolith
rockfalls
scarification
sediment
sheeting
sinkholes
slopes
soil creep
solution
spheroidal weathering
talus slope
weathering

Annotated Chapter Review Questions

1. Define geomorphology and describe its relationship with physical geography.

Geomorphology is a science that analyzes and describes the origin, evolution, form, and spatial distribution of landforms. It is an important aspect of the study of physical geography and the understanding of the spatial-physical aspects of landforms.

2. Define landmass denudation. What processes are included in the concept?

Denudation is a general term referring to all processes that cause reduction or rearrangement of landforms. The principal denudation processes affecting surface materials include weathering, mass movement, erosion, transportation, and deposition.

3. Give a brief overview of Davis's geomorphic cycle model. What was his basic thinking in setting up this cyclic model? What was Davis's principal contribution to the models of landmass denudation?

Davis theorized that a landscape goes through an initial uplift that is accompanied by little erosion or removal of materials and then enters a prolonged period of stability, an idea he later modified. The raised elevation of the landscape is such that streams begin flowing more rapidly downhill, cutting both headward (upstream) and downward. According to this cyclic model, the landscape eventually evolves into an old erosional surface (detailed further in Chapter 14). But Davis's theory, although it helped launch the science of geomorphology and was innovative at the time, was too simple and did not account for the processes being observed as systems theory entered geomorphology. Although not generally accepted today, his thinking about the evolution of landscapes is still influential.

4. What are the principal considerations in the dynamic equilibrium model?

The balancing act between tectonic uplift and reduction by weathering and erosion, between the resistance of rocks and the ceaseless attack of weathering and erosion, is summarized in the dynamic equilibrium model. A dynamic equilibrium demonstrates a trend over time. According to current thinking, landscapes in a dynamic equilibrium show ongoing adaptations to the ever-changing conditions of rock structure, climate, local relief, and elevation. Endogenic events (such as earthquakes and volcanic eruptions), or exogenic events (such as heavy rainfall or forest fire), may provide new sets of relationships for the landscape.

5. Describe conditions on a hillslope that is right at the geomorphic threshold. What factors might push the slope beyond this point?

As changing conditions provide new sets of relationships for the landscape, the system eventually arrives at a geomorphic threshold, or that point at which the system breaks through to a new set of equilibrium relationships and rapidly realigns landscape materials accordingly. Slopes, as parts of landscapes, are open systems and seek an angle of equilibrium among the forces described here. Conflicting forces work together on slopes to establish an optimum compromise incline that balances these forces. A geomorphic threshold (change point) is reached when any of the conditions in the balance is altered. Many factors could alter the hillside's equilibrium, such as an earthquake, saturation of the regolith, or the building of a house or dam (adding mass). All the forces on the slope then compensate by adjusting to a new dynamic equilibrium.

6. Given all the interacting variables, do you think a landscape ever reaches a stable, old-age condition? Explain.

In reality, a landscape behaves as an open system, with highly variable inputs of energy and materials. In response to this input of energy and materials, landforms constantly adjust toward a condition of equilibrium. As physical factors fluctuate, the surface constantly responds in search of an equilibrium condition, with every change producing compensating actions and reactions. The balancing act between tectonic uplift and the reduction rates of weathering and erosion, and between the resistance of crust materials and the ceaseless attack of denudation processes, is summarized in the dynamic equilibrium model. A dynamic equilibrium is different from a steady-state equilibrium, which fluctuates around an average. Instead, according to current thinking, landscapes adapt to the ever-changing conditions of rock structure, climate, local relief, and elevation. Various endogenic episodes, such as faulting, or exogenic episodes, such as a heavy rainfall, may episodically provide new sets of relationships for the landscape. According to Davis' evolutionary cyclic model, the landscape eventually evolves into an old erosional surface that he called a peneplain. In reality, a peneplain has never been satisfactorily identified in nature; landscapes do not hold still for such prolonged periods of time.

7. What are the general components of an ideal slope?

Figure 13-3b illustrates basic slope components that vary with conditions of rock structure and climate. Slopes generally feature an upper waxing slope near the top. The convex surface curves downward and grades into the free face below. The presence of a free face indicates an outcrop of resistant rock that forms a steep scarp or cliff.

Downslope from the free face is a debris slope, which receives rock fragments and materials from above. The debris slope grades into a waning slope, a concave surface along the base of the slope. This surface of erosional materials gently slopes at a continuously decreasing angle to the valley floor.

A slope is an open system seeking an angle of equilibrium. Conflicting forces work simultaneously on slopes to establish an optimum compromise incline that balances these forces.

8. Relative to slopes, what is meant by an "angle of equilibrium"? Can you apply this to the photograph in Figure 13-2?

Slopes, as parts of landscapes, are open systems and seek an angle of equilibrium among the forces described in the answer to Question 7, above. The recently disturbed hillslope in Figure 13-2 is in the midst of compensating adjustment. The disequilibrium was created by the failure of saturated slopes, which a day earlier were underwater. A dam break downstream, rapidly emptied the reservoir exposing these slopes, and leading to their ultimate failure.

9. Describe weathering processes operating on an open expanse of bedrock. How does regolith develop? How is sediment derived?

Rocks at or near Earth's surface are exposed to both physical and chemical weathering processes. Weathering encompasses a group of processes by which surface and subsurface rock disintegrates into mineral particles or dissolves into minerals in solution. Weathering does not transport the weathered materials; it simply generates these raw materials for transport by the agents of wind, water, and gravity. In most areas, the upper surface of bedrock is partially weathered to broken-up rock called regolith. In some areas, regolith may be missing or undeveloped, thus exposing an outcrop of unweathered bedrock. Loose surface material comes from further weathering of regolith and from transported and deposited regolith. This unconsolidated sediment and weathered rock forms the parent material from which soil evolves.

When rock is broken and disintegrated without any chemical alteration, the process is called physical weathering or mechanical weathering. By breaking up rock, physical weathering greatly increases the surface area on which chemical weathering may operate. Chemical weathering refers to actual decomposition and decay of the constituent minerals in rock due to chemical alteration of those minerals. A familiar example of chemical weathering is the eating away of cathedral facades and etchings on tombstones caused by increasingly acid precipitation. Water is essential to chemical weathering and, as shown in Figure 13-7, the chemical breakdown becomes more intense as both temperature and precipitation increase.

10. Describe the relationship between climatic conditions and rates of weathering activities.

Important in determining weathering rates are climatic elements, including the amount of precipitation, overall temperature patterns, and any freeze-thaw cycles. Figure 13-7 shows the relationship between annual precipitation and temperature and the physical (mechanical) or chemical weathering processes in an area. You can see that physical weathering dominates in drier, cooler climates, whereas chemical weathering dominates in wetter, warmer climates. Extreme dryness reduces most weathering to a minimum, as is experienced in desert climates (BW). (Refer to Chapter 10 for the key to these climate designations.) In the hot, wet-tropical and equatorial rainforest climates (Af), most rocks weather rapidly, and the weathering tends to be deep below the surface.

11. What is the relation among parent rock, parent material, regolith and soil?

Bedrock is the parent rock from which weathered regolith and soils develop. While a soil is relatively youthful, its parental rock is traceable through similarities in composition. This unconsolidated fragmental material, known as sediment, combines with weathered rock to form the parenta material from which soil evolves. See Figure 13-5.

12. What role do joints play in the weathering process? Give an ex-

ample from one of the illustrations in this chapter.

Joints are fractures or separations in rock that occur without displacement of the sides (as would be the case in faulting). The presence of these usually plane (flat) surfaces greatly increases the surface area of rock exposed to both physical and chemical weathering. An example is found in Figure 13-8, which portrays joint-block separation, a form of physical weathering.

13. What is physical weathering? Give an example.

Physical, or mechanical, weathering is the term used when rock is broken and disintegrated without any chemical alteration. By breaking up rock, physical weathering greatly increases the surface area on which chemical weathering can take place. An example of physical weathering is frost action. When water freezes, its volume expands. This creates a powerful mechanical force, which can exceed the tensional strength of rock. Repeated freezing and thawing of water break rocks apart. The work of ice begins in small openings, such as existing joints and fractures, gradually expanding until rocks are split apart.

14. What is the interplay between the resistance of rock structures and weathering variabilities?

Weathering is greatly influenced by the character of the bedrock: hard or soft, soluble or insoluble, broken or unbroken. The differing resistance of rock, coupled with these variations in the intensity of weathering, result in differential weathering.

15. Why is freezing water such an effective physical weathering agent?

Water expands by as much as 9% of its volume as it freezes (see Chapter 7). This expansion creates a powerful mechanical force that can exceed the tensional strength of rock. In a weathering action called frost-wedging, ice crystals grow in preexisting cracks in rock and push the sides apart along joints or fractures. Figure 13-8 shows this action, on blocks of rock, that causes joint-block separation, although cracking and breaking can be in any shape. The work of ice probably begins in small openings, gradually expanding until rocks are cleaved. Softer supporting rock underneath the slabs already has weathered physically–an example of differential weathering.

16. What weathering processes produce a granite dome? Describe the sequence of events.

As the tremendous weight of overburden is removed from a granitic pluton, the pressure of deep burial is relieved. The granite responds with a slow but enormous heave, and in a process known as pressure-release jointing, layer after layer of rock peels off in curved slabs, thinner toward the top of the formation and thicker toward the sides. As these slabs weather they slip off in a process called spalling, and form an exfoliation dome. This exfoliation process creates archshaped and dome-shaped features on the exposed landscape (Figure 13-11). Such domes probably are the largest single weathering features on Earth (in areal extent).

17. What is chemical weathering? Contrast this set of processes to physical weathering.

Chemical weathering is the actual decomposition of minerals in rock. Chemical weathering involves reactions between air and water and minerals in rock. Minerals may combine with water in chemical reactions (such as carbonation), or carbon dioxide and oxygen from the atmosphere (such as oxidation). Although physical weathering may create greater surface area for further weathering to take place, chemical weathering can dissolve minerals throughout the rock. An example which demonstrates the difference between physical and chemical weathering is the absorption of water in rocks. In cold climates dominated by physical weathering, the process of hydration takes place. In hydration water present in the rock expands with freezing and cracks the rock into smaller pieces. In humid climates, where chemical weathering occurs, water percolates into the rock, a process called hydrolysis, and breaks down the silicate minerals in rock. Hydrolysis dissolves silicate materials leaving behind resistant minerals, such as quartz.

18. What is meant by the rock form: spheroidal weathering? How is this formed?

Spheroidal weathering is an example of the way chemical weathering attacks rock. The sharp edges and corners of rock are rounded as the alteration of minerals progresses through the rock. Joints in the rock offer more surfaces of opportunity for weathering. Water penetrates joints and fractures and dissolves the rock's weaker minerals or cementing materials. The resulting rounded edges are the basis for the name spheroidal weathering.

19. What is hydrolysis? How does it affect rocks?

When minerals chemically combine with water, the process is called hydrolysis. Hydrolysis is a decomposition process that breaks down silicate minerals in rocks. Water is not simply absorbed in hydrolysis but actively participates in chemical reactions to produce different compounds and minerals.

20. Iron minerals in rock are susceptible to which form of chemical weathering? What color is associated with this?

Iron minerals in rock are susceptible to oxidation. Oxidation is an example of chemical weathering which occurs when oxygen combines with certain metallic minerals to form oxides. The rusting of iron in rocks or soils produces a reddish-brown stain of iron oxide.

21. What kind of minerals do carbon compounds react with and under what circumstances does this occur? What is this weathering process called?

Carbon compounds react with carbonic acid, created when water vapor dissolves carbon dioxide. Carbonic acid is strong enough to react with many minerals, especially limestone, in a process known as carbonation. When rainwater attacks formations of limestone, the constituent minerals are dissolved and wash away with the mildly acidic rainwater.

22. Describe the development of limestone topography. What is the name applied to such landscapes? From what was this name derived?

Limestone is so abundant on Earth that many landscapes are composed of it. These areas are quite susceptible to chemical weathering. Such weathering creates a specific landscape of pitted, bumpy surface topography, poor surface drainage, and well-developed solution channels underground. Remarkable labyrinths of underworld caverns also may develop. These are the hallmarks of karst topography, originally named for the Krs Plateau in Yugoslavia, where these processes were first studied. Approximately 15% of Earth's land area has some developed karst, with outstanding examples found in southern China, Japan, Puerto Rico, Cuba, the Yucatán of Mexico, Kentucky, Indiana, New Mexico, and Florida.

23. Differentiate among sinkholes, karst valleys, and cockpit karst. Within which form is the radio telescope at Arecibo, Puerto Rico?

See Figures 13-15, 13-16, and 13-18. The Arecibo radio telescope operated by Cornell University takes advantage of the symmetrical form of a doline in a complex cockpit topography of numerous dolines.

24. Generally, how would you characterize the region southwest of Orleans, Indiana?

The region southwest of Orleans, Indiana has over 1000 sinkholes in just

2.6 km^2 (1mi^2). In this area the Lost River, a "disappearing stream," flows more than 12.9 km (8mi) underground before it resurfaces; its flow diverted from the surface through sinkhole and solution channels. See Figure 13-14.

25. What are some of the unique erosional and depositional features you find in a limestone cavern?

See Figures 13-13, 13-15, 13-16, 13-17, and 13-18 for examples.

26. Define the role of slopes in mass movements–angle of repose, driving force, resisting force, and geomorphic threshold.

All mass movements occur on slopes. The steepness of a slope determines where loose material comes to rest, depending on the size and texture of the grains; this is called the angle of repose. This angle represents a balance of driving and resisting forces.

The driving force in mass movements is gravity, working in conjunction with the weight, size, and shape of the grains or surface material, the degree to which the slope is oversteepened, and the amount and form of moisture available–whether frozen or fluid. The greater the slope angle, the more susceptible the surface material is to mass movement. The resisting force is the shearing strength of slope material, that is, its cohesiveness and internal friction working against mass movement. To reduce shearing strength is to increase shearing stress, which eventually reaches the point at which gravity overcomes friction.

27. What events occurred in the Madison River Canyon in 1959?

In the Madison River Canyon near West Yellowstone, Montana, on the Wyoming border, a blockade of dolomite (a magnesium-rich rock in the carbonate group) had held back a deeply weathered and oversteepened slope (with a 40°- 60° slope angle) for untold centuries. Then, shortly after midnight on 17 August 1959, an earthquake measuring 7.5 on the Richter scale broke the dolomite structure along the foot of the slope and released 32 million m^3 (1.13 billion ft^3) of mass, which moved downslope at 95 kmph (60 mph), creating gale force winds through the canyon. The material continued more than 120 m (390 ft) up the opposite canyon slope, entombing campers beneath about 80 m (260 ft) of rock and debris.

The mass of material also effectively dammed the Madison River and created a new lake, dubbed Quake Lake. The landslide debris dam established a new temporary base level for the canyon.

28. What are the classes of mass movement? Describe each briefly and differentiate among these classes.

Four basic classifications of mass movement are used: fall, slide, flow, and creep. Each involves the pull of gravity working on a mass until the critical shearing strength is reduced to the point that the mass falls, slides, flows, or creeps downward. The Madison River Canyon event was a type of slide, whereas the Nevado del Ruiz lahar mentioned earlier was a flow. A rockfall is simply a quantity of rock that falls through the air and hits a surface. During a rockfall, individual pieces fall independently, and characteristically form a pile of irregular broken rocks called a talus cone at the base of a steep cliff.

29. Name and describe the type of mudflow associated with a volcanic eruption.

The hot eruption of Nevado del Ruiz in central Colombia, South America, melted about 10% of the ice on the mountain's snowy peak, liquefying mud and volcanic ash, and sending a hot mudflow downslope. Such a flow is called a lahar, an Indonesian word referring to flows of volcanic origin. The Mount Saint Helens eruption also produced a lahar in the Toutle River valley.

30. Describe the difference between a landslide and what happened on the slopes of Nevado Huascarán.

A debris avalanche is a mass of falling and tumbling rock, debris, and soil. It is differentiated from a debris slide or landslide by the tremendous velocity achieved by the onrushing materials. These speeds often result from ice and water that fluidize the debris. The extreme danger of a debris avalanche results from these tremendous speeds and lack of warning.

31. What is scarification, and why is it considered a type of mass movement? Give several examples of scarification. Why are humans a significant geomorphic agent?

Large open-pit strip mines–such as the Bingham Copper Mine west of Salt Lake City, the Berkeley Pit in Butte, Montana, and the extensive strip mining for coal in the East and West, are examples of human-induced mass movements, generally called scarification. At the Bingham Copper Mine, a mountain literally was removed. The disposal of tailings and waste material is a significant problem with such large excavations because the tailing piles prove unstable and susceptible to further weathering, mass wasting, or wind dispersal.

Overhead Transparencies

As an adopter you are provided with the following figures for overhead projector use.

- Figure 13-3: Slope mechanics and form (2parts)
- Figure 13-14: Features of karst topography
- Figure 13-20: Mass movement classes

14

River Systems and Landforms

Overview

Earth's rivers and waterways form vast arterial networks that both shape and drain the continents, transporting the by-products of weathering, mass movement, and erosion. To call them Earth's lifeblood is not an exaggerated metaphor, inasmuch as rivers redistribute mineral nutrients important for soil formation and plant growth. Not only do rivers provide us with essential water supplies, but they also receive, dilute, and transport wastes, and provide critical cooling water for industry. Rivers have been of fundamental importance throughout human history. This chapter discusses the dynamics of river systems and related landforms that streams produce.

I choose to begin setting the stage with a discussion of the drainage basin–a basic hydrologic unit. With this established, we move through streamflow characteristics, gradient, and deposition as water cascades through the hydrologic system. The human component is irrevocably linked to streams, with so many settlements along river banks and on floodplains.

The *Student Study Guide* presents 23 "Learning Objectives" to guide the student in reading the chapter. The *Applied Physical Geography* lab manual has portions of one exercise that involve fluvial landscapes.

New to the Third Edition

(Note: This section highlights major changes, new features, and additions in the third edition. This does not describe all the rewrite and recast of the text.)

1. A list of key learning concepts begins the chapter.

2. Discussion of base level of streams has been moved from Chapter 13.

3. Drainage patterns have been organized into a bullet format for greater clarity.

4. Figure 14-3 illustrates the drainage divides that separate drainage basins.

5. New photos of fluvial erosional and depositional features are included in Figures 14-6, 14-8, 14-21, and 14-23.

6. News Report #1: "Scouring the Grand Canyon for New Beaches and Habitats" describes a new approach to recreating beach areas. This report chronicles an experiment completed in 1996 used to remedy the problems associated with river dams, such as stream channel becoming filled with sediment and downstream beaches being depleted of sand. The dam released water for 7 days increasing the volume of the channels, allowing them to scour existing sediments and deposit beach areas downstream.

7. The appropriateness of rivers as political boundaries is debated in News

Report #2: "Rivers Make Poor Political Boundaries."

8. News Report #3: "Niagara Falls Closed for Inspection," discusses the process of nickpoint retreat causing Niagara Falls to recede more than 6.8 miles in the last 12,000 years. The Niagara Falls are often shut down to reinforce these nickpoints.

9. Figure 14-9 illustrates the retreat of Niagara Falls.

10. Description of the functional model of dynamic equilibrium.

11. Elaborate description of the Mississippi River delta and its impact upon the city of New Orleans.

12. News Report #4: " The 1993 Midwest Floods," correlates the floods of the upper Mississippi and lower Missouri Rivers with increased precipitation patterns during 1993.

13. Description of the methods used to assess flood risk.

14. News Report #4: "The Nile Delta Is Disappearing." This report discusses the impact of the Aswan Dam and extensive canal systems upon the Nile delta and flow of the Nile River.

15. A new summary and review section ends the chapter.

Key Learning Concepts

1. *Define* the term fluvial and *outline* the fluvial processes: erosion, tranportation, and deposition.

2. *Construct* a basic drainage basin model and *identify* different types of drainage patterns, with examples.

3. *Describe* the relation among velocity, depth, width, and discharge and *explain* the various ways that a stream erodes and transports its load.

4. *Develop* a model of a meandering stream, including point bar, undercut bank, and cutoff, and *explain* the role of stream gradient in these flow characteristics.

5. *Define* a floodplain and *analyze* the behavior of a stream channel during a flood.

6. *Differentiate* the several types of river deltas and detail each.

7. *Explain* flood probability estimates and *review* strategies for mitigating flood hazards.

Expanded Outline Discussion

The following headings (boldfaced) match some of the first, second, and third order headings in Chapter 14. The narrative under each heading contains information, sources, and anecdotal facts relating to portions of the chapter. Not all text headings are discussed.

FLUVIAL PROCESSES AND LANDSCAPES

Stream-related processes are termed fluvial, from the Latin *fluvius*, meaning river. So in this chapter we refer to fluvial processes, actions, and features. A stream is a more generic term for all bodies of running water under the influence of gravity that move to relatively lower levels. The term river is a general term for a stream of permanent or seasonal flow, or any large stream, bigger than a creek or brook, such as a major trunk stream. You will note that stream is used throughout *GEOSYSTEMS* for general reference, whereas river is used when specific streams or water resources are referenced.

Several hydrology texts may provide additional background for this chapter. See: S. Lawrence Dingman, *Physical Hydrology*, New York: Macmillan Publishing Company, 1994; C.W. Fetter, *Applied Hydrology*, 3rd ed., New York: Macmillan Publishing Company, 1994; R.G. Kazman, *Modern Hydrology*, 3rd ed., Dublin, OH: The National Water Well Association, 1988; John C. Manning, *Applied Principles of Hydrology*, 2nd ed., New York: Macmillan Publishing Company, 1992.

The Drainage Basin System

This section sets the stage for our discussion of streams by defining and illustrating the drainage basin–a basic organizational unit. It might be effective to follow a drop of water in a familiar drainage basin from the time it hits the ground until it exits the mouth, as I did in the text with the Allegheny-Ohio-Mississippi system. The USGS publishes small pamphlets for each of the major drainage basins in the United States. The drainage basin maps in these pamphlets are similar to the Delaware River drainage map in Figure 14-5.

Figure 14-4 presents the continental divides that separate the major drainage basins of the United States and Canada. Data are in million m^3 for Canada and millions of acre-feet per year for the United States. These form basic water resources regions. Figure 14-4 presents the major drainage basins themselves for both countries.

The USGS launched a study of the Delaware River basin in 1988 to research the potential impact of future climate change, namely global warming, on water resources. The study includes changes in streamflow, irrigation demand, reduction in soil moisture storage, possible salt water intrusion near the ocean, and problems associated with sea-level rise. This study is through USGS, National Research Program, Water Resources Division, Denver Federal Center, Lakewood, CO 80225. See: D.M. Wolock, M.A. Ayers, L.E. Hay, and G.J. McCabe, "Effects of Climate Change on Watershed Runoff," *Hydraulic Engineering '89 Proceedings National Conference on Hydraulic Engineering*, August 14-18, 1989.

The California River Assessment (CARA) provides a comprehensive inventory and evaluation of California's river resources. Contact them at: http://ice.ucdavis.edu/California_Rivers_Assessment.

Drainage Density and Patterns

You may have access to topographic maps that illustrate each type of drainage pattern. These might be helpful to use if this is a subject you emphasize.

I often take a class day and have the students trace a river system, such as the Nile or Amazon, using their atlases. The students can then identify the drainage pattern, the drainage basin, bedrock materials and the volume of these rivers as a method of determining their erosive and transportational capabilities. I have also added deltas to this list to give students a hands-on look at arcuate, bird-foot, and estuarian deltas.

Streamflow Characteristics

This section presents stream erosion, stream transport, flow characteristics, and channel patterns, beginning with the Leopold and Maddock graphs depicting the relationship of stream velocity, depth, and width to discharge. The classic stream channel cross-section of the San Juan River near Bluff, Utah is presented in Figure 14-9.

Discharge is calculated by multiplying the velocity of the stream by its width and depth for a specific cross section of the channel as stated in the simple expression:

$$Q = wdv$$

where:
Q = discharge
w = channel width
d = channel depth
v = stream velocity

As Q increases, some combination of channel width, depth, or stream velocity increases. Discharge is expressed either in cubic meters per second (m^3/sec) or cubic feet per second (ft^3/sec).

Stream Gradient; Nickpoints

Take care in comparing and contrasting gradient and graded concepts as discussed in the text.

The Davis Geomorphic Cycle

William Morris Davis's overall body of work and accomplishment represents an incredibly rich contribution to geomorphology and the understanding of landscape. His ideal geomorphic cycle, although supplanted by more recent models, demonstrates the value of theory, the scientific method, and the research it stimulates.

Landscapes do not, as Davis proposed, rapidly uplift with little or no erosion. Landscapes do not hold still while they evolve through an orderly sequence of aging. Press and Siever (*Earth*, 1986, W.H. Freeman, p. 145) pose an interesting analogy using chemical engineering: Davis's cycle is akin to the *batch process* in which the process starts and stops periodically, as opposed to a *continuous-flow process* in which all parts of the process are going on simultaneously as in the dynamic equilibrium approach to landscape development.

Davis's model was hypothetical since no observational evidence of a complete cycle in nature was available. The model does not recognize the open system characteristics actually found in nature, making a description of nature through Davisian eyes, at best, a vast "multicyclic" array of landforms. Landscapes are really combinations of each stage in the cycle at any one time–a point that Davis himself recognized. The character of any region is really a product of the conditions presently in effect and operating, not a progression of factors initiated in the past. Some feel that the influence of past landscapes is very important. The problem again is the time-scale involved. So why take the time and space to present such cyclic, sequential concepts? The historical importance and contribution of Davis' work to our understanding is undeniable. There is a value in having individual terms to apply to individual landscapes at a point in time, even though they are not part of a progressive cycle. Streams do operate toward a graded course, slopes do retreat and reduce, and lands do undergo landmass denudation. And if properly limited in application, there is an advantage to better understanding various landform groupings with these described features in mind.

Pay particular attention to the section in the text that states: "However, as suggested by Schumm and Lichty, two modern geomorphologists, the validity of cyclic or functional landscape models may depend on the time frame selected....understanding could be a function of the time scale selected."

Stream Deposition

Deposition, which produces floodplains, terraces, and deltas, logically follows as the next step in the discussion of fluvial processes. Floodplains are introduced in this section with the human-related aspects of hazards, zoning, and preventative constructions presented later. An idealized floodplain illustration is presented in Figure 14-21.

The floods pictured in Figure 14-22 occurred throughout the Midwest during 1993. Many of the concepts discussed in the text under "Floods and River Management" and Focus Study 14-1 "Floodplain Management Strategies," apply directly to these floods and the experiences of the many cities affected. See: Mary F. Myers and Gilbert F. White, "The Challenge of the Mississippi Flood," *Environment*, 35, no. 10, December 1993: 2-9+. This article presents a chronology of "Significant Events in the Development of U.S. Flood Control Policy (1825-1993)."

An extensive portrait by Alan Mairson, "The Great Flood of '93," appears in *National Geographic*, 185, no. 1, January 1994: 42-81. Included are remotely sensed orbital images before and during the flood showing surface water and soil moisture. Also: a summary of two dozen wire-service stories with maps and photographs that appeared in *Storm–The World Weather Magazine*, 1, no. 1, August 1993: 6-16 (1-708-406-8330).

The USGS produced a map of 1993 peak flood stages and discharges as compared to previous maximums experienced in the Mississippi River basin. See: C. Parrett, N.B. Melcher, and R.W. James, Jr., *Flood Discharges in the Upper Mississippi Basin*, Circular 1120-A, Washington, DC: USGS, 1993. A map in the circular shows that discharges were higher than previous maximums at 154 stations.

Much of the floodplain in the Sacramento Valley of central and northern California is susceptible to such periodic flooding despite the dams, bypasses and weirs, and reservoirs that are in operation. The aging system of levees is vulnerable to failure. This will become more severe in the lower parts of the valley with a rising sea level, land subsidence due to further groundwater removal, soil compaction, and peat oxidation in the delta.

A significant share of the world's population lives along rivers and is affected by fluvial processes. Those living along the Nile Delta have for centuries taken advantage of the fresh delivery of soil nutrients brought downstream by annual floods. Only since the construction of the Aswan High Dam and Lake Nasser in the late 1950s has this process been disrupted, for since that time alluvial materials have been trapped behind that new local base level formed by the dam. The delivery of fresh sediment to delta farms was curtailed and the chemistry of the eastern Mediterranean Sea is changing as a result of the dam. The management of rivers, watersheds, and floodplains represents an important aspect of applied physical geography.

The 2-volume set edited by Martha Lou Shirley, *Deltas In Their Geologic Framework*, Houston: Houston Geological Society, 1966, is extensive and includes many examples of each type of deltaic formation. The set is still available from the HGS.

FLOODS AND RIVER MANAGEMENT

Case study: Hazard perception, planning, and zoning (or lack of same), are issues of growing significance. In the region in which I live–the Sacramento Valley–urban limit lines had protected floodplain from settlement and preserved its use as prime-agricultural land, although this protective political structure eventually failed. The rest is history: The floodplain is now developed with warehousing, commerce, light industry, a fast-growing residential population, and a multipurpose sports/entertainment arena. The federal government assisted the exodus to the floodplain with a four-year moratorium of floodplain enforcement–they actually erased the label from restrictive maps. The area now is the site of a kind of land rush–all protected by an aging levee system. The last major levee break occurred in February 1986 north of Sacramento. This is placed in perspective when compared with the lower Mississippi River and the ongoing struggle between the river, the Atchafalaya River, the Old Control Structure, further tampering with channel controls, rising sea level, and an isostatically subsiding coastal area. These are amazing times when it comes to elementary aspects of hazards, business, politics, and human activities.

Hydrographs

Rainfall and snowmelt convert to streamflow at a rate based on the characteristics of the basin: soil permeability and infiltration rates, amount and type of vegetative cover, percentage of the basin under cultivation, and degree of urbanization and settlement. These various characteristics influence the pattern of the flood wave as it moves downstream.

Additional Flood Considerations

A reservoir, dam, and surrounding watershed constitute an open system. The essential input is precipitation, including both rain and snow meltwater. The system outputs are evaporation from the reservoir, evaporation and transpiration from soils and plants within the watershed, water absorbed by the ground that holds the reservoir, and water discharged through the dam's spillways. During a drought, less rainfall results in a reduction of runoff, which cuts back on the overall erosion in the watershed and resultant sediment load into the reservoir. However, balancing against that action is the reduction of vegetative cover because of the dieback of plants during a drought, which creates more bare ground and thus a potential for increased runoff, making soil erosion more rapid during whatever moisture events do take place. During the drought event, reservoir managers close down outflow penstocks, or spillways, to increase water in storage as much as possible, allowing lower, but regular releases to maintain downstream river flows for necessary uses below the dam.

The water held back in a reservoir is released to produce hydroelectric power on demand, known as peak load power. One reservoir strategy to extend the water resource involves installation of reverse-storage turbines so that during off-peak hours, water is pumped from afterbays through the turbines back into the reservoir for reuse.

Lake Powell, behind Glen Canyon Dam at the Utah-Arizona border, is silting in rapidly. Hydrologists generally agree that Lake Powell could be filled with sediment during the next century, perhaps in as little as 75 years. The 216 m [710 ft] dam might someday become an elaborate nickpoint on the river and the site of a spectacular waterfall, as the river channel flows across the sediment-filled reservoir, dropping on the buildings below.

The focus study discusses the James Bay Project in central and northern Québec and the uncompleted and controversial Great Whale project.

To supplement your materials for water resources please get a copy of a first-ever supplement to National Geographic: *Water–The Power, Promise, and Turmoil of North America's Fresh Water*, A National Geographic Special Edition, v. 184, no. 5A, 1993. Page 120 of this supplement presents an entire page of many resources that would be useful in teaching. Sections include: "Supply–Sharing the Wealth of Water," and "California: Desert in Disguise," "Development–When Human's Harness Nature's Forces," and "James Bay: Where Two Worlds Collide," "Pollution–Troubled Waters Run Deep," and "The Mississippi River Under Siege," "Restoration–New Ideas, New Understanding, New Hope." The issue includes a giant map supplement of water resources, surface and ground, for the United States. Call 1-800-638-4077 to order.

Glossary Review for Chapter 14 (in alphabetical order)

abrasion
aggradation
alluvial terraces
alluvium
backswamp
base level
bed load
braided stream
continental divides
delta
deposition
dissolved load
drainage basins
drainage density
drainage pattern
erosion
estuary

flood
floodplain
fluvial
graded stream
gradient
hydraulic action
hydrograph
meandering stream
natural levees
nickpoint
oxbow lake
peneplain
point bar
saltation
sheet flow
suspended load
traction
transport
undercut bank
watershed
yazoo tributary

Annotated Chapter Review Questions

1. What role is played by rivers in the hydrologic cycle?

Earth's rivers and waterways form vast arterial networks that both shape and drain the continents, transporting the by-products of weathering, mass movement, and erosion. To call them Earth's lifeblood is not an exaggerated metaphor, inasmuch as rivers redistribute mineral nutrients important for soil formation and plant growth. Not only do rivers provide us with essential water supplies, but they also receive, dilute, and transport wastes and provide critical cooling water for industry. Rivers have been of fundamental importance throughout human history.

2. Define the term fluvial. What is a fluvial process?

Stream-related processes are termed fluvial (from the Latin *fluvius*, meaning river). Insolation is the driving force of fluvial systems, operating through the hydrologic cycle and working under the influence of gravity. Denudation by water dislodges, dissolves, or removes surface material as erosional fluvial processes. Thus, streams supply weathered and wasted sediments for transport to new locations, where they are laid down in a process known as deposition.

3. What is the sequence of events that takes place as a stream dislodges material?

Water dislodges, dissolves, or removes surface material in the process called erosion. Streams produce fluvial erosion, in which weathered sediment is picked up for transport to new locations. Thus, a stream is a mixture of water and solids, the solids are carried by solution, suspension, and by mechanical transport. Materials are then laid down by another process, deposition.

4. Explain the base level concept. What happens to a local base level when a reservoir is constructed?

Base level is a level below which a stream cannot erode its valley further (Figure 14-2). The hypothetical absolute or ultimate base level is sea level (which is the average level between high and low tides). You can imagine base level as a surface extending inland from sea level, inclined gently upward, under the continents. Ideally, this is the lowest practical level for all denudation processes. Although base level is a very useful concept, no satisfactory working definition has yet been agreed upon. A local, or temporary, base level may control a regional landscape and the lower limit of local streams. That local base level might be a river, a lake, a hard and resistant rock structure, or a human-made dam. In arid landscapes, with their intermittent precipitation, local control is provided by valleys, plains, or other low points.

A reservoir and dam structure interrupt the gradient of a stream, producing a local base level that controls the upstream behavior and profile of the stream. The top of the dam is the precise location of the local base level. The load carried by the stream is deposited in the reservoir, since the stream loses velocity as it enters the body of water. If the dam should break, the stream would rapidly scour a channel through these deposits in response to a new downstream base level, forming terraces on either side of the stream through the former reservoir.

5. What is the spatial geomorphic unit of an individual river system? How is it determined on the landscape? Define the key relevant terms used.

Streams are organized into areas or regions called drainage basins. A drainage basin is the spatial geomorphic unit occupied by a river system. A drainage basin is defined by ridges that form drainage divides, i.e., the ridges are the dividing lines that control into which basin precipitation drains. Drainage divides define watersheds, the catchment areas of the drainage basin. The United States and Canada are divided by several continental divides; these are extensive mountain and highland regions that separate drainage basins, sending flows either to the Pacific, or to the Gulf of Mexico and the Atlantic, or to Hudson Bay and the Arctic Ocean.

6. On Figure 14-4, follow the Allegheny River to the Gulf of Mexico, analyze the pattern of tributaries, and describe the channel. What role do continental divides play in this drainage?

Rainfall in north-central Pennsylvania generates the streams that flow into the Allegheny River. The Allegheny River then joins with the Monongahela River at Pittsburgh to form the Ohio River. The Ohio flows southwest and connects with the Mississippi River at Cairo, Illinois, and eventually flows on past New Orleans to the Gulf of Mexico. Each contributing tributary adds its discharge and sediment load to the larger river. In our example, sediment weathered and eroded in north-central Pennsylvania is transported thousands of kilometers and accumulates as the Mississippi delta on the floor of the Gulf of Mexico.

7. Describe drainage patterns. Define the various patterns that commonly appear in nature. What drainage patterns exist in your home town? Where you attend school?

A drainage basin is the spatial geomorphic unit occupied by a river system. A drainage basin is defined by ridges that form drainage divides, i.e., the ridges are the dividing lines that control into which basin precipitation drains. Drainage basins are open systems whose inputs include precipitation, the minerals and rocks of the regional geology, and both the uplift and subsidence provided by tectonic activities. System outputs of water and sediment leave through the mouth of the river. Change that occurs in any portion of a drainage basin can affect the entire system as the stream adjusts to carry the appropriate load relative to discharge and velocity. Seven principal drainage patterns are shown in Figure 14-7.

8. What was the impact of flood discharge on the channel of the San Juan River near Bluff, Utah? Why did these changes take place?

Figure 14-9 shows changes in the San Juan River channel in Utah that occurred during a flood. The increase in discharge increased the velocity and therefore the carrying capacity of the river as the flood progressed. As a result, the river's ability to scour materials from its bed was enhanced. Such scouring represents a powerful clearing action.

9. How does stream discharge do its erosive work? What are the processes at work on the channel?

Several types of erosional processes are operative. Hydraulic action is the work of turbulence in the water–the eddies of motion. Running water causes friction in the joints of the rocks in a stream channel. A hydraulic squeeze-and-release action works to loosen and lift rocks. As this debris moves along, it mechanically erodes the streambed further through the process of abrasion, with rock particles grinding and carving the streambed.

10. Differentiate between stream competence and stream capacity.

Competence, which is a stream's ability to move particles of specific size, is a function of stream velocity. The total possible load that a stream can transport is its capacity.

11. How does a stream transport its sediment load? What processes are at work?

Eroded materials are transported by four processes: solution, suspension, saltation, and traction. Solution refers to the dissolved load of a stream, especially the chemical solution derived from minerals such as limestone or dolomite or from soluble salts. The suspended load consists of fine particles physically held aloft in the stream, with the finest particles not deposited until the stream velocity slows to near zero. The bed load refers to those coarser materials that are dragged along the bed of the stream by traction or are rolled and bounced along by saltation (from the Latin *saltim*, which means "by leaps or jumps").

12. Describe the flow characteristics of a meandering stream. What is the pattern of the flow in the channel, the erosional and depositional features, and the typical landforms created?

A meandering channel pattern is common for a stream that slopes gradually, a sinuous form weaving across the landscape. The outer portion of each meandering curve is subject to the greatest erosive action and can be the site of a steep bank called a cut bank (Figure 14-14). On the other hand, the inner portion of a meander receives sediment fill and forms a deposit called a point bar. As meanders develop, these scour-and-fill features gradually work their way downstream. If the load in a stream exceeds the capacity of the stream, sediments accumulate as an aggradation in the stream channel (the opposite of degradation) as the channel builds up through deposition. With excess sediment, a stream becomes a maze of interconnected channels laced with sediments that form a braided pattern.

13. Explain these statements: (a) all streams have a gradient, but not all streams are graded; and (b) graded streams may have ungraded segments.

Every stream has a degree of inclination or gradient, which is the rate of decline in elevation from its headwaters to its mouth, generally forming a concave-shaped slope (Figure 14-15). Theoretically, a stream gradient becomes graded when the load carried by the stream and the landscape through which it flows become mutually adjusted, forming a state of dynamic equilibrium among erosion, transported load, deposition, and the stream's capacity. Attainment of a graded condition does not mean that the stream is at its lowest gradient, but rather that it represents a balance among erosion, transportation, and deposition over time along a specific portion of the stream. One problem with applying the graded stream concept in an absolute sense, however, is that an individual stream can have both graded and ungraded portions and may have graded sections without having an overall graded slope. In fact, variations and interruptions in a graded profile of equilibrium occur as a rule rather than an exception, making a universally acceptable definition difficult.

14. Why is Niagara Falls an example of a nickpoint? Without human intervention what do you think would eventually take place at Niagara Falls?

At Niagara Falls on the Ontario-New York border, glaciers advanced and receded over the region, exposing resistant rock strata underlain by less-resistant shales. As the less-resistant material continued to weather away, the overlying rock strata collapsed, allowing the falls to erode farther upstream toward Lake Erie. In fact, the falls have retreated more than 11 km (6.8 mi) from the steep face of the Niagara escarpment (long cliff) during the past 12,000 years (see Figure 14-19).

15. What is meant by "the validity of cyclic or equilibrium models depends on the time frame being considered"? Explain and discuss.

A complete model has yet to replace the cyclic model of Davis, now abandoned, although landscapes are generally viewed as operating in a state of dynamic equilibrium. However, as suggested by Schumm and Lichty, two modern geomorphologists, the validity of cyclic or functional landscape models may depend on the time frame selected. Over the long span of geologic time, cyclic models of evolutionary development might explain, for example, aspects of the disappearance of entire mountain ranges through denudation. But these generalizations would not be applicable to "steady time," the adjustments ongoing in a portion of a drainage basin. In between these two time frames lies the realm of "graded time," or the conditions of dynamic equilibrium.

16. Describe the formation of a floodplain. How are natural levees, oxbow lakes, backswamps, and yazoo tributaries formed?

The low-lying area near a stream channel that is subjected to recurrent flooding is a floodplain. It is formed when the river leaves its channel during times of high flow. Thus, when the river channel changes course or when floods occur, the floodplain is inundated with water. When the water recedes, alluvial deposits generally mask the underlying rock. Figure 14-21 illustrates a characteristic floodplain, with the present river channel embedded in the plain's alluvial deposits. The former meander scars form water-filled (often ephemeral) loops on the floodplain called oxbow lakes (known as a billabong in Australia, an aboriginal term meaning *dead river* that is only intermittently filled with water).

On either bank of the river are natural levees, which are by-products of flooding. When flood waters arrive, the river overflows its banks, loses velocity as it spreads out, and drops a portion of its sediment load to form the levees. Larger sand-sized particles drop out first, forming the principal component of the levees, with finer silts and clays deposited farther from the river. Successive floods increase the height of the levees and may even raise the overall elevation of the channel bed so that it is perched above the surrounding floodplain.

Notice on Figure 14-21 an area labeled backswamp and a stream called a yazoo tributary. The natural levees and elevated channel of the river prevent this tributary from joining the main channel, so it flows parallel to the river and through the backswamp area.

17. Can you identify any of these features on the Philipp, Mississippi, topographic quadrangle in Figure 14-21b?

See Figure 14-21b. Students should recognize a meandering stream, oxbox lakes, meander scars, point bars and undercut banks, yazoo rivers and other features that are illustrated in Figure 14-21a.

18. Are there any floodplains where you live or where you go to college? Have you seen any of these features?

Personal analysis and response.

19. What is a river delta? What are the various deltaic forms? Give some examples.

The mouth of a river marks the point where the river reaches a base level. Its forward velocity rapidly decelerates as it enters a larger body of standing water, with the reduced velocity causing its transported load to be in excess of its capacity. Coarse sediments drop out first, with finer clays being carried to the extreme end of the deposit.

This depositional plain formed at the mouth of a river is called a delta, named after the triangular shape of the Greek letter delta, which was perceived by Herodotus in ancient times to be similar to the shape of the Nile River delta. See the discussion of deltaic forms in the text.

20. How might life in New Orleans change in the next century? Explain.

Due to the dynamic character of the Mississippi River delta, the main channel of the delta persists in its present location because of much effort and expense directed to maintain an artificial levee system. Compaction and tremendous weight of the sediments in the Mississippi River create isostatic adjustment in Earth's crust. This is causing the entire region of the delta to subside, placing tremendous stress on natural and artificial levees along the lower Mississippi.

The city of New Orleans is now almost entirely below river level, with some sections of the city below sea level. Severe flooding is a certainty for existing and planned settlements unless further intervention or urban relocation occurs. The building of multiple flood-control structures and extensive reclamation efforts by the U.S. Army Corps of Engineers apparently have only delayed the peril, as demonstrated by recent flooding.

An additional problems for the lower Mississippi Valley is the possibility that the river could break from its existing channel and seek a new route to the Gulf of Mexico. The obvious alternative route for the Mississippi is along the Atchafalaya River. If the Mississippi would bypass New Orleans, the threat of flooding would be reduced, yet, it would be a financial disaster for New Orleans since the port would silt in and sea water would intrude the fresh water resources.

21. Describe the Ganges River delta. What factors upstream explain its form and pattern? Assess the consequences of settlement on this delta.

The Ganges River delta features an intricate braided pattern of distributaries. Alluvium carried from deforested slopes upstream provides excess sediment that forms the many deltaic islands.

Catastrophic floods continue to be a threat. In Bangladesh, intense monsoonal rains and tropical cyclones in 1988 and 1991 created devastating floods over the country's vast alluvial plain (130,000 km^2 or 50,000 mi^2). One of the most densely populated countries on Earth, Bangladesh was more than three-fourths covered by floodwaters. Excessive forest harvesting in the upstream portions of the Ganges-Brahmaputra River watersheds increased runoff and added to the severity of the flooding. Over time, the increased load carried by the river was deposited in the Bay of Bengal, creating new islands. These islands, barely above sea level, became sites of new settlements and farming villages. When the recent floodwaters finally did recede, the lack of freshwater–coupled with crop failures, disease, and pestilence–led to famine and the death of tens of thousands. About 30 million people were left homeless and many of the alluvial-formed islands were gone.

22. What is meant by the statement, "the Nile River delta is disappearing"?

The Nile delta is disappearing due to the building of the Aswan Dam and the extensive network of canals which have been built in the delta to augment the natural distributary system. Yet, as the river enters the network of canals, flow velocity is reduced, stream competence and capacity are lost, and seidment load is deposited far short of where the delta reaches the Mediterranean Sea. River flows no longer reach the sea! The Nile delta is receding from the coast at an alarming 50 to 100 m per year. Seawater is intruding farther inland in both surface water and groundwater.

23. Specifically, what is a flood? How are such flows measured and tracked?

A flood is a high water level that overflows the natural (or artificial) banks along any portion of a stream. Understanding flood patterns for a drainage basin is as complex as understanding the weather, for floods and weather are equally variable, and both include a level of unpredictability. The key is to measure streamflow–the height and discharge of a stream. A staff gauge, a pole placed in a stream bank and marked with water heights, is used to measure stream level. With a fully measured cross section, stream level can be used to determine discharge. A stilling well is sited on the stream bank and a gauge is mounted in it to measure stream level. A portable current meter can be used to sample velocity at various locations. Approximately 11,000 stream gauge stations are used in the United States (an average of over 200 per state). Of these, 7000 have continuous stage and discharge records operated by the U.S. Geological Survey. Many of these stations automatically telemeter data to satellites, from which information is retransmitted to regional centers. Environment Canada's Water Survey of Canada maintains more than 3000 gauging stations.

24. Differentiate between a hydrograph from a natural terrain and one from an urbanized area.

A graph of stream discharge over a time period for a specific place is called a hydrograph. The hydrograph in Figure 14-28 shows the relationship between stream discharge and precipitation input. During dry periods, at low water stages, the flow is described as base flow and is largely maintained by contributions from the local water table. When rainfall occurs in some portion of the watershed, the runoff collects and is concentrated in streams and tributaries. The amount, location, and duration of the rainfall episode determines the peak flow. Also important is the nature of the surface in a watershed; for example, a hydrograph for a specific portion of a stream changes after a forest fire or urbanization of the watershed.

Human activities have enormous impact on water flow in a basin. The effects of urbanization are quite dramatic, both increasing and hastening peak flow as shown in the same figure. In fact, urban areas produce runoff patterns quite similar to those of deserts. The sealed surfaces of the city drastically reduce infiltration and soil moisture recharge, behaving much like the hard, nearly barren surfaces of the desert.

25. What do you see as the major consideration regarding floodplain management? How would you describe the general attitude of society toward natural hazards and disasters?

Throughout history, civilizations have settled floodplains and deltas, especially since the agricultural revolution that occurred some 8000 years B.P. when the fertility of floodplain soils was discovered. Early villages were generally built away from the area of flooding, or on stream terraces, because the floodplain was the location of intense farming. However, as the era of commerce grew, sites near rivers became important for transportation: port and dock facilities and river bridges to related settlements were built. Also, because water is a basic industrial raw material used for cooling and for diluting and removing wastes, waterside industrial sites became desirable.

Human activities on vulnerable flood-prone lands require planning to reduce or avoid disaster. Essentially, relative to all natural disasters, including floodplains, human societies appear to be unwilling, unable, or incapable of perceiving hazards in a familiar environment. The development and urbanization of the Sacramento River floodplain presented in the discussion a few pages earlier exemplifies these attitudes.

26. What do you think the author of the article, "Settlement Control

Beats Flood Control," meant by the title? Explain your answer, using information presented in the chapter.

As suggested in an article, "Settlement Control Beats Flood Control" (Walter Kollmorgen, *Economic Geography* 29, no. 3, July 1953: 215.), there are other ways to protect populations than with enormous, expensive, sometimes environmentally disruptive projects. Strictly zoning the floodplain is one approach, but flat, easily developed floodplains near pleasant rivers might be perceived as desirable for housing, and thus weaken political resolve. This strategy would set aside the floodplain for farming or passive recreation, such as a riverine park, golf course, or plant and wildlife sanctuary, or for other uses that are not hurt by natural floods. Kollmorgen's study concludes that "urban and industrial losses would be largely obviated by set-back levees and zoning and thus cancel the biggest share of the assessed benefits which justify big dams." Focus Study 14-1 takes a closer look at floodplain management.

Overhead Transparencies

As an adopter you are provided with the following figures for overhead projector use.

- Figure 14-4: Continental divides and major drainage basins for the United States and Canada
- Figure 14-15: Meandering stream development (photo and 4 arts)
- Figure 14-21: A floodplain (illustration and topographic map segment)

15

Eolian Processes and Arid Landscapes

Overview

Wind is an agent of erosion, transportation, and deposition. Its effectiveness has been the subject of much debate; in fact, wind at times was thought to produce major landforms. Presently, wind is regarded as a relatively minor exogenic agent, but it is significant enough to deserve our attention. In this chapter we examine the work of wind, its associated processes, and resulting landforms. In addition, we'll look at desert landscapes, where water remains the major erosional force but where an overall lack of moisture and stabilizing vegetation allows wind processes to operate. Evidence of this fact include extensive sand seas and sand dunes of infinite variety. Beach and coastal dunes, which form in many climates, also are influenced by wind and are discussed in the next chapter.

The *Student Study Guide* presents 20 "Learning Objectives" to guide the student in reading the chapter. The *Applied Physical Geography* lab manual has portions of one exercise that involve arid landscapes.

New to the Third Edition

(Note: This section highlights major changes, new features, and additions in the third edition. This does not describe all the rewrite and recast of the text.)

1. A list of key learning concepts begins the chapter.

2. News Report #1: "War Tears Up the Pavement" examines the impact of the Persian Gulf War upon desert pavement, and the resultant increase of particles that were loosened for deflation.

3. A new photo complements Figure 15-1, illustrating sand movement and wind velocity.

4. A photo supports the illustration of sand transportation in Figure 15-4, depicting sand grains saltating along the Stovepipe Wells Dune Field.

5. Some yardangs are large enough to be detected on satellite imagery. News Report #2: "Yardangs From Mars" discusses the use of remote sensors to study the Martian surface, locating wind-blown sediments as proof of eolian processes on Mars.

6. Figure 15-3 is a new photograph of a yardang.

7. Figure 15-14 has been updated with photos of arid landforms found in the the Sonoran, the Taklimakan, the Atacama, and the Kalahari deserts.

8. News Report #3: "The Dust Bowl," discusses the deflation and wind transport of loess soils that created tremendous soil erosion in the Great Plains during the 1930's. Land management strategies used at that time left the Great Plains vunerable to such erosion.

9. A schematic illustration accompanies Figure 15-19, a photograph of the Mitten Buttes and the Merrick Butte along the Utah-Nevada border, to demonstrate the

mass amount of material that has been eroded and transported from this area through eolian processes.

10. More elaborate discussion of Basin and Range topography.

11. A photograph of Death Valley as an example of a Basin and Range Province, Figure 15-23.

12. A new summary and review section ends the chapter.

Key Learning Concepts

1. *Characterize* the unique work accomplished by wind and eolian processes.

2. *Describe* eolian erosion, including deflation, abrasion, and the resultant landforms.

3. *Describe* eolian transportation; *explain* saltation and surface creep.

4. *Identify* the major classes of sand dunes and *present* examples within each class.

5. *Define* loess deposits, their origins, locations, and landforms.

6. *Portray* desert landscapes and *locate* these regions on a world map.

Expanded Outline Discussion

The following headings (boldfaced) match some of the first, second, and third order headings in Chapter 15. The narrative under each heading contains information, sources, and anecdotal facts relating to portions of the chapter. Not all text headings are discussed.

THE WORK OF WIND

I include a quote from Ralph Bagnold who is credited with the first important arid-land and sand-sea research. While stationed with the British army in Egypt, he explored the western deserts of Egypt. Much of his research time had to be sandwiched in on available leaves from his post. Bagnold took Henry Ford at his word that a Model-T Ford could handle the most difficult terrain. He drove a Model-T all over the desert, carrying sections of chickenwire for areas where support was needed under the wheels. Additional wind-tunnel tests completed key aspects of his research on eolian processes (aeolian is the European spelling).

I recommend that you obtain a copy of McKee, Edwin D. *A Study of Global Sand Seas*, U.S. Geological Survey Professional Paper 1052. Washington: Government Printing Office, 1979. You can get this through any one of the USGS regional bookstores. This oversized book is an excellent comprehensive treatment of erg desert landscapes and forms. The dune illustrations in Figure 15-9 were adapted from this publication.

Eolian Erosion Deflation (Persian Gulf War effects)

In addition to the discussion in the text, be sure to mention the importance of intermittent flows of water in the formation of desert pavement, or lag desert deposits. Water delivers materials to help cement the cobbles and pebbles of the surface.

The disruption of desert pavement in the recent Persian Gulf War represents a *spatial* consequence of war activities in an arid environment, especially when the expenditure of munitions and movement of heavy vehicles are so concentrated in time and area (p. 451). Several recent references are available, including Constance Holden's "Kuwait's Unjust Deserts: Damage to Its Desert," *Science* Vol. 251 (8 March 1991): 1175–subtitled "Among the less heralded consequences of the Gulf war could be a massive movement of sand engulfing roads and towns"; also, see the survey article by Thomas Y. Canby in the August 1991 *National Geographic* (Vol. 180, No. 2, pp. 2-35). The environment is overlooked in both theater and strategic military planning, yet with modern firepower and mobilization prowess, the environment should be considered.

A recent study completed by Dr. Farouk El-Baz, Director of the Center for Remote Sensing at Boston University, reported on the impact of the Persian Gulf War in Kuwait. Using *Landsat* and *SPOT* images and ground-based confirmation studies he found that 5446 km^2 of Kuwait's total land-surface area suffered war related damage–*that's 30.6% of Kuwait*! Impacts were classified as: land mines and post-war cleanup of mines and unexploded ordinance, land blackened by oil and soot, destroyed desert vegetation, land altered by the construction of berms, oil lakes, oil-well fire fighting alteration of the land, and vehicle tracks. The overall destruction produced by these impacts destroyed the desert pavement. Freed fine-grained materials were then available for deflation and transport, covering farms, roads, and encroaching on cities. Groundwater resources are being watched closely for oil contamination as it percolates to the water table. Dr. El-Baz presented these results at the October 1993 meeting of the Geological Society of America (GSA) in Boston.

These are issues that geographers are equipped to handle. With adequate inputs and a GIS-type work up on the potential impacts of the war. I wonder what weight such a GIS approach to the environment would have in encouraging the economic-sanction alternative to going to war. Also see "Burning Questions" and "Why are Data from Kuwait Being Withheld?" in *Scientific American* 265, No. 1 (July 1991) and "Up in Flames" and "U.S. Gags Discussion of War's Environmental Effects," in *Scientific American* 264, No. 5 (May 1991): 17-24; and, Sir Frederick Warner's "The Environmental Consequences of the Gulf War," *ENVIRONMENT* 33, No. 5 (June 1991): 7-9, 25-26.

As an example of a proactive statement about war, please consult Association of American Geographers' "Statement of the Association of American Geographers on Nuclear War." Adopted by AAG, 6 May 1986, Minneapolis.

Abrasion

Understanding of the overall effects of abrasion and deflation processes on the landscape has gone through a transformation of thought during this century. Earlier researchers imagined that these eolian processes had shaped the valleys of the Basin and Range province in the western United States. Eolian processes are currently believed to contribute little to the formation of erosional landforms.

Eolian Transportation

Note the difference between fluvial saltation and eolian saltation, i.e., hydraulic lift vs. aerodynamic lift, elastic bounce, and impact (Figure 14-11 compared to Figure 15-4).

Eolian Depositional Landforms

Figure 15-9 is designed to assist students in understanding the classes of dunes and representative types. If you have ever visited a dune field somewhere in the world, perhaps you have sensed the beauty of the shifting sand. If Earth wrote poems I think it is a toss-up between clouds and sand seas as to the most poetic forms.

Loess Deposits

The term loess is from the German word *löss*, for the original name of a fine loam in northern Germany. In the text, I suggest the pronunciation, "luss" which I learned from several soil scientists at the Soil Conservation Service in Washington and from several professors at universities in the Midwest. I assume that either "luss" or "lers" are acceptable pronunciations.

Two types of loess are mentioned in the text–glacial loess, or true loess, and desert loess. The latter is subject to more conjecture as to origins, although China and central Asia possess enormous deposits of apparent desert loess. Also, there is new evidence referred to in the text of wind-blown silt deposited in the Amazon region that originates from African deserts, thus enriching Oxisols of South America.

OVERVIEW OF DESERT LANDSCAPES

The 1992 revival of the movie *Lawrence of Arabia* in a restored form allows us to sense the awesome scale and spectacle of erg and reg deserts. Many shots were made with David Lean's patient camera remaining fixed on vast scenic panoramas, unlike the more hectic editing in modern film.

Relative to form and process at the Grand Canyon, the discussion in Chapter 13 of this resource manual may be a useful description to accompany your slides and discussion. Many aspects of arid-landscape processes are in evidence at the Grand Canyon of the Colorado in northern Arizona, where a grand river valley is being carved out of horizontal rock strata. Continuous adjustments with each storm and passing season are occurring in the landscape.

The exposed bedrock of the canyon walls reveals 20 distinct layers and records an incredible history, covering a time span of almost 2 billion years. The Kaibab limestone, which makes up the plateau and rim at the top of the canyon (the light-colored strata you see along the rim in the Part 3 opening photo, p. 309, and the Chapter 8 opening photo, pp. 208-09), is dated at 250 million years. At least 14 distinct layers that once sat atop the Kaibab are missing, long since removed by erosion and transportation. These layers are still in place in Utah, north of the canyon; thus, the Grand Canyon is understood to be at the bottom of what was once a huge sequence of strata. Other bedrock layers that are missing in the canyon are seen as unconformities in the sequence of strata, or a gap in the geologic record. The roots of old mountains are represented by metamorphic rocks in the deepest inner gorge.

Desert Fluvial Processes

The experience portrayed in Figure 15-16 and also in 11-11 was a once-in-a-lifetime chance to witness the extreme contrasts of fluvial processes in an arid landscape. I waited at the roadblock for the flooded playas to retreat and went into Death Valley as soon as access was permitted. Few people were there because of the road conditions around the monument and heavy snows that had fallen in mountains to the north and west. This was in spring 1983 during the heavy precipitation associated with the 1982-83 El Niño. As I walked across the Stovepipe Wells sand dune field, I heard the sound of running water. As I topped one ridge there was a wide, shallow, swift river cutting through the sand. The clays being deposited were finer than face powder. The 2.57 cm (1.01 in.) rainfall in a 24-hour period equaled 55% of the annual average of 4.65 cm (1.83 in.), producing the runoff through the dune field. The temporal river flowed to the lowest point in that section of the valley. One month later I returned with a battery-powered slide viewer in an effort to match the comparative shots.

Alluvial Fans

The sorting that occurs from the mouth of a canyon to the salt pan of the playa is remarkable. The top of the alluvial fan–in Death Valley these are almost 1500 m (5000 ft) higher than the valley floor–features large rocks and boulders. As you walk toward the center of the valley the individual pieces grow smaller and smaller in a graded fashion. Finally toward the center of the valley the silts give way to clays as fine as powder covered during dry times with a light crust of borated salts. Before the water evaporates the playa lake can be several kilometers wide and about 10 kilometers long, yet only a few centimeters in depth. The clay surface beneath these shallow waters bears linear striations in the direction of flow toward the lowest point.

Desert Landscapes

Perhaps not since Joseph Wood Krutch has a writer so captured the feeling of arid lands as Edward Abbey, in his *Desert Solitaire* (New York: McGraw-Hill Book Co., 1968).

Focus Study: The Colorado River: A System Out of Balance

I hope discussion and dialog result from this focus study since the budget presented in Table 1 demonstrates a river overbudgeted by some 3.5 million acre-feet per year. Present understanding of Colorado River flows seems to be based on the exceptionally high river discharges that occurred before the Colorado River Compact was signed in 1922.

The floods generated by the El Niño event of 1982-83 produced near-record discharges in the river. Federal regulators, under the leadership of the Secretary of the Interior, apparently saw an opportunity to have all the reservoirs in the system full for the first time in the history of reclamation activities on the river. In essence this meant that the government would invert the reclamation law that demands flood control as the prime function of these multipurpose facilities. Instead, water for irrigation and hydroelectric interests were placed first. The resultant flood described in the focus study occurred–a human-caused flood along the most regulated river in the world!

A few dry years in a row will certainly stress the river's water budget, especially with the operation of the new Central Arizona Project (CAP) and increasing demands by the upper basin states for their allotted shares. Perhaps the most telling fact of irrigating the desert occurs in the Wellton-Mohawk irrigation area (east of Yuma, Arizona). It would have been cheaper for the government to buy the land at a fair market price and take it out of production than it cost to subsidize the irrigation and to construct a desalinization plant for the runoff drainage from the irrigated farmland.

For graphs that demonstrate the reduction of discharge in the Colorado River, see Andrew Goudie, *The Human Impact*, Oxford Press, 1994.

Glossary Review for Chapter 15 (in alphabetical order)

abrasion
alluvial fan
bajada
Basin and Range Province
blowout depressions
bolson
deflation
desert pavement
dunes
eolian
erg desert
flash flood
loess
playa
sand sea
slipface
surface creep
ventifacts
wash
yardangs

Annotated Chapter Review Questions

1. Who was Ralph Bagnold? What was his contribution to eolian studies?

A British major, Ralph Bagnold was stationed in Egypt in 1925. Bagnold was an engineering officer who spent much of his time in the deserts west of the Nile, where he measured, sketched, and developed hypotheses about the wind and desert forms. His often-cited work, *The Physics of Blown Sand and Desert Dunes*, was published in 1941 following the completion of wind-tunnel simulations in London. For example, Bagnold studied the ability of wind to transport sand over the surface of a dune. Figure 15-1 shows that a wind of 50 kmph (30 mph) can move approximately one-half a ton of sand per day over a one-meter-wide section of dune. The graph also demonstrates how rapidly the amount of transported sand increases with wind speed.

2. Explain the term eolian and its application in this chapter. How would you characterize the ability of the wind to move material?

Wind-eroded, wind-transported, and wind-deposited materials are called eolian (also spelled aeolian; named for Aeolus, the ruler of the winds in Greek mythology). The actual ability of wind to move materials is small compared with that of other transporting agents such as water and ice, because air is so much less dense than these other media.

3. Describe the erosional processes associated with moving air.

Two principal wind-erosion processes are deflation, the removal and lifting of individual loose particles, and abrasion, the grinding of rock surfaces with a "sandblasting" action by particles captured in the air.

4. Explain deflation and the evolutionary sequence that produces desert pavement.

Deflation literally blows away unconsolidated or noncohesive sediment. After wind deflation and sheetwash do their work on an arid landscape, a desert pavement is formed from the concentrated pebbles and gravels left behind. That pavement, which resembles a cobblestone street, protects underlying sediment from further deflation (Figure 15-2). Desert pavements are so common that many provincial names have been used for them–for example, gibber plain in Australia, gobi in China, and in Africa, lag gravels or serir (or reg desert if some fine particles remain). Water also should be considered in the formation of lag desert plains that wind deflation has modified.

5. Differentiate between a dust storm and a sand storm.

Only the finest dust particles travel significant distances, and consequently the finer material suspended in a dust storm is lifted much higher than the coarser particles of a sand storm, which may be lifted only about 2 m (6.5 ft).

6. How are ventifacts and yardangs formed by the wind?

Rocks exposed to eolian abrasion appear pitted, grooved, or polished, and usually are aerodynamically shaped in a specific direction, according to the flow of airborne particles. Rocks that bear such evidence of eolian erosion are called ventifacts. On a larger scale, deflation and abrasion are capable of streamlining rock structures that are aligned parallel to the most effective wind direction, leaving behind distinctive, elongated ridges called yardangs. These can range from meters to kilometers in length and up to many meters in height (Figure 15-3).

7. What is the difference between eolian saltation and fluvial saltation?

The term saltation was used in Chapter 14 to describe movement of particles along stream beds. The term saltation also is used in eolian processes to describe the wind transport of grains along the ground, grains usually larger than 0.2 mm (0.008 in.). About 80% of wind transport of particles is accomplished by this skipping and bouncing action (Figure 15-4). In comparison with fluvial transport, in which saltation is accomplished by hydraulic lift, eolian saltation is executed by aerodynamic lift, elastic bounce, and impact (compare Figures 15-4 and 14-11).

8. Explain the concept of surface creep.

Wind exerts a drag or frictional pull on surface particles. Bagnold studied the relationship between wind velocity and grain size, determining the fluid threshold (minimum wind speed) required for initial movement of grains of various sizes. A slightly lower wind velocity suffices if the particle already has been set into motion by the impact of a saltating grain. Bagnold termed this lesser velocity the impact threshold. Once in motion, particles continue to be transported by lower wind velocities. As you can see from the graph, the impact threshold is less than the fluid threshold.

9. What is the difference between an erg and a reg desert? Which type is a sand sea? Are all deserts covered by sand? Explain.

A common assumption is that most deserts are covered by sand. Instead, desert pavements predominate across most subtropical arid landscapes; only about 10% of desert areas are covered with sand. Sand grains generally are deposited as transient ridges or hills called dunes. A dune is a wind-sculpted accumulation of sand. An extensive area of dunes, such as that found in North Africa, is characteristic of an erg desert, which means sand sea. Most desert landscapes are not covered with sand but are desert pavements, which are so common that many provincial names have been used for them–for example, gibber plain in Australia, gobi in China, and in Africa, lag gravels or serir or reg desert.

10. What are the three classes of dune forms? Describe the basic types of dunes within each class. What do you think is the major shaping force for sand dunes?

We can simplify dune forms into three classes–crescentic, linear, and star dunes. See Figure 15-9 for classes of dunes, their descriptions, and an illustration of each major type.

11. Which form of dune is the mountain giant of the desert? What are the characteristic wind patterns that produce such dunes?

Star dunes are the mountain giants of the sandy desert. They form in response to complicated, changing wind patterns and have multiple slipfaces. They are rosette-shaped or pinwheel-shaped, with several radiating arms rising and joining to form a common central peak or crest. The best examples of star dunes are in the Sahara, where they approach 200 m (650 ft) in height (Figure 15-10).

12. How are loess materials generated? What form do they assume when deposited?

Pleistocene glaciers advanced and retreated in many parts of the world, leaving behind large glacial outwash deposits of fine-grained clays and silts (<0.06 mm or 0.0023 in.). These materials were blown great distances by the wind and redeposited in unstratified, homogeneous deposits named loess. Loess deposits form some complex weathered badlands and some good agricultural land.

13. Name a few examples of significant loess deposits.

See Figure 15-13. In Europe and North America, loess is thought to be derived mainly from glacial and periglacial sources. The vast deposits of loess in China, covering more than 300,000 km^2 (115,800 mi^2), are thought to be derived from desert rather than glacial sources. Accumulations in the Loess Plateau of China exceed 300 m (984 ft) of thickness.

14. Characterize desert energy and water balance regimes. What are the significant patterns of occurrence for arid landscapes in the world?

The daily surface energy balance for El Mirage, California, presented in Figure 4-21, highlights the high sensible heat conditions and intense ground heating in the desert. Such areas receive a high input of insolation through clear daylight skies and experience high radiative heat losses at night. A typical desert water balance is given in Figure 9-13, for Phoenix, Arizona, showing high potential evapotranspiration demand, low precipitation supply, and prolonged summer deficits. Fluvial processes in the desert generally are dominated by intermittent running water, with hard, sparsely vegetated desert pavement yielding high runoff during rainstorms. The spatial distribution of these dry lands is related to several factors: subtropical high-pressure cells between 15° and 35° N and S, rain shadows to the lee of mountain ranges, and great distance from moisture-bearing air masses. Figure 15-14, portrays this distribution according to the modified Köppen climate classification used in this text.

15. How would you describe the water budget of the Colorado River? What was the basis for agreements regarding distribution of the river? Why has thinking about the river's discharge been so optimistic?

Exotic stream flows are highly variable, and the Colorado is no exception. In 1917, the discharge measured at Lees Ferry totaled 24 million acre-feet (maf), whereas in 1934 it reached only 5.03 maf. In 1977, the discharge dropped to 5.02 but rose to an all-time high of 24.5 maf in 1984. In addition to this variability, approximately 70% of the year-to-year discharge occurs between April and July.

The average flows between 1906 and 1930 were almost 18 maf a year, with averages dropping to 13 maf during the last 50 years. As a planning basis for the Colorado River Compact, the government used average river discharges from 1914 up to the treaty signing in 1923, an exceptionally high 18.8 maf. That amount was perceived as more than enough for the upper and lower basins each to receive 7.5 maf and, later, enough for Mexico to receive 1.5 maf in the 1944 Mexican Water Treaty.

16. Describe a desert bolson from crest to crest. Draw a simple sketch with the components of the landscape labeled.

Figure 15-22 illustrates a typical bolson, which is a slope-and-basin area between the crests of two adjacent ridges in a dry region. Basin-and-range relief is abrupt, and rock structures are angular and rugged. As the ranges erode, the transported materials accumulate to great depths in the basins, gradually producing extensive desert plains.

17. Where is the Basin and Range Province? Briefly describe its appearance and character.

The basins average 1220-1525 m (4000-5000 ft) above sea level, with mountain crests rising higher by some 915-1525 m (3000-5000 ft). Death Valley is an extreme, being the lowest of these basins, with an elevation of –86 m (–282 ft). However, to the west of the valley, the Panamint Range rises to 3368 m (11,050 ft) at Telescope Peak–over 3 vertical kilometers (2 mi) of desert relief!

Overhead Transparencies

As an adopter you are provided with the following figures for overhead projector use.

- Figure 15-9: Major dune forms (large multi-part illustration and table)
- Figure 15-22: Basin and Range Province of the western U. S.

16

Coastal Processes and Landforms

Overview

Coastal regions are unique and dynamic environments. Most of Earth's coastlines are relatively new and are the setting for continuous change. The land, ocean, and atmosphere interact to produce waves, tides, erosional features, and depositional features along the continental margins. The interaction of vast oceanic and atmospheric masses is dramatic along a shoreline. At times, the ocean attacks the coast in a stormy rage of erosive power; at other times, the moist sea breeze, salty mist, and repetitive motion of the water are gentle and calming.

You will find coastal processes in this chapter organized in a system of specific inputs, actions, and outputs. This should permit you to conduct the students through a logical progression of subjects. We conclude with a look at the considerable human impact on coastal environments.

The *Student Study Guide* presents 22 "Learning Objectives" to guide the student in reading the chapter. The *Applied Physical Geography* lab manual has portions of one exercise that involve coastal landscapes.

New to the Third Edition

(Note: This section highlights major changes, new features, and additions in the third edition. This does not describe all the rewrite and recast of the text.)

1. A list of key learning concepts begins the chapter.

2. Description of the chemical composition of seawater and the physical structure of the ocean have been incorporated from Chapter 7.

3. A satellite image mapping ocean topography and sea level data is included as Figure 16-5.

4. News Report #1: "Sea Level Varies Along U.S. Coastline," describes the sea level differences along North American shores, identifying the major factors that create such variation.

5. Inputs to the coastal system are presented in a bullet format for greater clarity.

6. The Great Lakes are used as an example of coastal environments (a good connection to Focus Study 19-1).

7. Tidal Bores and tsunamis are often confused; News Report #2: "Tidal Bores Are True Tidal Waves," clarifies these differences.

8. Figure 16-11 illustrates tsunami travel times to Honolulu, Hawaii from various points along the Pacific Rim.

9. News Report #3: "Killer Waves Strike Beachcombers," compares the destructive capacity of waves which are in-phase with each other, aligning to create an increased wave height, vs. waves which are out-of-phase with each other, reducing the height of the crest and destructive abilities.

10. Updated statistics from Scripps concerning ocean warming and new predictions from the IPCC concerning the rise of sea levels between 1995 and 2100.

11. New photos of coastal landscapes are included in Figures 16-10, 16-12, and 16-16.

12. News Report #4: "Engineers Nourish a Beach," describes how Miami and Dade County, Florida have rebuilt their beaches, costing $70 million since the 1970s. The environmental consequences of such actions are addressed.

13. A new summary and review section ends the chapter.

Key Learning Concepts

1. *Describe* the chemical composition of seawater and the physical structure of the ocean.

2. *Identify* the components of the coastal environment and *list* the physical inputs to the coastal system, including tides and mean sea level.

3. *Describe* wave motion at sea and near shore and *explain* coastal straightening as a product of wave refraction.

4. *Identify* characteristic coastal erosional and depositional features.

5. *Describe* barrier islands and their hazards as they relate to human settlement.

6. *Assess* living coastal environments: corals, wetlands, salt marshes, and mangroves.

7. *Construct* an environmentally sensitive model for settlement and land use along the coast.

Expanded Outline Discussion

The following headings (boldfaced) match some of the first, second, and third order headings in Chapter 16. The narrative under each heading contains information, sources, and anecdotal facts relating to portions of the chapter. Not all text headings are discussed.

GLOBAL OCEANS AND SEAS

Oceanography is the integrative discipline that studies the ocean. Oceanography, much like geography, draws on many academic disciplines: geology, geophysics, geochemistry, physics, biology, meteorology, and climatology. From these fields, specialties in chemical, biological and physical oceanography are derived; also, marine geology, marine engineering, and public policy disciplines involving the law of the sea are important today. Figure 16-1 lists with reference numbers each of Earth's principal oceans and seas. The physical structure of the oceans is portrayed in an integrative illustration in Figure 16-3 with plots of temperature, salinity, dissolved carbon dioxide, and oxygen levels, with increasing depth.

The comparison of land and ocean hemispheres in Figure 7-2 can be demonstrated to your class using a globe.

Table 7-3 compares the four oceans as to area, volume, depth, and deepest point. The southern portions of the Indian, Atlantic, and Pacific is sometimes called the Southern Ocean. See: Institute of Oceanographic Sciences, Deacon Laboratory. *The FRAM (Fine Resolution Antarctic Model) Atlas of the Southern Ocean.* Wormley, Surrey: Natural Environment Research Council, 1991, for detailed description of currents and extent of the Southern Ocean.

Chemical Composition of Seawater

This section presents at least four ways of expressing the average worldwide value of 35‰. Water as the universal solvent has dissolved an enormous quantity of materials, which is expressed as its salinity. Figure 16-2 shows the variation of salinity with latitude. The following table lists the concentration of some of the elements in a cubic kilometer (and cubic mile) of seawater.

Concentration of elements in km^3 of seawater.

Element	Tons* / km^3	Tons / mi^3
Hydrogen (H) + Oxygen (O) = H_2O	991,000,000	4,546,000,000
Chlorine(C)	19,600,000	89,500,000
Sodium (Na)	10,900,000	49,500,000
Magnesium (Mg)	1,400,000	6,125,000
Sulfur (S)	920,000	4,240,000
Calcium (Ca)	420,000	1,880,000
Potassium (K)	390,000	1,790,000
Bromine (Br)	67,000	306,000
Carbon (C)	29,000	132,000
Strontium (St)	8,300	38,000

COASTAL SYSTEM COMPONENTS

The littoral zone is an excellent place to contemplate the interaction of the spheres: atmosphere, hydrosphere, lithosphere, biosphere–the four parts of this textbook. Vast systems drive the processes that produce a dynamic geomorphological environment. Processes that take many decades inland appear accelerated along the coast. This chapter can serve as an overview of the endogenic and exogenic systems in this text. The quote from Rachel Carson's *The Edge of the Sea* eloquently refers to many of these systems.

The Coastal Environment

Sea level is described in this section as an input to the coastal system, whereas ongoing sea-level change is discussed as an action along the coast. Changes in sea level and the daily fluctuation of the tides adds to the dynamism of the littoral zone, subjecting varying elevations of the shore to erosion, transport, and deposition.

COASTAL SYSTEM ACTIONS

Tides; Tidal Power

I chose to include a section on tides for several reasons including the point just made about the importance to coastal geomorphology of changes in sea level. For many students, physical geography is perhaps their only science class so this becomes their only exposure to the worldwide physical phenomena of tides. This is also an opportunity to present the inertial explanation for the tidal phases as opposed to the cumbersome non-inertial centrifugal explanations. I hope you find this discussion and Figure 16-6 a useful presentation.

Predicting specific tides from a model is extremely difficult because of the number of interacting components. Tide tables that list the times and variation of flood and ebb tides for each day of the year, and annual spring and neap tides, are based on actual empirical measurements of past-tide experience along a specific portion of coast. Tide tables allow ships to time their passages with high tide. Some bays empty of water completely, leaving vessels "parked" on the mud flats awaiting the return of the tides to again float (Figure 16-7). Knowing the timing of spring tides is particularly important if the weather is stormy and storm surges are high. A stream tributary into a bay, estuary, or coastal location will flood upstream if storm runoff coincides with a spring tide, for the stream discharge will flood further inland. Such was the case with Tropical Storm Keith in late November 1988 as it passed across Florida. The storm's arrival coincided with exceptionally high tides and was thus magnified in coastal flood effects. Between 1635 and 1976, it is estimated that approximately 100 storms affecting North

America correlated with high tides in this way.

The tidal pattern is influenced by many other variables. To name a few: the Moon spends half of the month north of the equator and half of the month below the equator; the Moon takes 29.5 days to revolve around Earth; the distance between the Moon and Earth varies 11 percent during the month, which makes an overall 35 percent difference in tidal bulge height (the Sun and Earth distance varies 3.2 percent during the year); and finally, the plane of the Moon's orbit and that of Earth's orbit is out of alignment by 5°, so that the Moon ranges between 18.5° and 28.5° in tilt compared with Earth's 23.5° tilt, which produces a "wobble" between the two planes so the exact same alignment position occurs every 18.6 years. This is why a 19-year record is necessary to determine mean sea level at a site, for only after that length of time does the full cycle of tidal patterns repeat!

Although very site specific, tidal power remains one underdeveloped alternative for the generation of electricity. And, if economics and true cost were really considered as a basis for energy choices, a Canadian study found that tidal power could be competitive with fossil fuels, given that certain environmental considerations are resolved.

There is a web site devoted to marine weather observations. This site will provide you with information on water temperature, wave height and frequency, and wind conditions. Contact them at:

http://thunder.met.fsu.edu/Ÿnws/buoy/.

Waves

The text describes the production of waves as truly an ocean-wide, indeed, global, phenomenon. As you observe the pulse of the waves along the coast, you are seeing energy patterns that could have been imparted to the water at a generating region in another hemisphere. The interference or amplification of various wave trains produces the variable beat of groups of waves. Geomorphological processes increase in intensity when storm-driven waves work the littoral zone. A distant storm will be reflected several days after it occurs on a far away shoreline.

The discussion of the transformation from waves of transition to waves of translation and Figure 16-8 may help present the mechanism of breaker formation.

Wave Refraction

Note that the principle of wave refraction continues away from the coast as affected by submarine topography–headlands and coves often lead to ridges and valleys offshore.

Tsunami, or Seismic Sea Wave

Note in the text the effect of sea-floor topography on tsunamis because of their great wavelength, for these waves can be refracted by rises and ridges on the ocean floor. Strangely, people have been known to show up at the predicted site of a tsunami with surf board in hand ready for the monster wave. Observers are sometimes swept away by the surge of water.

Sea Level Changes

In researching this text I was told by a prominent United States government scientist that various research projects that were involved with analyzing the present sea-level rise were either cut or threatened with cuts. It seems, according to this contact, that any increase in sea level would act as further compelling proof of global warming and verification of the greenhouse warming theory. Principal studies of sea-level rise are being conducted by other nations, particularly those on islands, and by lowland countries such as the Netherlands and a growing leadership from the European Community. As the scientist I spoke to said, "Sea-level increases really represent a 'smoking gun' of evidence to go along with the existing instrumental record of higher temperatures."

See one of the IPCC reports for a treatment of "Sea Level Rise" (Houghton, J.T., Jenkins, G.J. and J.J. Ephraums, editors. *Climate Change–The IPCC Scientific Assessment*, Working Group I, World Meteorological Organization/United Nations Environment Programme. Port Chester, N.Y.: Cambridge University Press, 1991, pp. 261-81). Headings include: sea level rise, factors affecting sea level; has sea level been rising over the last 100 years?; possible contributing factors to past and future sea level rise; how might sea level change in the future? A full section of over 100 references follows on pp. 279-81. Mean sea level (MSL) is not only rising over the past 100 years, but it is doing it at an accelerated rate faster than the previous two centuries.

Figure 9.8, p. 278 in the IPCC report, demonstrates that if anthropogenic greenhouse forcing ceased by A.D. 2030, additional sea level rise would still occur throughout the remainder of the century. This certainly represents a challenge to science and society to analyze and understand this spatial phenomena and act to lessen the increasing inundation of the world's coastlines.

The impact of rising sea level on the barrier islands of our coastlines is discussed by Ruth Flanagan in "Beaches on the Brink," *Earth*, 2, no. 6, November 1993: 24-33. The article states that "Retreat of the shoreline is nature's response to a rising sea level. Recent storms have renewed the debate over attempts to hold it in place."

Note in the suggested readings section for Chapter 16 the reference to an article H.R. Wanless, "The Inundation of Our Coastline." Also, the California Coastal Commission (631 Howard Street, 4th Floor, San Francisco, CA 94105; 415-543-8555) produced an 85-page report titled *Planning for an Accelerated Sea Level Rise along the California Coast*, June 26, 1989. The report divides the topic among 28 headings and includes an extensive 10-page list of sources and references in a bibliography.

COASTAL SYSTEM OUTPUTS Erosional and Depositional Coastal Processes

Given the actions of the ocean currents, tides, waves, and sea level changes, we can now discuss the outputs, or processes that are a result and consequence of the above inputs and actions. Outputs of consequence include the essential form of the coastline as emerged or submerged relative to sea level, erosional and depositional features such as beaches, spits, barriers, sandbars, and certain organic processes that lead to the formation of coral reefs and islands.

Beaches

You might want to consider that the beach acts in a manner similar to floodplain alluvium, acting to absorb energy even in storm situations. However, development right on the beach or disruption in sand replenishment through construction will eliminate this energy-absorbing advantage. The dwelling will then absorb the destructive aspects of the sea. The third edition includes new material on beach nourishment, especially focused on efforts by the Army Corps in south Florida.

Near Pacific Grove, California, heavy pedestrian traffic had almost eliminated the primary dunes, making the highway and residences susceptible to wave erosion. In response, Conservation Corps work crews stabilized the dunes, planted native varieties of plants, and built several boardwalks to direct erosive foot traffic and prevent further loss.

Relative to replenishment of artificial beach sands, a further disadvantage is that unforeseen environmental impacts may accompany the addition of new and foreign sand types. If the new sands do not match the natural varieties, untold disruption of coastal marine life is possible. Along the Florida coast, sand derived from foreign parent materials was introduced and it proceeded to adversely affect indigenous organisms and coral colonies offshore.

Barrier Forms

There is not complete agreement on the exact way barrier islands form. In the text, I present a synthesis and simplification of a basic hypothesis for barrier formation. At least four ideas have been proposed as likely explanations of such formations and their relationship to sea level. You may want to consult one of the geomorphology texts referenced in this resource manual, where these hypotheses are described.

Organic Processes: Coral Formations; Coral Reefs

Coral reef colonies are the most biologically productive of natural communities. A coral colony consists of a coral component– coelenterates, animals whose body cavity is for both circulatory and digestive functions (such as jelly fish and anemones)–and an algae component. The algae member of this system permits the colony to thrive near the surface where light is more available. The algae process waste from the coral and provide critical "cementation" between the corals, essential in building large coral fossil rock structures. Growth rates for corals approach 2 m (6.7 ft) per 1,000 years, which under average conditions can keep pace with an increasing sea level. Corals appeared some 500 million years ago and have been forming limestone deposits ever since. Such limestone formations as the Permian reef of southeastern New Mexico and west Texas are more than 1300 m (400 ft) thick. Those corals lived during times of higher sea level in the interglacial periods and are now exposed far above sea level.

Think of these large coral colonies as home to many forms of marine life, all benefiting from their mutual adaptations. Various experiments with cinder blocks and other artificial coral-like formations placed in tropical waters show how rapidly, and with what species, marine plants and animals colonize. The volume of a single coral reef exceeds by a thousand times the largest human-made pyramid. Some solitary forms occur farther from the equator than noted in the text, but these solitary forms do not build reefs.

Coral Bleaching

Coral bleaching is a phenomenon of increasing severity world wide. The 1992 IPCC update referred to this on pp. 145-46 as possibly related to increasing sea-surface temperatures (SSTs). (Houghton, J.T., Jenkins, G.J. and J.J. Ephraums, editors. *Climate Change 1992–The Supplementary Report to the IPCC Scientific Assessment*. World Meteorological Organization/United Nations Environment Programme. New York: Cambridge University Press, 1992.) See the reference in the suggested readings section of Chapter 16 in *GEOSYSTEMS* by B.E. Brown, "Coral Bleaching."

Other studies focus on pollution, such as sewage and agricultural runoff, as the culprit. See: Maria Burke, "Phosphorus Fingered as Coral Killer," *Science*, 263, no. 5150, February 25, 1994: 1086. The nitrates and phosphates from these sources threaten coastal reefs. A team of Australian scientists are applying this hypothesis to studies of the Great Barrier Reef in a project called Effect of Nutrient Enrichment on Coral Reefs (ENCORE).

Attention by other scientists is focused on the effect of increasing ultraviolet light (UV-B) on organisms such as corals and phytoplankton in the photic layer of the ocean. Increasing mortality is identified in tropical and subpolar oceans and seas. The increase in surface UV-B radiation is directly related to the global thinning of stratospheric ozone.

An interactive laser disc has been produced by National Geographic, entitled *The Great Ocean Rescue*. If you have the extra time in class or if you have a Physical Geography Lab, try spending 30 minutes with this resource. *The Great Ocean Rescue* has students playing a variety of roles, from geologist to oceanographer, trying to decipher the cause of destruction to coral reefs on a global basis. This laser disc has four different crises that it addresses, and challenges students to identify the cause of coral reef destruction and select appropriate solutions. The disc is very good at stimulating the students ability to see interrelationships among Earth systems.

Salt Marshes and Mangrove Swamps

Many tracts of wetlands will probably be exposed to new threats of development, encroachment, and loss to pollution and damaging activities because of the actions of the Bush Administration despite a 1988 campaign promise of "no net loss of wetlands." This removal of protection for millions of hectares of wetlands follows an unusual bipartisan effort in 1987 among environmentalists, politicians from both parties, and industrialists to forge a new understanding for the protection of these valuable lands.

Now, of the 38.5 million hectares (95 million acres) of identified wetland in the contiguous 48 states, some 13 million hectares (32.1 million acres) will be reclassified as non-wetlands and therefore will be available for industry, drilling, landfills, pollution discharges, and development. One effect of this is that cities upstream from the newly developed and closed wetlands will have to treat wastes to a greater degree, for the systems they depend on for natural cleansing will be lost or made less effective. The pollution control replacement cost of losing 13 million hectares of wetlands exceeds $100 billion (1991 dollars). For example, approximately one-half of the Florida Everglades, already at increasing risk and disruption from a century of redirected water flows, will not fit the new definition from the Bush Administration. Also, about a third of the wetlands in the lower Mississippi Valley will be lost–the wet, forested east shore region along the Chesapeake Bay could be logged for housing, and even sandy wetlands along the Rocky Mountains will be affected.

This is a story to follow over the third edition of *GEOSYSTEMS* because the rewording of the guide manual for wetlands represents the single greatest reversal of environmental regulation in American history. The elimination of the previous White House Council on Competitiveness in early 1992 will probably mean a reversal of many of these policies–and an increase in preservation efforts.

Even in the Netherlands, where living and surviving with the sea is a daily issue of levees, dikes, and pumping, further conversion of estuaries to farmland is quite controversial. And the people of the Netherlands live in fear of a potential rise in sea level over the next century.

HUMAN IMPACT ON COASTAL ENVIRONMENTS

Note the section added to the Focus Study 16-1, "An Environmental Approach to Shoreline Planning," discussing the approach in South Carolina.

Remember, only 28 of 280 barrier islands identified before World War II remain undeveloped, and less than a fourth of original natural wetlands remain undeveloped to some degree. Focus Study 16-1, "An Environmental Approach to Shoreline Planning," discusses the basics for preparing a "Do" map and a "Don't" map for a portion of the New Jersey shore. Ian McHarg developed the "McHargian" style of planning with meticulous hand-rendered maps of various spatial components. These components were then synthesized onto a single master map identifying regions for protection or development, lowest impact routes for a transmission line, areas for preservation due to unique attributes or environmental characteristic, etc. One of his books is still available in bookstores: *Design with Nature*, Garden City, NY: Doubleday, 1969. Lewis Mumford wrote the introduction to *Design with Nature* in these words, "...I would put it on the same shelf that contains as yet only a handful of works in a similar vein....classics such as those by Henry Thoreau, George Perkins Marsh, Patrick Geddes, Carl Sauer, Benton MacKaye, and Rachel Carson." The roots of today's GIS revolution resides in McHarg's work as described in this book. For instance, examine the thorough geographic analysis and work up of Staten Island (pp. 102-113). Many of you will find a 1970 film by McHarg, produced by WGBH Boston with assistance from Narendra Juneja, called "Multiply and Subdue the Earth" (2 reels), in your local educational film library.

For a depth report of Boston Harbor and coastal environs, its environmental history, polluted condition, and future, see: Eric Jay Dolin, "Boston Harbor's Murky Political Waters," *Environment*, 34, no. 6, July/August 1992: 6-11+. The adventure with its sewage problem began during the 19th Century and continues to this day.

Marco Island, Florida: An Example of Impact

Among several sources, I derived the material on the development of Marco Island, Florida, from Samuel G. Patterson, *Mangrove Community Boundary Interpretation and Detection of Areal Changes in Marco Island, Florida: Application of Digital Image Processing and Remote Sensing Techniques*. Biological Report 86 (10) for the U.S. Fish and Wildlife Service, August 1986. The digital and photographic information was used to determine the impact on this formerly, varied mangrove community (Figure 16-22). The development of 2175 hectares (5300 acres) of subtropical mangrove habitat, susceptible to tropical cyclones, is chronicled in this GIS study. Imagine the analysis capability geographers possess to monitor the impact of the proposed changes in wetland designations.

I recently went to a conference of our regional AAG group (the Association of Pacific Coast Geographers) and attended a panel discussion featuring five geographers who make a living in the public and private sectors doing local, regional, and problematic GIS work. The career potential for undergraduate and graduate geography majors was stressed by each of the participants. Every one of the professionals emphasized and encouraged the students in attendance. The AAG has several excellent brochures mentioning these career fields. Also, see the listings in the GIS section in Chapter 1 of this resource manual for information and publications.

Glossary Review for Chapter 16 (in alphabetical order)

barrier beaches
barrier islands
barrier spit
bay barrier
beach
beach drift
brackish
breaker
brine
coral
ebb tide
flood tide
lagoon
littoral zone
longshore current
mean sea level
neap tide
salinity
salt marshes
spit
spring tide
swells
tides
tombolo
tsunami
wave-cut platform
wave refraction
wave
wetland

Annotated Chapter Review Questions

1. Describe the salinity of seawater: its composition, amount, and distribution.

Water acts as a solvent, dissolving at least 57 of the elements found in nature. In fact, most natural elements and the compounds they form are found in the seas as dissolved solids, or solutes. Thus, seawater is a solution, and the concentration of dissolved solids is called salinity. Seven elements comprise more than 99% of the dissolved solids in seawater: chlorine (Cl), sodium (Na), magnesium (Mg), sulfur (S), calcium (Ca), potassium (K), and bromine (Br). Seawater also contains dissolved gases (such as carbon dioxide, nitrogen, and oxygen), solid and dissolved organic matter, and a multitude of trace elements. Salinity worldwide normally varies between 34‰ and 37‰; variations are attributable to atmospheric conditions above the water and to the quantity of freshwater inflows. The term brine is applied to water that exceeds the average of 35‰ salinity, whereas brackish applies to water that is less than 35‰.

2. Analyze the latitudinal distribution of salinity shown in Figure 16-2. Why is salinity less along the equator and greater in the subtropics?

Generally, oceans are lower in salinity near landmasses, because of river discharges and runoff. Extreme examples include the Baltic Sea (north of Poland and Germany) and the Gulf of Bothnia (between Sweden and Finland), which average 10‰ or less salinity because of heavy freshwater runoff and low evaporation rates. On the other hand, the Sargasso Sea, within the North Atlantic subtropical gyre, averages 38‰, and the Persian Gulf is at 40‰ as a result of high evaporation rates in an almost-enclosed basin. Deep pockets near the floor of the Red Sea register a very salty 225‰. In equatorial water, precipitation is high throughout the year, diluting salinity values to slightly lower than average (34.5‰). In subtropical oceans–where evaporation rates are high due to the influence of hot, dry subtropical high-pressure cells–salinity is more concentrated, increasing to 36.5‰.

3. What are the three general zones relative to physical structure within the ocean? Characterize each by temperature, salinity, dissolved oxygen, and dissolved carbon dioxide.

The ocean's surface layer is warmed by the Sun and is wind-driven. Variations in water temperature and solutes are blended rapidly in a mixing zone that represents only 2% of the oceanic mass. Below this is the thermocline transition zone, a region of strong temperature gradient that lacks the motion of the surface. Friction dampens the effect of surface currents, with colder water temperatures at the lower margin tending to inhibit any convective movements. Starting at a depth of 1-1.5 km (0.62-0.93 mi) and going down to the bottom, temperature and salinity values are quite uniform. Temperatures in this deep cold zone are near 0°C (32°F); but, due to its salinity, seawater freezes at about –2°C (28.4°F). The coldest water is at the bottom except near the poles, where cold water may be near or at the surface. Refer to graphs on right side of Figure 16-3.

4. What are the key terms used to describe the coastal environment?

See Figure 16-4. The coastal environment is called the littoral zone. (Littoral comes from the Latin word for shore.) The littoral zone spans both land and water. Landward, it extends to the highest water line that occurs on shore during a storm. Seaward, it extends to the point at which storm waves can no longer move sediments on the seafloor (usually at depths of approximately 60 m or 200 ft). The specific contact line between the sea and the land is the shoreline, and adjacent land is considered the coast.

5. Define mean sea level. How is this value determined? Is it constant or variable around the world? Explain.

Mean sea level is a calculated value based on average tidal levels recorded hourly at a given site over a period of at least 19 years, which is one full lunar tidal cycle. Mean sea level varies spatially from place to place because of ocean currents and waves, tidal variations, air temperature and pressure differences, and ocean temperature variations.

6. What interacting forces generate the pattern of tides?

Earth's orientation to the Sun and the Moon and the reasons for the seasons are discussed in Chapters 2 and 4. The same astronomical relationships produce the pattern of tides, the complex daily oscillations in sea level that are experienced to varying degrees around the world. Tides also are influenced by the size, depth, and topography of ocean basins, by latitude, and by shoreline configuration. Tides are produced by the gravitational pull exerted on Earth by both the Sun and the Moon. Although the Sun's influence is only about half that of the Moon (46%) because of the Sun's greater distance from Earth, it is still a significant force. Figure 16-6 illustrates the relationship among the Moon, the Sun, and Earth and the generation of variable tidal bulges on opposite sides of the planet.

7. What characteristic tides are expected during a new Moon or a full Moon? During the first-quarter and third-quarter phases of the Moon? What is meant by a flood tide? An ebb tide?

Figure 16-6a shows the Moon and the Sun in conjunction (lined up with Earth–new Moon or full Moon), a position in which their gravitational forces add together. The combined gravitational effect is strongest in the conjunction alignment and results in the greatest tidal range between high and low tides, known as spring tides. (Relative to tides, spring means to "spring forth"; it has no relation to the season of the year.) Figure 16-6b shows the other alignment that gives rise to spring tides, when the Moon and Sun are at opposition. When the Moon and the Sun are neither in conjunction nor in opposition, but are more-or-less in the positions shown in c and d (first- and third-quarter phases), their gravitational influences are offset and counteract each other somewhat, producing a lesser tidal range known as neap tide. Every 24 hours and 50 minutes, any given point on Earth rotates through two bulges as a direct result of this rotational positioning. Thus, every day, most coastal locations experience two high (rising) tides known as flood tides, and two low (falling) tides known as ebb tides. The difference between consecutive high and low tides is considered the tidal range.

8. Is tidal power being used anywhere to generate electricity? Explain briefly how such a plant would utilize the tides to produce electricity. Are there any sites in North America?

The fact that sea level changes daily with the tides suggests an opportunity that these could be harnessed to produce electricity. The bay or estuary under consideration must have a narrow entrance suitable for the construction of a dam with gates and locks, and it must experience a tidal range of flood and ebb tides large enough to turn turbines, at least a 5 m (16 ft) range. About 30 locations in the world are suited for tidal power generation, although at present only two of them are actually producing electricity–an experimental 1 megawatt station in Russia at Kislaya-Guba Bay, on the White Sea, since 1969, and a facility in the Rance River estuary on the Brittany coast of France since 1967.

According to studies completed by the Canadian government, the present cost of tidal power at ideal sites is economically competitive with that of fossil fuels, although certain environmental concerns must be addressed. Among several favorable sites on the Bay of Fundy, one plant is in operation. The Annapolis Tidal Generating Station was built in 1984 to test electrical production using the tides. Nova Scotia Power Incorporated operates the 20 megawatts plant (Figure 16-7c).

9. What is a wave? How are waves generated, and how do they travel across the ocean? Does the water travel with the wave? Discuss the process of wave formation and transmission.

Undulations of ocean water called waves travel in wave trains, or groups of waves. Storms around the world generate large groups of wave trains. A stormy area at sea is the generating region for these large waves, which radiate outward from their formation center. As a result, the ocean is crisscrossed with intricate patterns of waves traveling in all directions. The waves seen along a coast may be the product of a storm center thousands of kilometers away. Water within such a wave is not really migrating but is transferring energy through the water in simple cyclic undulations, which form waves of transition. In a breaker, the orbital motion of transition gives way to waves of translation, in which both energy and water move forward toward shore as water cascades down from the wave crest (Figure 16-8).

10. Describe the refraction process that occurs when waves reach an irregular coastline. Why is the coastline straightened?

Generally, wave action is a process which results in coastal straightening. As waves approach an irregular coast, they bend and focus around headlands, or protruding landforms generally composed of more resistant rocks (Figure 16-9). Thus, headlands represent a specific point of wave attack along a coastline. Waves tend to disperse their energy in coves and bays on either side of the headlands. This wave refraction (wave bending) along a coastline redistributes wave energy so that different sections of the coastline are subjected to variations in erosion potential.

11. Define the components of beach drift and the longshore current and longshore drift.

Particles on the beach are moved along as beach drift, or littoral drift, shifting back and forth between water and land in the effective wind and wave direction. These dislodged materials are available for transport and eventual deposition in coves and inlets and can represent a significant volume. Beach drift is transported by the longshore current. A longshore current is generated only in the surf zone and works in combination with wave action to transport large amounts of sand, gravel, sediment and debris along the shore as longshore drift.

12. Explain how a seismic sea wave attains such tremendous velocities. Why is it given a Japanese name?

Tsunamis, often incorrectly referred to as tidal waves, are formed by sudden and sharp motions in the seafloor, such as those caused by earthquakes, submarine landslides, or eruptions of undersea volcanoes. Usually, a solitary wave of great wavelength, or sometimes a small group of two or three long waves, is stimulated. These waves generally exceed 100 km (60 mi) in wavelength but are only a meter or so in height. Because of their great wavelength, tsunamis are affected by the topography of the deep-ocean floor and are refracted by rises and ridges. They travel at great speeds in deep-ocean water–velocities of 600 to 800 kmph (375-500 mph) are not uncommon–but often pass unnoticed because their great length makes the slow rise and fall of water hard to observe. Throughout history Japan has suffered greatly from these seismic waves. *Tsunami* is Japanese for "harbor wave," named for its devastating effect in harbors.

13. What is meant by an erosional coast? What are the expected features of such a coast?

The active margins of the Pacific along the North and South American continents are characteristic coastlines affected by erosional landform processes. Erosional coastlines tend to be rugged, of high relief, and tectonically active, as expected from their association with the leading edge of a drifting lithospheric plate. Sea cliffs are formed along a coastline by the undercutting action of the sea. As indentations are produced at water level, such a cliff becomes notched, leading to subsequent collapse and retreat of the cliff. Other erosional forms evolve along cliff-dominated coastlines, including sea caves, sea arches, and sea stacks. As erosion continues, arches may collapse, leaving isolated stacks out in the water. The coasts of southern England and Oregon are prime examples of such erosional landscapes. See Figure 16-12.

14. What is a depositional coast? What are the expected features of such a coast?

Depositional coasts generally are located near onshore plains of gentle relief, where sediments are available from many sources. Such is the case with the Atlantic and Gulf coastal plains of the United States, which lie along the relatively passive, trailing edge of the North American lithospheric plate. A spit consists of sand deposited in a long ridge extending out from a coast; it partially crosses and blocks the mouth of a bay. Classic examples include Sandy Hook, New Jersey, and Cape Cod, Massachusetts. The spit becomes a bay barrier if it completely cuts the bay off from the ocean and forms an inland lagoon. Spits and bars are made up of materials that have been eroded and transported by littoral drift; for much sand to accumulate, offshore currents must be weak. A tombolo occurs when sand deposits connect the shoreline with an offshore island or sea stack.

15. Why do people attempt to modify littoral drift? What strategies are used? What are the positive and negative impacts of these actions?

Particles on the beach are moved along as *beach drift*, shifting back and forth between water and land in the effective wind and wave direction with each *swash* and *backwash* of surf. Individual sediment grains trace arched paths along the beach. You have perhaps stood on a beach and heard the sound of myriad grains and seawater especially in the backwash of surf. These dislodged materials are available for transport and eventual deposition in coves and inlets and can represent a significant volume.

Figure 16-16, illustrates several of the common approaches: jetties to block material from harbor entrances, groins to slow drift action along the coast, and a breakwater to create a zone of still water near the coastline. However, interrupting the coastal drift that is the natural replenishment for beaches may lead to unwanted changes in sand distribution downcurrent. In addition, enormous energy and materials must be committed to counteract the enormous and relentless energy that nature invests along the coast.

16. Describe a beach–its form, composition, function, and evolution.

A beach is that place along a coast where sediment is in motion. Material from the land temporarily resides there while it is in active transit along the shore. The beach zone ranges, on average, from 5 m (16 ft) above high tide to 10 m (33 ft) below low tide, although specific definition varies greatly along individual shorelines. Beaches are dominated by quartz (SiO_2) because it is the most abundant mineral on Earth, resists weathering, and therefore remains after other minerals are removed. In volcanic areas, beaches are derived from wave-processed lava. A beach acts to stabilize a shoreline by absorbing wave energy, as is evident by the amount of material that is in almost constant motion. Some beaches are continuous and stable. Others cycle seasonally; they accumulate during the summer only to be removed by winter storm waves, and are again replaced the following summer. During quiet, low-energy periods, both backshore and foreshore areas experience net accumulations of sand and are well defined. However, a storm brings increased wave action and high-energy conditions to the beach, usually reducing its profile to a sloping foreshore only. Sediments from the backshore area do not disappear but may be moved by wave action to the nearshore (offshore) area, only to be returned by a later low-energy surf.

17. What success has Miami had with beach replenishment? Is it a practical strategy?

The city of Miami, Florida and surrounding Dade County have spent almost $70 million rebuilding their beaches, and, needless to say, the effort is continuous. As of 1989, much of the beach-replenishment work done in recent years already had been washed away by winter storms. A further disadvantage is that unforeseen environmental impacts may accompany the addition of new and foreign sand types to a beach.

Beach nourishment refers to the artificial placement of sand along a beach. Through such efforts, a beach that normally experiences a net loss of sediment will instead show a net gain. In contrast to hard structures this hauling of sand to replenish a beach is considered "soft" shoreline protection.

Enormous energy and material must be committed to counteract the relentless energy that nature invests along the coast. Years of human effort and expense can be erased by a single storm, such as when offshore islands were completely eliminated by Hurricane Camille along the Gulf Coast in 1969. In Florida, the city of Miami and surrounding Dade County have spent almost $70 million since the 1970s rebuilding their beaches, and needless to say, the effort is continuous. To maintain a 200 m-wide beach (660 ft), net sand loss per year is determined and a schedule of needed replenishment is set. For Miami Beach an eight-year replenishment cycle is maintained. During Hurricane Andrew in 1992 the replenished Miami Beach is estimated to have prevented millions of dollars in shoreline structural damage.

Unforeseen environmental impact may accompany the addition of new and foreign sand types to a beach. If the new sands do not match the existing varieties, untold disruption of coastal marine life is possible. The U.S. Army Corps of Engineers, which operates the Miami replenishment program, is running out of "borrowing areas" for sand that matches the natural sand of the beach. A proposal to haul a different type of sand from the Bahamas is being studied as to possible environmental consequences.

18. Based on the information in the text and any other sources at your disposal, do you think barrier islands and beaches should be used for development? If so, under what conditions? If not, why not?

(The first portion of this question asks the student for an opinion as to hazard perception.) Offshore sand bars gradually migrate toward shore as the sea level rises. Because many barrier beaches evidence this landward migration today, they are an unwise choice for a homesite or commercial building. Nonetheless, they are a common choice, even though they take the brunt of storm energy and actually act as protection for the mainland. The hazard represented by the settlement of barrier islands was made graphically clear when Hurricane Hugo (1989) assaulted South Carolina. Beachfront houses, barrier beach developments, and millions of tons of sand were swept away; up to 95% of the single-family homes in Garden City alone were destroyed.

19. After the Grand Strand off South Carolina was destroyed by Hurricane Hazel in 1954, settlements were rebuilt, only to be hit by Hurricane Hugo 35 years later,

in 1989. Why do these recurring events happen to human populations? Compare the impact of the two storms.

Some homes that had escaped the last major storm, Hurricane Hazel in 1954, were lost to Hugo's peak winds of 209 kmph (130 mph). In comparable dollars, Hugo caused nearly $4 billion in damage, compared with Hazel's $1 billion. The increased damage partially resulted from expanded construction during the intervening years between the two storms. Due to increased development and real estate appreciation, each future storm can be expected to cause ever-increasing capital losses. The inability, unwillingness, or incapacity of humans to perceive hazards in a familiar environment is well-illustrated. This perception void allows the political-business sector to proceed with development and rezoning or nonzoning of high-risk areas. There are many examples of an informed citizenry blocking such efforts by business. The irony is that the conservative atmosphere that seems to prevail across the country relative to taxes does not in some way translate to proper hazard zoning. After all, the taxpayer pays for the reconstruction through disaster benefits and assistance.

20. How are corals able to construct reefs and islands?

A coral is a simple marine animal with a cylindrical, saclike body; it is related to other marine invertebrates, such as anemones and jellyfish. Corals secrete calcium carbonate ($CaCO_3$) from the lower half of their bodies, forming a hard external skeleton. Although both solitary and colonial corals exist, it is the colonial forms that produce enormous structures, varying from treelike and branching forms to round and flat shapes. Through many generations, live corals near the ocean's surface build on the foundation of older corals below, which in turn may rest upon a volcanic seamount or some other submarine feature built up from the ocean floor. An organically derived sedimentary formation of coral rock is called a reef and can assume one of several distinctive shapes; principally, a fringing reef, a barrier reef, or an atoll

21. Evaluate a trend in corals that is troubling scientists.

Scientists at the University of Puerto Rico and NOAA are tracking an unprecedented bleaching and dying-off of corals worldwide. The Caribbean, Australia, Japan, Indonesia, Kenya, Florida, Texas, and Hawaii are experiencing this phenomenon. The bleaching is due to a loss of colorful algae from within and upon the coral itself. Normally colorful corals have turned stark white as nutrient-supplying algae are expelled by the host coral. Exactly why the coral ejects its living partner is unknown. Possibilities include local pollution, disease, sedimentation, and changes in salinity.

Another possible cause is the 1 to 2C° (1.8–3.6F°) warming of sea-surface temperatures, as stimulated by greenhouse warming of the atmosphere. During the 1982-1983 ENSO (Chapter 10) areas of the Pacific Ocean were warmer than normal and widespread coral bleaching occurred. Coral bleaching worldwide is continuing as average ocean temperatures climb higher. Further bleaching through the decade will effect most of Earth's coral-dominated reefs and may be an indicator of serious and enduring environmental trauma. If present trends continuc most of the living corals on Earth could perish by A.D. 2000.

22. Why are the wetlands poleward of 30° N and S latitude different from those that are equatorward? Describe the differences.

In terms of wetland distribution, salt marshes tend to form north of the 30th parallel, whereas mangrove swamps form equatorward of that point. This is dictated by the occurrence of freezing conditions, which control the survival of mangrove seedlings. Roughly the same latitudinal limits apply in the Southern Hemisphere. Salt marshes usually form in estuaries and behind barrier beaches and sand spits. An accumulation of mud produces a site for the growth of halophytic (salt-tolerant) plants. Plant growth then traps additional alluvial sediments and adds to the salt marsh area. Sediment accumulation on tropical coastlines provides the site for growth of mangrove trees, shrubs, and other small trees. The adventurous prop roots of the mangrove are constantly finding new anchorages and are visible above the water line but reach below the water surface, providing a habitat for a multitude of specialized lifeforms. Mangrove swamps often secure and fix enough material to form islands.

23. Describe the present condition of Marco Island, Florida. Was a rational model used to assess the environment prior to development? What economic and political forces were involved?

In 1984, a GIS approach (Geographic Information System is described in Chapter 2) was used to study Marco Island, incorporating existing maps, remote-sensing techniques, and a knowledge of the ecology of an estuarine mangrove community. Using the three images in Figure 16-22, you can compare the 1952 and 1984 digitized inventory maps and a 1984 color-infrared aerial photograph. With these tools, it is possible to visually and quantitatively compare this island over time to see the impact of urban development. Note that the only remaining natural community is restricted to a very limited portion along the extreme perimeter.

The key to environmental planning and zoning is to allocate responsibility and cost in the event of a disaster. Development began on Marco Island in 1962, without consideration of tropical storms, such as Hurricane Donna which hit the island in 1960. Development included artificial landfill for housing sites and general urbanization of the entire island. Scientists regarded the island as a highly productive island and estuarine habitat, unique in its mixture of mangrove species, endangered bird species and protective aquatic environment. Numerous challenges and court actions attempted to halt development.

Appropriate planning would have assessed the level of hazard which exists on Marcos Island to avoid recurrent disasters. Such assessment would have addressed the island as an integrated ecosystem (similar to a large marine ecosystem, LME -see Chapter 20) in order to assess the damage any development would have upon the entire system.

24. What type of environmental analysis is needed for rational development and growth in a region like the New Jersey shore? Evaluate South Carolina's approach to coastal hazards and protection.

See the analysis completed by Ian McHarg as presented in Focus Study 16-1. Prior to an intensive study completed in 1962, no analysis had been completed outside academic circles. Thus, what is common knowledge to botanists, biologists, ecologists, and geographers in the classroom and laboratory still has not filtered through to the general planning and political processes. As a result, on the New Jersey shore and along much of the Atlantic and Gulf coasts, improper development of the fragile coastal zone led to extensive destruction during the storms of March 1962. The New Jersey shore provides an excellent example of how proper understanding of a coastal environment could have avoided problems encountered during major storms and from human development and settlement.

South Carolina enacted their Beach Management Act of 1988 (modified 1990) to apply some of the principles McHarg described in his analysis of the New Jersey shore. Vulnerable coastal areas are protected from new construction or rebuilding. Now guided by law are structure size, replacement limits for damaged structures, and placement of structures on lots, although liberal interpretation is allowed. The Act sets standards for different coastal forms:

beach and dunes, eroding shorelines, or a hypothetical baseline along a coast that lacks dunes or has unstabilized inlets. In the first two years more than 70 lawsuits protested the Act as an invalid seizure of private property without compensation. Implementation has been difficult, political pressure intense, and results mixed.

Overhead Transparencies

As an adopter you are provided with the following figures for overhead projector use.

- Figure 16-1: Principal oceans and seas of the world
- Figure16-4: The littoral zone
- Figure 16-6: What causes tides? (Sun, Moon and Earth relationships)
- Figure 16-12: Erosional coastal landforms
- Figure 16-14: Characteristic depositional coastal features

17

Glacial and Periglacial Processes and Landforms

Overview

A large measure of the freshwater on Earth is frozen, with the bulk of that ice sitting restlessly in just two places–Greenland and Antarctica. The remaining ice covers various mountains and fills alpine valleys. More than 29 million km^3 (7 million mi^3), or about 77% of all freshwater, is tied up as ice. These deposits of ice, laid down over several million years, provide an extensive frozen record of Earth's climatic history and perhaps some clues to its climatic future. The inference is that rather than distant frozen places of low population, these frozen lands are dynamic and susceptible to change, just as they have been over Earth's past history. The changes in the ice mass worldwide signal vast climatic change and further glacio-eustatic increases in sea level. As you study this chapter, keep this significance in mind.

As an example, please consult Mark F. Meier (U.S. Geological Survey, Tacoma, Washington), "Contribution of Small Glaciers to December 1984): 1416-1421. He states in his summary that

> **Observed long-term changes in glacier volume and hydrometeorological mass balance models yield data on the transfer of water from glaciers, excluding those in Greenland and Antarctica, to the oceans....These glaciers appear to account for a third to half of observed rise in sea level, approximately that fraction not explained by thermal expansion of the ocean.**

The *Student Study Guide* presents 20 "Learning Objectives" to guide the student in reading the chapter. The *Applied Physical Geography* lab manual has portions of one exercise that involve glacial landscapes.

New to the Third Edition

(Note: This section highlights major changes, new features, and additions in the third edition. This does not describe all the rewrite and recast of the text.)

1. A list of key learning concepts begins the chapter.

2. New photos of glaciation, cirque glaciers in southeastern Alaska (figure 17-2) and a peidmont glacier in Alaska reaching the sea causing ice calving (Figure 17-3).

3. A description of glacial ice mass, and the cause of glacial surges from new evidence accumulated from Greenland ice cores.

4. News Report #1: "South Cascade Glacier Loses Mass, "describes the net mass balance of the South Cascade Glacier in Washington State, including updated statistics and discussion of reasons for its loss of mass. News Report Figure 1 illustrates the net balance of the South Cascade Glacier between 1955 and 1995 in graphic form.

5. Use of bullet format to list erosional features.

6. Depositional features, such as erratics, are what early researchers, such as

Louis Agassiz, used to validate their ideas concerning continental glaciation. The importance of erratics is described in News Report #2: "Glacial Erratic Marks a Famous Grave."

7. Many photos have been added to illustrations of glacial erosion and deposition, including; Figures 17-9, 17-10, 17-13, 17-15, and Focus Study Figures 5 and 6.

8. An illustration of a roche moutonnée has been added to Figure 17-14.

9. Paleoclimatology was removed from the Focus Study, yet is still discussed within the chapter.

10. Description of the modern research techniques which are used to study past climates through oxygen isotopes, coral growth, and ice cores from Greenland and Antarctica.

11. News Report #4: "GRIP and GISP-2: Boring Ice for Exciting Activity" summarizes the findings of ice core drilling projects in the Greenland Ice Cap, for such information on past atmospheric content, such as gas concentrations, greenhouses gases and pollutants from volcanic eruptions. Figure 1 in News Report #4 is a map locating the GRIP and GISP-2 cores.

12. Description of the formation of the Great Lakes, as illustrated in Figure 17-18.

13. News Report #3: "Great Salt Lake Floods" describes how increased precipitatio in the Salt Lake Region during the 1980's caused great changes in the lake's surface levels. The Great Salt Lake is an ancient paleolake, which lacks natural drainage, allowing us to study the impact of climate change upon water levels.

14. Description of the Medieval Warm Period and the Eriksson Little Ice Age, both effecting the growth and decay of Greenland Viking colonies. Evidence of both periods has been found in the Greenland ice cores.

15. News Report #5: "An Arctic Ice Sheet?" questions the nature of the Arctic Ocean during the Pleistocene. Recent sonar images of the ocean floor taken in 1990 show deep gouges and grooves in bottom sediments on the ocean floor between Greenland and Norway, supporting the existence of super ice-bergs scarring the ocean floor.

16. Description of periglacial environments has been moved to the Focus Study entitled "Periglacial Landscapes."

17. The changes in Earth's orbit which affect climate change are put into a bullet format.

18. A new summary and review section ends the chapter.

Key Learning Concepts

1. *Differentiate* between alpine and continental glaciers and *describe* their principal features.

2. *Describe* the process of glacial ice formation and *portray* the mechanics of glacial movement.

3. *Describe* characteristic erosional and depositional landforms created by alpine glaciation and continental glaciation.

4. *Explain* the Pleistocene ice age epoch, and related glacials and interglacials; and *describe* some of the methods used to study paleoclimatology.

Expanded Outline Discussion

The following headings (boldfaced) match some of the first, second, and third order headings in Chapter 17. The narrative under each heading contains information, sources, and anecdotal facts relating to portions of the chapter. Not all text headings are discussed.

RIVERS OF ICE

The point is established in Chapter 7 and Table 7-1 that there are presently 29,180,000 km^3 (7,000,000 mi^3) of ice, amounting to 77% of all the freshwater on Earth, with 54% of this ice residing in Greenland and Antarctica. Antarctica is a full 89% of this amount.

The U.S. Geological Survey has initiated a major project of professional papers surveying all Earth's glaciers. The first of these is listed in the chapter references (Swithinbank, Charles. *Antarctica*. U.S. Geological Survey Professional Paper 1386-B. Washington, D.C.: Government Printing Office, 1988). The second professional paper in the series is now available: Williams, Richard S., Jr. and Jane G. Ferrigno eds., *Glaciers of Irian Jaya, Indonesia and New Zealand*, Professional Paper 1386-H, Washington, D.C.: Government Printing Office, 1989. These are beautiful publications on high quality paper and present heretofore unpublished, high-resolution Navy trimetrogon aerial photographs from the 1950s.

Types of Glaciers; Alpine Glaciers; Continental Glaciers

An aspect of glaciers to consider is whether they are confined or unconfined, e.g., the valley glacier is confined and becomes an unconfined piedmont glacier as it emerges from the valley onto a plain, whereas ice caps or ice sheets are unconfined. Figures 17-1, 17-3, 17-4, and the topographic map in 17-2 present the student with satellite imagery, high altitude photography, and a topographic map depiction of these glacier types. Perhaps you have your own favorite topographic map to illustrate these types of glaciers.

GLACIAL PROCESSES

The three parts of Figure 17-5 are meant to give the student a chance to compare the mass balance relationships of a glacier with a cross-section illustration of typical alpine glacier and an aerial photograph of a glacier. This is a good example of an open system with easily determined inputs and outputs and equilibrium line.

Formation of Glacial Ice

The formation of glacial ice is analogous to formation of a metamorphic rock with sediments (snow and firn) pressured and recrystallized into a dense metamorphic rock (glacial ice). The time taken for such a process is dependent on climatic conditions as noted in the text. A dry climate such as that found in Antarctica may produce glacial ice in 1000 years, whereas a wetter climate may take only a few years to complete the formation process.

The term névé is a French term for a mass of hardened snow at the head or in the accumulation area of the glacier. When you see this term used interchangeably with firn, that stems from an earlier English application. The American Geological Institute *Glossary of Geology* (3rd edition) recommends that we restrict the usage of névé to refer only to the geographic area covered with perennial snow, or the accumulation area at the head of the glacier. Firn on the other hand is representative of the intermediate stage between snow and glacial ice.

Glacial Mass Balance

In data provided to me by the Ice and Climate Project, University of Puget Sound, Tacoma, Washington 98416, by Robert M. Krimmel, U.S. Geological

Survey, 1989, it is interesting to note the net mass balance for the South Cascade Glacier, Washington for the period 1955 through 1992. To update this analysis a new report is now available: Robert M. Krimmel, *Mass Balance, Meteorological, and Runoff Measurements at South Cascade Glacier, Washington, 1992 Balance Year*, Open-File Report 93-640, Tacoma, WA: U.S. Geological Survey, 1993.

You may want to have your students plot these positive and negative values on a graph, and determine the algebraic sum of the net balance values. The glacier has gone through a net wastage at lower and middle elevations. (Note that this is presented in the *Student Study Guide* and in the *Applied Physical Geography* lab manual.) These statistics are graphically illustrated in News Report #1 documenting the net mass balance for the South Cascade Glacier in Washington state between 1955 and 1995.

The glacier has gone through a net wastage between 1955 and 1992 at lower and middle elevations. In just one year (9/1991 to 10/1992) the terminus retreated 38m and resulted in other major changes in the glacier. This represents more than a 2% loss in the glacier's mass in that time frame. Net mass balance is specified as cm of water equivalent spread over the entire glacier.

South Cascade Glacier Net Mass Balance Data (cm)

Year	Net balance
1955	30.00
1956	20.00
1957	–20.00
1958	–330.00
1959	70.00
1960	–50.00
1961	–110.00
1962	20.00
1963	–130.00
1964	120.00
1965	–17.00
1966	–103.00
1967	–63.00
1968	1.00
1969	–73.00
1970	–120.00
1971	60.00
1972	143.00
1973	–104.00
1974	102.00
1975	–5.00
1976	95.00
1977	–130.00
1978	–38.00
1979	–156.00
1980	–102.00
1981	–84.00
1982	8.00
1983	–77.00
1984	12.00
1985	–120.00
1986	–71.00
1987	256.00
1988	–164.00
1989	–71.00
1990	–73.00
1991	–20.00
1992	–201.00
1993	-100.00
1994	-158.00
1995	-70.00

Glacial Movement
Glacial Surges

A subject to follow with a personal clippings file is the occurrence of documented glacial surge episodes, which at this time are evidently on an increase. At least three moving ice streams are now being monitored along the West Antarctic ice sheet, although the three do not reach the sea and are blocked by the Ross Ice Shelf from entering the ocean. Research is ongoing to determine the cause of these surges and any relationship this may have to global warming trends.

Glacial Erosion

A visit to areas of active glaciation reveals to the eye a scene of extensive excavation, much like a massive earth-moving construction project. Areas of former glaciation reveal the scars and polish of the abrasive tools used by the mass of the glacier in carving the landscape.

Glacial Landforms
Erosional and Depositional Landforms

Figure 17-9 is adapted from a drawing by William Morris Davis published in 1909. The familiar sequence is fully labeled to assist you with this discussion. This color illustration is in your overhead transparency packet. I included Table 17-1 to allow an easy differentiation between glacially produced features as produced by either alpine or continental glaciation. Figure 17-13 presents an illustration and three inset photographs showing some common depositional landforms.

Periglacial Landscapes

In addition to journals in the field I used the American Geological Institute's *Glossary of Geology*, 3rd edition, edited by Robert L. Bates and Julia A. Jackson, 1987, to verify definitions and key term usage in *GEOSYSTEMS.* A valuable resource is by Peter J. Williams and Michael W. Smith, *The Frozen Earth–Fundamentals of Geocryology*, Cambridge: Cambridge University Press, 1989. This text contains a twenty-two page bibliography of useful references. A large volume titled *Geomorphology from Space–A Global Overview of Regional Landforms*, edited by Nicholas M. Short and Robert W. Blair, Jr. (Washington, D.C.: Scientific and Technical Information Branch, National Aeronautics and Space Administration, 1966, available from the Superintendent of Documents, U.S. GPO), has a chapter on "Glaciers and Glacial Landforms," by Richard S. Williams, Jr., pp. 521-96. Other sources for this section are listed in the credit lines in captions in the chapter and in the suggested readings section at the end of Chapter 17.

The term *solifluction* is often applied to all processes of soil flow in a generic sense and inclusive of such processes in periglacial regions. The term *gelifluction* specifically refers to downslope, slow soil flows in frozen landscapes whereas solifluction may apply to these same processes in temperate and tropical landscapes as well these periglacial regions. Geomorphology texts appropriately distinguish these terms. I have attempted to cover both the proper and common usage in the text.

THE PLEISTOCENE ICE AGE EPOCH

The text describes the traditional four major glacials and three interglacials previously used by scientists (and still appearing in some texts) to describe the episodes in North America. Figure 17-26 presents a more valid representation of glacial and interglacial cycles as published in 1976 through the G.S.A. Note the two named glacials and one named interglacial and 20 other stages during the last 900,000 years that have replaced the discarded sequence.

You may obtain a complete set of the *CLIMAP* Project presentation at a nominal cost (Figures 17-17 and 17-19). The many maps that accompany the explanatory booklet are sent in a tube so they are not folded. There are 18 maps in the full series; *CLIMAP* Project Members' "Seasonal Reconstructions of the Earth's Surface at the Last Glacial Maximum"; and *CLIMAP* Project, compiled by Andrew McIntyre-Leader in the LGM Project and Senior Research Scientist, Geological Society of America Map and Chart Series MC-36, Lamont-Doherty Geological Observatory of Columbia University, Palisades, NY 10964: Contribution No. 3153, 1981. It is also presented and overviewed by the *CLIMAP* Project Members in *Science* Vol. 191 (19 March 1976): 1131-1140.

A recent article by Wallace S. Broecker and George H. Denton, "What Drives Glacial Cycles?" that appeared in *Scientific American* (January 1990): 49-56, discusses glacial cycles and the massive reorganizations that occur in the ocean-atmosphere system. They link changes to Earth's orbit and add more detail to the points made in the focus study.

Milutin Milankovitch, like Alfred Wegener, died before seeing his life's work on the mechanisms that produce ice ages authenticated and elevated to acceptance by the scientific community. We can always wonder what the impact of modern computers would have been on the numbers that Milankovitch calculated by hand.

The U.S. Geological Survey distributes a 21-page booklet for $3 called "Glaciers: Clues to Future Climate?" by Richard S. Williams, Jr., 1983.

Numerous articles appeared in *Science* and *Nature* during the summer and fall of 1993 reporting on the GRIP and GISP 2 Greenland ice drilling projects– they are approximately 28 km apart. Rather than list all of them here I refer you to the indexes of these journals or to the electronic indexing systems in your library. GRIP is for Greenland Ice Core Project and GISP 2 stands for Greenland Ice Sheet Project II. The GRIP Project members reported in *Nature*, 364, July 15, 1993: 203-07; and Dansgaard, *et al.*, in that issue pp. 218-20. An overview of these reports appears on page 186 of the same issue. Also see reports by K.C. Taylor, *et al.*, and another by P.M. Grootes, *et al.*, *Nature*, 366, December 9, 1993: 549-554, for comparative analysis of the two cores that are sampling ice back some 150,000 years.

Resources concerning paleoclimatology are also available on the Internet. Paleoenvironmental records of past climate change can be accessed at: http://www.ngdc.noaa.gov/paleo/paleo.html.

Pluvial Lakes (Paleolakes)

In response to the increasing levels of the Great Salt Lake, the Utah state legislature funded a project to create an additional system output for the lake; enormous pumps were installed on the west side of the lake to move excess water to a nearby desert basin. At full ideal capacity the pumps lowered the desert lake by about 50.8 cm (20 in.) a year. In 1986, however, the lake level increased to a high of 1284 m (4212 ft) above sea level and by 1989 had dropped just to 1282 m (4206 ft), with only about 61 cm (24 in.) of that reduction related to the pumping. Fortunately, the summers of 1986 and 1987 had high evaporation rates, and precipitation dropped to 50 percent below average. Predictions indicate that lake levels are going to rise again to even greater levels in the 1990s. See Adler, William J., and Gerald Williams, "The West Desert Pond or Another Mini-Great Salt Lake," Western Region Technical Attachment no. 88-05, Salt Lake City: National Oceanic and Atmospheric Administration, National Weather Service, 26 January 1988. Only time will verify the accuracy of the models being used.

ARCTIC AND ANTARCTIC REGIONS

Annual precipitation averages only 178 mm (7 in.) at the U.S. McMurdo station in Antarctica. Resolute (74° N), Cornwallis Island, N.W. Territories, receives 130 mm (5.1 in.) of precipitation. These polar regions are deserts, in that precipitation values are extremely low. Snow and ice accumulate because evapotranspiration rates are extremely small.

The quantity of ice and dimension of the Antarctic and Greenland ice sheets is mentioned earlier in this unit. Greenland is only about 12% the size of Antarctica. Until the space age placed satellites in orbit, the mapping of these frozen worlds was poor. There remain portions of Antarctica that are not as accurately mapped as is the other side of the Moon! Today, as a result of the *Landsat* series, and other spacecraft, actual measurement and mapping is greatly improved. This precision is important because the status of these ice volumes is such an important indicator of Earth's climatic trends.

You can now contact the Antarctica Resource Guide on the Internet. The following site will link you to other Internet resources regarding Antartica, inlcuding reaseach reports and news articles,

http://http2.sils.umich.edu/Antartica /Bibliography/Bib.html

The Antarctic Ice Sheet

A resource from the USGS is *Antarctica*. U.S. Geological Survey Professional Paper 1386-B. Washington: Government Printing Office, 1988. The entire September 1962 issue of *Scientific American* was devoted to many aspects of Antarctica. Uwe Radok's "The Antarctic Ice," appeared in the August 1985 *Scientific American*, pp. 98-105. From the studies in Polar Research Series: George Deacon, *The Antarctic Circumpolar Ocean*, Cambridge University Press, 1984. And, J. Jouzel, *et al.*, "Extending the Vostok ice-core record of paleoclimate to the penultimate glacial period," *Nature*, 364, July 29, 1993: 407-12.

On 4 October 1991, 24 countries signed an historic treaty to ban oil and mineral exploration in Antarctica for at least 50 years. Also included were provisions to protect wildlife, prevent waste disposal and marine pollution of various forms, and to continue scientific monitoring and research. This agreement is technically a protocol to the original 1959 Antarctic Treaty. This is regarded as a virtual ban on any economic exploitation of this, Earth's last above sea-level frontier. Think of Antarctica as a vast new international park dedicated to scientific research and peace.

Glossary Review for Chapter 17 (in alphabetical order)

ablation
abrasion
alpine glacier
arête
cirque
col
continental glacier
crevasses
drumlin
esker
firn
firn line
fjord
glacial drift
glacial ice
glacial surge
glacier
horn
ice age
ice cap
ice field
ice sheet
icebergs
kame
kettle
lateral moraine
medial moraine
moraine
outwash plain
paleolakes
palsa
paternoster lakes
periglacial
permafrost
roche moutonnée
snowline
stratified drift
talik
tarn
terminal moraine
till
till plain

Annotated Chapter Review Questions

1. Describe the location of most of the freshwater on Earth today?

A large measure of the freshwater on Earth is frozen, with the bulk of that ice sitting restlessly in just two places–Greenland and Antarctica. The remaining ice covers various mountains and fills alpine valleys. More than 29 million km^3 (7 million mi^3) of water, or about 77% of all freshwater, is tied up as ice.

2. What is a glacier? What is implied about existing climate patterns in a glacial region?

A glacier is a large mass of perennial ice, resting on land or floating shelf-like in the sea adjacent to land. Glaciers form by the accumulation and recrystallization of snow. They move under the pressure of their own mass and the pull of gravity. Today, about 11% of Earth's land area is dominated by these slowly flowing ice streams. During colder episodes in the past, as much as 30% of continental land was covered by glacial ice because below-freezing temperatures prevailed at lower latitudes, allowing snow to accumulate. Relative to elevation, in equatorial mountains, the snowline is around 5000 m (16,400 ft); on midlatitude mountains, such as the European Alps, snowlines average 2700 m (8850 ft); and in southern Greenland snowlines are down to 600 m (1970 ft).

3. Differentiate between an alpine glacier and a continental glacier.

With few exceptions, a glacier in a mountain range is called an alpine glacier, or mountain glacier. It occurs in several subtypes. One prominent type is a valley glacier, an ice mass constricted within the confines of a valley. Such glaciers range in length from only 100 m (325 ft) to over 100 km (62 mi). The snowfield that feeds the glacier with new snow is at a higher elevation. As a valley glacier flows slowly downhill, the mountains, canyons, and river valleys beneath its mass are profoundly altered by its passage. A continuous mass of ice is known as a continental glacier and in its most extensive form is called an ice sheet. Two additional types of continuous ice cover associated with mountain locations are designated as ice caps and ice fields. Most glacial ice exists in the snow-covered ice sheets that blanket 80% of Greenland (1.8 million km^3, or 0.43 million mi^3) and 90% of Antarctica (13.9 million km^3, or 3.3 million mi^3).

4. Name the three types of continental glaciers. What is the basis for dividing continental glaciers into types? Which type covers Antarctica?

The three types are ice sheet, ice cap, and ice field. Both ice caps and ice sheets completely bury the underlying landscape, although an ice cap is somewhat circular and covers an area of less than 50,000 km^2 (19,300 mi^2). Antarctica alone has 91% of all the glacial ice on the planet as an enormous ice sheet, or more accurately several ice sheets acting in concert. An ice field is the smallest of the three.

5. Trace the evolution of glacial ice from fresh fallen snow.

Snow that survives the summer and into the following winter begins a slow transformation into glacial ice. Air spaces among ice crystals are pressed out as snow packs to a greater density. The ice crystals recrystallize under pressure and go through a process of re-growth and enlargement. In a transitional step to glacial ice, snow becomes firn, which has a granular texture. As this process continues, many years pass before denser glacial ice is produced. Formation of glacial ice is analogous to formation of metamorphic rock: sediments (snow and firn) are pressured and recrystallized into a dense metamorphic rock (glacial ice).

6. What is meant by glacial mass balance? What are the basic inputs and outputs underlying that balance?

A glacier is fed by snowfall and is wasted by losses from its upper and lower surfaces and along its margins. A snowline called a firn line is visible across the surface of a glacier, indicating where the winter snows and ice accumulation survived the summer melting season. A glacier's area of excessive accumulation is, logically, at colder, higher elevations. The zone where accumulation gain ends and loss begins is the equilibrium line. This area of a glacier generally coincides with the firn line, except in subpolar glaciers, where frozen surface water occurs below the firn line. Glaciers achieve a positive net balance of mass–grow larger–during colder periods with adequate precipitation. In a glacier's lower elevation, losses of mass occur because of surface melting, internal and basal melting, sublimation, wind removal by deflation, and the calving, or breaking off, of ice blocks. In warmer times, the equilibrium line migrates to a higher elevation and the glacier retreats–grows smaller–due to its negative net balance.

7. What is meant by a glacial surge? What do scientists think produces surging episodes?

Some glaciers will lurch forward with little or no warning in a glacial surge. This is not quite as abrupt as it sounds; in glacial terms, a surge can be tens of meters per day. The Jakobshavn Glacier in Greenland, for example, is known to move between 7 and 12 km (4.3 and 7.5 mi) a year. The exact cause of such a glacial surge is still being studied. Some surge events result from a buildup of water pressure in the basal layers of the glacier. Sometimes that pressure is enough to actually float the glacier slightly during the surge. As a surge begins, icequakes are detectable, and ice faults are visible along the margins that separate the glacier from the surrounding stationary terrain.

8. How does a glacier accomplish erosion?

Glacial erosion is similar to a large excavation project, with the glacier hauling debris from one site to another for deposition. As rock fails along joint planes, the passing glacier mechanically plucks the material and carries it away. There is evidence that rock pieces actually freeze to the basal layers of the glacier and, once embedded, allow the glacier to scour and sandpaper the landscape as it moves, a process called abrasion. This abrasion and gouging produces a smooth surface on exposed rock, which shines with glacial polish when the glacier retreats. Larger rocks in the glacier act much like chisels, working the underlying surface to produce glacial striations parallel to the flow direction.

9. Describe the evolution of a V-shaped stream valley to a U-shaped glaciated valley. What kinds of features are visible after a glacier has retreated?

W. M. Davis characterized the stages of a valley glacier in a set of drawings published in 1906 and redrawn here in Figure 17-9. Illustration (a) shows a typical river valley with characteristic V-shape and stream-cut tributary valleys that exist before glaciation. Illustration (b) shows that same landscape during a later period of active glaciation. Glacial erosion and transport are actively removing much of the regolith (weathered bedrock) and the soils that covered the preexisting valley landscape. Illustration (c) shows the same landscape at a later time when climates have warmed and ice has retreated. The glaciated valleys now are U-shaped, greatly changed from their previous stream-cut form. You can see the oversteepened sides, the straightened course of the valley, and the presence of hanging valleys and waterfalls. The physical weathering associated with a freeze-thaw cycle has loosened much rock along the steep cliffs, falling to form talus cones along the valley sides during the postglacial period. See Figure 17-9c for labeled details of the features that are formed as a result of the formation, growth, passage, and retreat of an alpine glacier.

10. How is an iceberg generated?

Where a glacier ends in the sea, large pieces break off and drift away as icebergs. These are portions of a glacier at drift in the sea. When large pieces of an ice shelf break off, such as the portions of the Ross Ice Shelf in the west Antarctic area have done during the past few years, enormous tabular islands are formed as a type of iceberg.

11. Differentiate between two forms of glacial drift–till and outwash.

Where the glacier melts, debris accumulates to mark the former margins of the glacier–the end and sides. Glacial drift is the general term for all glacial deposits. Direct deposits appear unstratified and unsorted and are called till. In contrast, sorted and stratified glacial drift, characteristic of stream-deposited material, is called outwash and forms an outwash plain of glacio-fluvial deposits across the landscape.

12. What is a morainal deposit? What specific forms of moraines are created by alpine and continental glaciers?

Glacial till moving downstream in a glacier can form a marginal unsorted deposit known as a moraine–Figure 17-5 and 17-11. A lateral moraine forms along each side of a glacier. If two glaciers with lateral moraines join, their point of contact becomes a medial moraine. Eroded debris that is dropped at the glacier's farthest extent is called a terminal moraine. However, there also may be end moraines, formed wherever a glacier pauses after reaching a new equilibrium. If a glacier is in retreat, individual deposits are called recessional moraines. And finally, a deposition of till generally spread across a surface is called a ground moraine.

13. What are some common depositional features encountered in a till plain?

With the retreat of the glaciers, many relatively flat plains of unsorted coarse till were formed behind terminal moraines. Low, rolling relief and deranged drainage patterns are characteristic of these till plains. As the glacier melts, this unsorted cargo of *ablation till* is lowered to the ground surface, sometimes covering the clay-rich *lodgement till* deposited along the base. The rock material is poorly sorted and is difficult to cultivate for farming, but the clays and finer particles can provide a basis for soil development.

14. Compare a roche moutonnée and a drumlin regarding appearance, orientation and the way each forms.

Two landforms created by glacial action are streamlined hills, one erosional (called a roche moutonnée) and the other depositional (called a drumlin). A roche moutonnée is an asymmetrical hill of exposed bedrock. Its gently sloping upstream side (stoss side) has been polished smooth by glacial action, whereas its downstream side (lee side) is abrupt and steep where rock was plucked by the glacier (Figure 17-14). A drumlin is deposited till that has been streamlined in the direction of continental ice movement, blunt end upstream and tapered end downstream. Multiple drumlins (called swarms) occur in fields in New York and Wisconsin, among other areas. Sometimes their shape is that of an elongated teaspoon bowl lying face down (see Figure 17-13, and topo map in Figure 17-15).

15. What is paleoclimatology? Describe Earth's past climatic patterns. Are we experiencing a nominal climate pattern in this era or have scientists noticed any significant trends?

Paleoclimatology is the science of past climates. The most recent episode of cold climatic conditions began about 1.65 million years ago, launching the Pleistocene epoch. At the height of the Pleistocene, ice sheets and glaciers covered 30% of Earth's land area, amounting to more than 45 million km^2 (17.4 million mi^2). The Pleistocene is thought to have been one of the more prolonged cold periods in Earth's history. At least 18 expansions of ice occurred over Europe and North America, each obliterating and confusing the evidence from the one before. The term ice age is applied to any such extended period of cold, even though an ice age is not a single cold spell. Instead, it is a period of generally cold climate, called a glacial, interrupted by brief warm spells, known as interglacials. There is a worldwide retreat of alpine glaciers, higher snowlines in Greenland, and at least three accelerating ice streams on the West Antarctic ice sheet. The reduction in alpine glacial mass balances is particularly true of low and middle elevation glaciers. This trend in ice mass reduction may be attributed to the present century-long increase in mean global air temperatures. Additionally, over the past ten years, we have experienced the eight warmest years in instrumental history.

16. Define an ice age. When was the most recent? Explain "glacial" and "interglacial" in your answer.

The term ice age is applied to any extended period of cold, even though an ice age is not a single cold spell. Instead, it is a period of generally cold climate, called a glacial, interrupted by brief warm spells, known as interglacials. Traditionally, four major glacials and three interglacials were acknowledged for the Pleistocene epoch. (The glacials were named the Nebraskan, Kansan, Illinoian, and Wisconsinan.) In Europe, similar episodes coincided with those in North America, but were given different names. Modern techniques have opened the way for a new chronology and understanding. Currently, glaciologists acknowledge the Illinoian glacial and Wisconsinan glacial, and the Sangamon interglacial between them. These span the past 300,000 years (Figure 17-16). The Illinoian is believed to have had two glacials (designated stages 6 and 8 in the figure), as did part of the Wisconsinan (stages 2 and 4), which is dated at 10,000 to 35,000 years ago. The stages on the chart are numbered back to stage 23, at approximately 900,000 years ago. The Holocene (past 10,000 years) is regarded as either an interglacial or a post-glacial epoch.

17. Overview what science has learned about the various causes of ice ages by listing and explaining at least four possible climate change factors.

The mechanisms that bring on an ice age are the subject of much research and debate. Because past occurrences of low temperatures appear to have followed a pattern, researchers have looked for causes that also are cyclic in nature. They have identified a complicated mix of interacting variables that appear to influence long-term climatic trends, including galactic and Earth-Sun relationships, solar variability, geophysical factors, and geographical-geological factors.

18. Describe the role of ice cores in deciphering past climates. What record do they preserve? Where were they drilled?

Ice core analysis has opened the way for a new chronology and understanding of glaciation. Chemical and physical properties of the atmosphere and snow that accumulated each year are frozen into place. Locked into the ice cores are the air bubbles from past atmospheres, which indicate ancient gas concentrations, such as greenhouse gases, carbon dioxide and methane. For examples, during cold periods, high concentrations of dust are present, brought by winds from distant dry lands, acting as condensation nuclei. Past volcanic eruptions are recorded in this manner. The presence of ammonia indicates ancient forest fires at lower latitudes and the ratio between stable forms of oxygen is measured with each snowfall.

Greenland has been the location of two ice cores, the Greenland Ice Core Project (GRIP) which began in 1989, and more recently, the Greenland Ice Sheet Project (GISP-2) began in 1990. Each project was successful in reaching bedrock. The GRIP catalogued 250,000 years of history, drilling a 3030m deep core, and the GISP-2 core reached bedrock at a 2700m, and records 115,000 years of past climatic history.

19. What criteria define the Arctic and Antarctic regions? Is there any coincidence in these criteria and the distribution of Northern Hemisphere forests on the continents?

Climatologists use environmental criteria to define the Arctic and the Antarctic regions (Figure 17-25). For the Arctic area, the 10°C (50°F) isotherm for July is used; it coincides with the visible treeline, which is the boundary between the northern forests and tundra climates, i.e., a temperature below which boreal forests cannot survive.

Annotated Focus Study Review Questions

1. In terms of climatic types, describe the areas on Earth where periglacial landscapes occur. Include both higher latitude and higher altitude climate types.

Periglacial regions occupy over 20% of Earth's land surface (Focus Study 1, Figure 1). The areas are either near permanent ice or are at high elevation, and have ground that is seasonally snow free. Under these conditions, a unique set of periglacial processes operate, including permafrost, frost action, and ground ice.

Climatologically, these regions are in *Dfc*, *Dfd subarctic*, and *E polar* climates (especially *ET tundra* climate). Such climates occur either at high latitude (tundra and boreal forest environments) or high elevation in lower-latitude mountains (alpine environments).

2. Define two types of permafrost and differentiate their occurrence on Earth. What are the characteristics of each?

When soil or rock temperatures remain below 0°C (32°F) for at least two years, a condition of *permafrost* develops. An area that has permafrost but is not covered by glaciers is considered periglacial. Note that this criterion is based solely on temperature and not on whether water is present. Other than high latitude and low temperatures, two other factors contribute to permafrost: the presence of fossil permafrost from previous ice-age conditions and the insulating effect of snow cover or vegetation that inhibits heat loss.

Permafrost regions are divided into two general categories, continuous and discontinuous, that merge along a general transition zone. *Continuous permafrost* describes the region of the most severe cold and is perennial, roughly poleward of the –7°C (19°F) mean annual temperature isotherm. Continuous permafrost affects all surfaces except those beneath deep lakes or rivers in the areas shown in Focus Study 1, Figure 1. Continuous permafrost may exceed 1000 m in depth (over 3000 ft) averaging approximately 400 m (1300 ft).

3. Describe the active zone in permafrost regions and relate the degree of development to specific latitudes.

The *active layer* is the zone of seasonally frozen ground that exists between the subsurface permafrost layer and the ground surface. The active layer is subjected to consistent daily and seasonal freeze-thaw cycles. This cyclic melting of the active layer affects as little as 10 cm depth in the north (Ellesmere Island, 78° N), up to 2 meters in the southern margins (55° N) of the periglacial region, and 15 m in the alpine permafrost of the Colorado Rockies (40° N).

4. What is a talik? Where might you expect to find taliks and to what depth do they occur?

A *talik* (derived from a Russian word) is an unfrozen portion of the ground that may occur above, below, or within a body of discontinuous permafrost or beneath a body of water in the continuous region. Taliks are found beneath deep lakes and may extend to bedrock and noncryotic soil under large deep lakes (Focus Study 1, Figure 2). Taliks form connections between the active layer and groundwater, whereas in continuous permafrost groundwater is essentially cut off from water at the surface. In this way, permafrost disrupts aquifers and causes water supply problems.

5. What is the difference between permafrost and ground ice?

In regions of permafrost, subsurface water that is frozen is termed *ground ice*. The moisture content of areas with ground ice may vary from nearly absent in regions of drier permafrost to almost 100% in saturated soils. From the area of maximum energy loss, freezing progresses through the ground along a *freezing front*, or boundary between frozen and unfrozen soil. The presence of frozen water in the soil initiates geomorphic processes associated with *frost action* and the expansion of water volume as it freezes (Chapter 7 and 13).

6. Describe the role of frost action in the formation of various landform types in the periglacial region.

The 9% expansion of water as it freezes produces strong mechanical forces that fracture rock and disrupt soil at and below the surface. Frost-action shatters rock, producing angular pieces that form a *block field*, or *felsenmeer*, accumulating as part of the arctic and alpine periglacial landscape, particularly on mountain summits and slopes.

If sufficient water undergoes the phase change to ice, the soil and rocks embedded in the water are subjected to *frost-heaving* (vertical movement) and *frost-thrusting* (horizontal motions). Boulders and slabs of rock generally are thrust to the surface. Soil horizons may appear disrupted as if stirred or churned by frost action, a process termed *cryoturbation*. Frost action also produces a contraction in soil and rock, opening up cracks for ice wedges to form. Also, there is a tremendous increase in pressure in the soil as ice expands, particularly if there are multiple freezing fronts trapping unfrozen soil and water between them.

7. Relate some of the specific problems humans encounter in developing periglacial landscapes.

Human populations in areas that experience frozen ground phenomena encounter various difficulties. Because thawed ground in the active layer above the permafrost zone frequently shifts in periglacial environments, the maintenance of road beds and railroad tracks is a particular problem. In addition, any building placed directly on frozen ground will begin to melt itself into the defrosting soil. Thus, the melting of permafrost can create subsidence in structures and complete failure of building integrity (Focus Study 1, Figure 7).

Overhead Transparencies

As an adopter you are provided with the following figures for overhead projector use.

- Figure 17-5: A retreating alpine glacier and mass balance
- Figure 17-9a, b, and c: The geomorphic handiwork of alpine glaciers (before, during, and after a valley glacier occupies a formerly stream cut valley)
- Figure 17-13: Continental glacier depositional features
- Figure 17-17: Pleistocene glaciation
- Figure 17-18: Late stages of Great Lakes formation
- Figure 17-25: Illustrations of the Arctic and Antarctic regions
- Focus Study 1, Figure 1: Periglacial environments

- Focus Study 1, Figure 1: Periglacial environments

PART FOUR:
Soils, Ecosystems, and Biomes

Overview–Part Four

Earth is the home of the only known biosphere in the Solar System–a unique, complex, and interactive system of abiotic and biotic components working together to sustain a tremendous diversity of life. Thus we begin Part Four of *GEOSYSTEMS* and an examination of the geography of the biosphere–soils, ecosystems, and terrestrial biomes. Remember the description of the biosphere from Chapter 1:

> The intricate web that connects all organisms with their physical environment is the biosphere. Sometimes referred to as the ecosphere, the biosphere is the area in which physical and chemical factors form the context of life. The biosphere exists in an area of overlap among the spheres. Life processes have also powerfully shaped the other spheres through various interactive processes. The biosphere has evolved, reorganized itself at times, faced extinction, gained new vitality, and managed to flourish overall. Earth's biosphere is the only known one in the solar system; thus, life as we know it is unique to Earth.

Part Four is a synthesis of many of the elements covered throughout the text.

18
The Geography of Soils

Overview–Chapter 18

Earth's landscape is generally covered with soil. Soil is a dynamic natural body which comprises fine materials in which plants grow, and which is composed of both mineral and organic matter. By their diverse nature, soils are a complex subject and pose a challenge for spatial analysis. This chapter presents an overview of the modern system of soil classification used in the United States, with mention of the Canadian system. The Soil Taxonomy system is often misnamed in texts, and anachronistic soil classification terminology is sometimes used. Since the system has been in constant use since 1975 and is integrated into all major pedology texts, I have attempted to present this system in an appropriate manner.

Please find Table 18-6 and utilize it to simplify this geographic study of soils. The table summarizes the 10 soil orders of the Soil Taxonomy, including general location and climate association, areal coverage estimate, and basic description. Also, you will find small locator maps for most soil orders included with a picture of that soil's profile. These locator maps are derived from the worldwide distribution map presented in Figure 18-9.

The *Student Study Guide* presents 21 "Learning Objectives" to guide the student in reading the chapter.

New to the Third Edition

(Note: This section highlights major changes, new features, and additions in the third edition. This does not describe all the rewrite and recast of the text.)

1. A list of key learning concepts begins the chapter.

2. New description of pedology appears in the introduction.

3. More elaborate description of A and B horizons.

4. Figure 18-2 is a typical soil profile with a photo of a Mollisol pedon in southeastern South Dakota.

5. A new photo illustrating the usefulness of the Munsell Color chart (Figure 18-3).

6. The varying conditions of soil consistence are clearly outlined in a bullet format.

7. News Report #1: "Geophagy: Eating Clay" gives students some factors behind this practice, its worldwide distribution, and American adaptations.

8. New statistics concerning American and worldwide soil loss, more elaborate descriptions of the effects of this loss and appropriate solutions that need to be taken.

9. The amount of farm land lost each year to mismanagement and conversion to nonagricultural uses is summarized in News Report #2: "Soil is Slipping Through Our Fingers".

10. Figure 1, News Report 2, illustrates a soil drainage canal which collects contaminated water from field drains and transports it directly into the Salton Sea.

11. Focus Study 1, "Selenium Concentration in Western Soils," discusses the problems of waterlogging and salinization of soils associated with agricultural irrigation. These problems are especially prevalent in arid lands that are poorly drained, and reduce the productivity of once fertile lands, such as the Indus River Valley, Tigris and Euphrates Valley and the Central Valley of California.

12. Focus Study 1, Figure 1, is a photo of Kesterson Wildlife Refuge in California's Central Valley, where selenium-tainted agricultural drainage has destroyed the refuge, now considered a toxic waste site.

13. News Report #3: "Drainage Tiles, But No Where To Go," describes methods used in California and Arizona agricultural valleys to reduce salinization and waterlogging. Drainage drains have been installed to carry away excess water, yet excess water flows directly into the Salton Sea (from Imperial and Coachella Valleys in California) and the Colorado River (in Arizona) creating political problems and causing great economic costs.

14. Types of soil horizons has are listed in a bullet format to clearly identify each horizon.

15. A new inset to Figure 18-9 illustrates soil taxonomy in the United States, a nice addition to the global soil taxonomy map.

16. New photos complement the global distribution maps for each soil type; Figure 18-14 Mollisols, Figure 18-17 Soils of the Midwest, Figure 18-20 Ultisols, Figure 18-21 Spodosols, and Figure 18-23 Aridosols.

17. A new summary and review section ends the chapter.

Key Learning Concepts

1. *Define* soil and soil science and *describe* a pedon, polypedon, and typical soil profile.

2. *Describe* soil properties of color, texture, structure, consistence, porosity, and soil moisture.

3. *Explain* basic soil chemistry, including cation-exchange capacity, and *relate* these concepts to soil fertility.

4. *Evaluate* principal soil formation factors, including the human element.

5. *Describe* the eleven soil orders of the Soil Taxonomy classification system and *explain* their general occurrence.

Expanded Outline Discussion

The following headings (boldfaced) match some of the first, second, and third order headings in Chapter 18. The narrative under each heading contains information, sources, and anecdotal facts relating to portions of the chapter. Not all text headings are discussed.

SOIL CHARACTERISTICS

Specific soil conditions determine soil fertility, which is the ability of soil to support plant productivity. Geographers should be concerned about the pattern of this fertility, or the patterns of human-induced loss of fertility. This precept is mentioned in the introduction to this chapter and here and there throughout the chapter because of its inherent importance to the geography of soils. After all, the main push that occurred following World War II to improve soil science was really aimed at increasing crop production worldwide.

Soil Profiles; Soil Horizons

Remember that the pedon is the basic *sampling unit* in soil surveys and that the polypedon is the essential soil individual, constituting an identifiable soil series, and as such, is the basic local *mapping unit.*

The profile photographs of soil horizon layers shown in Figures 18-10, 18-13, 18-15, 18-19, 18-20, 18-21, and 18-24 should prove helpful in identifying the physical characteristics of horizons O, A, E, B, C, and R. The text mentions that transition areas are present in the soil profile between horizons–this is the rule rather than an exception.

The lower portion of the A horizon grades into the E horizon, which is a bit more pale and is made up of coarse sand, silt, and resistant minerals. Clays and oxides of aluminum and iron are leached (removed) from the E horizon and migrate to lower horizons with water as it percolates through the soil. This process of rinsing through upper horizons and removing finer particles and minerals is termed *eluviation*; thus the designation E for this horizon. The greater the precipitation in an area, the higher the rate of eluviation that occurs in the E horizon.

Soil Properties: Color, Texture, Structure, Consistence, and Porosity

The physical properties of soil are key to the identification of specific soil series that form the basis of the Soil Taxonomy. The modern system of classification is based on actual observable soil properties seen in the field. This is in contrast to the 1938 system that recognized the importance of soil processes. These were used to establish pedogenic regimes characteristic of climatic conditions. Pedogenic regimes are presented in *GEOSYSTEMS* within specific orders of the

Soil Taxonomy. The 1938 system fell short in that it based soil classification on the uncertainty of such pedogenic inferences. In recognition of this judgment by soil scientists I elected to include the regimes within related orders rather than feature them as a basis for classification. Soil classification and mapping is a constantly evolving area of geography and soil science.

Soil Moisture Regimes

Soil Taxonomy establishes various moisture regimes (Table 18-3) on Thornthwaite's water balance principles and soil-moisture considerations. This gave me further reason to include the extended treatment of the water balance in Chapter 9. Students can be referred back to those sections to refresh their memories on soils moisture utilization and recharge, field capacity and wilting point (Chapter 9) and the operation of the balance for specific and selected stations.

Soil Chemistry; Soil Acidity and Alkalinity

The importance of colloidal particles and soil pH are described in the text on pages 554-55. This section ties together soil texture, structure, and moisture consideration. Note that an analysis of overall soil chemicals may have little to do with actual soil fertility; rather, what is important are ion exchange rates and mineral and chemical transformations–the focus of our brief overview.

Soil Formation and Management

There are many inferences that can be drawn about soils and history, such as the overgrazing and occupation of the Great Plains that led to the dust bowl of the 1930s. A soil scientist might view the American Civil War as one fought over the need for new soils when the slave-labor intensive, Ultisol-depleting, cotton crop destroyed the fertility of available soils in the South. As the movement of cotton farming headed west and north, with its awful cultural baggage of an enslaved work force, the rest of the country eventually said no. The southern agricultural interests were told that they could not move onto still fertile Mollisols and Alfisols of the north and west–that is, expand the geographic diffusion of slavery.

New paragraphs and a map present worldwide soil degradation (Figure 18-8). This map was to be a Robinson projection, however the decision was made to use a Mercator projection because of the difficulty in transferring data from the source map. Note the areas of serious concern through the central portion of North America, portions of Brazil and valleys in the Andes, the Sahel and portions of southern Africa, regions of Eastern Europe, Ukraine, India, and China.

A great source of information concerning soil erosion and land degradation is *Land Degradation and Society*, by Piers Blaikie and Harold Brookfield, Methuen, 1987. Blaikie has another book that discusses the political and economic impact of soil erosion, entitled, *The Political Economy of Soil Erosion.*

SOIL CLASSIFICATION: A Brief History of Soil Classification

Taxonomy is defined in Webster's as "Classification according to natural relationships...the systematic distinguishing, ordering, and naming of type groups within a subject field....a formal system of nomenclature." Additional background is added to this section concerning the history of soil classification efforts in Canada.

The first formal American system is credited to Dr. Curtis F. Marbut, who derived his approach from Dukuchaev. Published in the *1935 Atlas of American Agriculture* and the *Yearbook of Agriculture–1938*, Marbut's classification of great soil groups recognized the importance of soils as products of dynamic natural processes. However, as the database of soil information grew, the inadequacies of this system became apparent. Following World War II, there ensued efforts in many nations to increase agricultural output. As part of this, soil survey efforts were stimulated and the number of soil series identified increased to over 5500 by 1951. The inadequacies of the 1938 system were obvious to soil scientists at about this time. As the number of identified soil series grew, it became increasingly problematic to associate all the soil series to soil families, groups, and higher categories. The 1938 system simply did not contain detailed criteria for the higher categories of classification.

The work of developing a new classification system began about 1950 with a series of *approximations*, each tested in-progress by the Soil Survey Staff of the Soil Conservation Service in the United States. International cooperation was important, as evidenced by conferences at Ghent, Leopoldville, Paris, London, and Washington. The 3rd Approximation included many of the European suggestions. By 1955, the 4th Approximation was introduced with efforts to group soil series for possible identification of patterns in categories above the soil-family level. This produced the 5th Approximation at the 1956 6th Congress of the International Society of Soil Science in Paris. By 1957 the 6th Approximation was developed to provide the Soil Survey Staff with a guide for better grouping of subgroups and families. Also, through this stage the nomenclature was developed in consultation with specialists in language, the classics, and with particular assistance from Belgian scientists.

By 1960, the *7th Approximation* was completed and distributed at the 7th International Congress of Soil Science held at Madison, Wisconsin. At this time we were approaching some 10,000 soil series in active use in the United States. Many texts still use this 30-year old outdated name to describe the Soil Taxonomy system–a practice not duplicated in any soil science or pedology texts! In preparing *GEOSYSTEMS* I spoke to Dr. Henry D. Foth of Michigan State (*Fundamentals of Soil Science*, 8th ed., Wiley, 1990). He commented on this use of misnomers for Soil Taxonomy and that we should get in line with the related field of soil science. I have made an effort to do this.

Various supplements and updates were subsequently prepared to improve the system especially with regard to improving tropical soil classification. The effort is clearly international in scope and reflects work from many scientists worldwide. All this led to the publication of Agricultural Handbook No. 436 in 1975, the *Soil Taxonomy-A Basic System of Soil Classification for Making and Interpreting Soil Surveys.* Many aspects of Soil Taxonomy overlap with the world soil map published by the Food and Agriculture Organization. If it is not already on your shelf you might want to obtain a copy of this 750-page volume, along with Agricultural Handbook No. 18 the *Soil Survey Manual.* Also useful is the Soil Management Support Services Technical Monograph No. 6, the *Keys to Soil Taxonomy,* by the Soil Survey Staff (3rd ed., 1987).

Diagnostic Soil Horizons

I included a simplified version of surface (epipedons) and subsurface diagnostic horizons in Table 18-5 to accommodate a general perusal by students. Or, the table will assist you if you choose go into more depth with these topics if this is an area of special interest. The table will be useful as a reference for the balance of the chapter since several of these diagnostic horizons are mentioned in the description of the soil orders.

The Eleven Soil Orders of Soil Taxonomy

Soil Taxonomy-A Basic System of Soil Classification for Making and Interpreting Soil Surveys (Agricultural Handbook No. 436), by the Soil Survey Staff, will provide you with more in-depth material on each of the soil orders. You also might want to check several of the current soil science texts mentioned in the suggested readings in the text.

The integrative table presented as Table 18-6 will allow you to overview the eleven soil orders without proceeding through the rest of the chapter, and in a descriptive sense, it can serve as a summary for introducing Soil Taxonomy and geography to the introductory student. Or, for a more comprehensive approach, the table can act as an organizer for relating detailed information back to the overall classification system. This chapter is divided between a systematic treatment and a regional treatment separated by this table.

The newest soil order, the Andisols, was added in 1990 by taking suborders derived from volcanic glass and ash out of the Inceptisol and Entisol soil orders. The student can refer back to the igneous rock mineral Table 11-2 to see the position of the pyroxene mineral family and fine-grained basalt. Basalt has a low resistance to weathering so that weathering and mineral transformations in these soils occur quite rapidly.

THE CANADIAN SYSTEM OF SOIL CLASSIFICATION (CSSC)

The Canadian System of Soil Classification (CSSC) in Appendix C provides taxa for all soils presently recognized in Canada and is adapted to Canada's expanses of forest, tundra, prairie, frozen ground, and colder climates. As in the U. S. Soil Taxonomy System, the CSSC classifications are based on observable and measurable properties found in real soils rather than idealized soils that may result from the interactions of genetic processes. The system is flexible in that its framework can accept new findings and information in step with progressive developments in the soil sciences. The system is arranged in a nested, hierarchical pattern to allow generalization at several levels of detail. Elements of the horizon suffix descriptions are derived from the Soil Taxonomy although adapted to conditions in the Canadian environment.

Appendix C, Table 1 is a summary derived from several Canadian publications listed in the suggested readings of Chapter 18. I tried to summarize in an admittedly succinct form a large volume of information. Note that the percentage of land area classified under each soil order and the Soil Taxonomy equivalent is given with each characteristic's section.

Glossary Review for Chapter 18 (in alphabetical order)

- calcification
- cation-exchange capacity (CEC)
- eluviation
- epipedon
- gleization
- humus
- illuviation
- laterization
- loam
- pedogenic regimes
- pedon
- podzolization
- polypedon
- salinization
- soil
- soil colloids
- soil fertility
- soil horizon
- soil science
- Soil Taxonomy
  - Alfisols
  - Andisols
  - Aridisols
  - Entisols
  - Histosols
  - Inceptisols
  - Mollisols
  - Oxisols
  - Spodosols
  - Ultisols
  - Vertisols
- solum
- subsurface diagnostic horizon

Annotated Chapter Review Questions

1. Soils provide the foundation for animal and plant life and therefore are critical to Earth's ecosystems. Why is this true?

Soil is a dynamic natural body comprised of fine materials in which plants grow, and which is composed of both mineral and organic matter. Specific soil conditions determine soil fertility, which is the ability of soil to support plant productivity. Plants capture sunlight and fix carbon in organic compounds that sustain the biosphere.

2. What are the differences among soil science, pedology, and edaphology?

Soil science is interdisciplinary, involving physics, chemistry, biology, mineralogy, hydrology, taxonomy , climatology, and cartography. Pedology concerns the origin, classification, distribution, and description of soil. Pedology is at the center of learning about soils, yet is does not dwell on its practical uses. Edaphology focuses on soil as a medium for sustaining higher plants. Edaphology emphasizes plant growth, fertility, and the differences in productivity among soils. Pedology gives us a general understanding of soils and their classification, whereas edaphology reflects society's concern for food and fiber production and the management of soils to increase fertility and reduce soil losses.

3. Define polypedon and pedon, the basic units of soil.

A soil profile selected for study should extend from the surface to the lowest extent of plant roots, or to the point where regolith or bedrock is encountered. Such a profile, known as a pedon, is imagined as a hexagonal column encompassing from 1 m^2 to 10 m^2 in surface area (Figure 18-1). At the sides of the pedon, the various layers of the soil profile are visible in cross section. A pedon is the basic sampling unit in soil surveys. Many pedons together in one area comprise a polypedon, which has distinctive characteristics differentiating it from surrounding polypedons. These polypedons are the essential soil individuals, constituting an identifiable series of soils in an area. A polypedon has a minimum dimension of about 1 m^2 and no specified maximum size. It is the soil unit used in preparing local soil maps.

4. Characterize the principal aspects of each soil horizon. Where does the main accumulation of organic material occur? The formation of humus? Explain the difference between the eluviated layer and the illuviated layer. Which horizons comprise the solum?

Each layer exposed in a pedon is a soil horizon. A horizon is roughly parallel to the pedon's surface and has characteristics distinctly different from horizons directly above or below. The boundary between horizons usually is visible in the field, using the properties of color, texture, structure, consistence, porosity, the presence or absence of certain minerals, moisture, and chemical processes.

At the top of the soil profile is the O horizon, composed of organic material derived from plant and animal litter that was deposited on the surface and transformed into humus. Humus is a mixture of decomposed organic materials in the soil and is usually dark in color. At the bottom of the soil profile is the R horizon, representing either unconsolidated material or consolidated bedrock of granite, sandstone, limestone, or other rock. The A, B, and C horizons mark differing mineral strata between O and R; these middle layers are composed of sand, silt, clay, and other weathered by-products. In the A horizon, the presence of humus and clay particles is particularly important, for they provide essential chemical links between soil nutrients and plants.

The lower portion of the A horizon grades into the E horizon, which is a bit more pale and is made up of coarse sand, silt, and resistant minerals. Clays and oxides of aluminum and iron are leached (removed) from the E horizon and migrate to lower horizons with water as it percolates through the soil. This process of rinsing through upper horizons and removing finer particles and minerals is termed eluviation; thus the designation E for this horizon. The greater the precipitation in an area, the higher the rate of eluviation that occurs in the E horizon.

Materials are translocated to lower horizons by internal washing in the soil. This process of rinsing through upper horizons and removing finer particles and minerals is termed eluviation–an erosional process. In contrast to the A horizons, B horizons demonstrate an accumulation of clays, aluminum, iron, and possibly humus. These horizons are dominated by illuviation–a depositional process. The C horizon is weathered bedrock or weathered parent material, excluding the bedrock itself. This zone is identified as regolith.

The combination of the A horizon with its eluviation removals and the B horizon with its illuviation accumulations is designated the solum, considered the true soil of the pedon. The A and B horizons are most representative of active soil processes.

5. Soil color is identified and compared using what technique?

Color is important, for it sometimes reflects composition and chemical makeup. Soil scientists describe a soil's color by comparing it with a Munsell Color Chart (from the Munsell Color Company). These charts are in a loose-leaf binder, with 175 different colors arranged by hue (the dominant spectral color, such as red), value (degree of darkness or lightness), and chroma (purity and strength of the color saturation, which increases with decreasing grayness). Color is identified by a name and a Munsell notation, and checked at various depths within a pedon.

6. Define a soil separate. What are the various sizes of particles in soil? What is loam? Why is loam regarded so highly by agriculturalists?

Individual mineral particles are called soil separates; those smaller than 2 mm in diameter (0.08 in.), such as very coarse sand, are considered part of the soil, whereas larger particles are identified as pebbles, gravels, or cobbles. Figure 18-4 shows a diagram of soil textures with sand, silt, and clay concentrations. The figure includes the common designation loam, which is a mixture of sand, silt, and clay in almost equal shares. A sandy loam with clay content below 30% usually is considered ideal by farmers because of its water-holding characteristics and ease of cultivation.

7. What is a quick, hands-on method for determining soil consistence?

A corollary to texture and structure is soil consistence, which is the cohesion in soil and its resistance to mechanical stress and manipulation under varying moisture conditions. Wet soils are variably sticky when held between the thumb and forefinger, ranging from a little adherence to either finger, to sticking to both fingers, to stretching when the fingers are moved apart. Plasticity, the quality of being molded, is roughly measured by rolling a piece of soil between your fingers and thumb to see whether it rolls into a thin strand. Moist soil implies that it is filled to about half of field capacity, and its consistence grades from loose (noncoherent), to friable (easily pulverized), to firm (not crushable between thumb and forefinger).

8. Summarize the five soil moisture regimes common in mature soils.

Soil moisture regimes and their associated climate types shape the biotic and abiotic properties of the soil more than any other factor. Based on Thornthwaite's water-balance principles (Chapter 9), the U.S. Soil Conservation

Service recognizes five soil moisture regimes. Please see Table 18-3, for a description of each of these five regimes.

9. What are soil colloids? How do they relate to cations and anions in the soil? Explain cation-exchange capacity.

Soil colloids are important for retention of ions in soil. These tiny clay and organic particles carry a negative electrical charge and consequently are attracted to any positively charged ions in the soil. Clay colloids and organic colloids exhibit different levels of chemical activity. Individual clay colloids are thin and platelike, with parallel surfaces that are negatively charged. Colloids can exchange cations between their surfaces and the soil solution, a measured ability called cation-exchange capacity (CEC). A high CEC means that the soil colloids can store or exchange more cations from the soil solution, an indication of good soil fertility (unless there is a complicating factor, such as the soil being too acid). Cations attach to the surfaces of the colloids by adsorption; that is, the metallic cations are adsorbed by the soil colloids. A cation is a positively charged ion and an anion is a negatively charged ion.

10. What is meant by the concept of soil fertility?

Soil fertility is the ability of soil to sustain plants. Soil is fertile when it contains organic substances and clay mineral that absorb certain elements needed by plants.

11. Briefly describe the contribution of the following factors and their effect on soil formation: parent material, climate, vegetation, landforms, time, and humans.

The role of parent material in providing weathered minerals to form soils is important in establishing the basic mineral structure and character of the developing soil. At the bottom of the soil profile is the R horizon, representing either unconsolidated material or consolidated bedrock of granite, sandstone, limestone, or other rock. When bedrock weathers to form regolith, it may or may not contribute to overlying soil horizons.

Worldwide, soil types show a close correlation to climate types. The moisture, evaporation, and temperature regimes associated with varying climates determine the chemical reactions, organic activity, and eluviation rates of soils. Not only is the present climate important, but many soils also exhibit the imprint of past climates, sometimes over thousands of years.

The organic content of soil is determined in part by the vegetation growing in that soil, as well as by animal and bacterial activity. The chemical makeup of the vegetation contributes to the acidity or alkalinity of the soil solution. For example, broadleaf trees tend to increase alkalinity, whereas needleleaf trees tend to produce higher acidity.

Landforms also affect soil formation, mainly through slope and orientation. Slopes that are too steep do not have full soil development, but slopes that are slight may inhibit soil drainage. As for orientation, in the Northern Hemisphere, a southern slope exposure is warmest (slope faces the southern Sun), which affects water balance relationships.

All of these identified factors require time to operate. A few centimeters' thickness of prime farmland soil may require 500 years for maturation. Yet these same soils are being lost at a few centimeters per year to sheetwash and gullying produced by human abuse of the soil resource. The GAO estimates that between 3 and 5 million acres of prime farmland are being lost each year in the United States directly because of poor practices.

12. Explain some of the details supporting concern for the loss of our most fertile soils. What cost estimates have been placed on soil erosion?

Much effort and many dollars are expended to create fertile soil conditions, yet we live in an era when the future of Earth's most fertile soils is threatened. Soil erosion is increasing worldwide. Some 35% of farmland is losing soil faster than it can form–a loss exceeding 22.75 billion metric tons per year (25 billion tons). Increases in production resulting from artificial fertilizers and new crop designs partially mask this effect, but such compensations for soil loss are nearing an end. Soil depletion and loss are at record levels from Iowa to China, Peru to Ethiopia, and the Middle East to the Americas. The impact on society could be significant. One 1995 study tabulated the market value of lost nutrients and other variables at over $25 billions of a year in the United States and hundreds of billion dollars worldwide. The cost to bring soil erosion under control in the United States is estimated at approximately $8.5 billion, or about 30 cents on every dollar of damage and loss.

13. Summarize the brief history of soil classification described in this chapter. What lead soil scientists to develop the new Soil Taxonomy classification system?

In 1883, the Russian soil scientist V. V. Dukuchaev published a monograph that organized soils into rough groups based on observable properties, most of which resulted from climatic and biological soil-forming processes. The Russians contributed greatly to modern soil classification because they were first to consider soil an independent natural body with a definite individual soil genesis that was recognizable in an orderly global pattern. By contrast, American and European scientists were still considering soil as a geologic product, or simply a mixture of chemical compounds.

The Russians were first to demonstrate broad interrelationships among the physical environment, vegetation, and soils, at a time when a scientific revolution was beginning to sweep their country in various fields. The first formal American system is credited to Dr. Curtis F. Marbut, who derived his approach from Dukuchaev. Published in the 1935 *Atlas of American Agriculture* and the *Yearbook of Agriculture*–1938, Marbut's classification of great soil groups recognized the importance of soils as products of dynamic natural processes. In 1951, the U.S. Soil Conservation Service began researching a new soil classification system. That process went through seven exhaustive reviews, culminating in what is called the *Seventh Approximation* in 1960 and 1964.

Canadian efforts at soil classification began in 1914 with the partial mapping of soils in Ontario by A.J. Galbraith. Efforts to develop a taxonomic system spread across the country, anchored by academic departments at universities in each province. Regional differences emerged, hampered by a lack of specific soil details. By 1936 only 1.7% of Canadian soil, totaling 15 million hectares, was surveyed.

14. What is the basis of the Soil Taxonomy system? How many orders, suborders, great groups, subgroups, families, and soil series are there?

The current version of the Soil Taxonomy system was published in 1975: *Soil Taxonomy–A Basic System of Soil Classification for Making and Interpreting Soil Surveys*, is generally called simply Soil Taxonomy. Much of the information in this chapter is derived from that keystone publication.

The classification system divides soils into six categories, creating a hierarchical sorting system (Table 18-4). Each soil series (the smallest, most detailed category) ideally includes only one polypedon, but may include portions continuous with adjoining polypedons in the field. Soil orders = 10; suborders = 47; great groups = 230; subgroups = 1200; families = 6000; and soil series = 13,000.

15. Define an epipedon and a subsurface diagnostic horizon. Give a simple example of each.

Two diagnostic horizons may be identified in the solum: the epipedon and the subsurface. The epipedon (literally, over the soil) is the diagnostic horizon that forms at the surface and may extend through the A horizon, even including all or part of an illuviated B horizon. The epipedon is visibly darkened by organic matter and sometimes is leached of minerals. There are six recognized epipedons. See Table 18-5 for an example. A diagnostic horizon often reflects a physical property (color, texture, structure, consistence, porosity, moisture), or a dominant soil process in a pedon.

The second type of diagnostic horizon is the subsurface diagnostic horizon. It originates below the surface at varying depths and may include part of the A and/or B horizons. Many subsurface diagnostic horizons have been identified. See Table 18-5 for an example.

In the Canadian System soil horizons are named and standardized as diagnostic in the classification process. Several mineral and organic horizons and layers are used in the CSSC. Three *mineral horizons* are recognized by capital letter designation, followed by lowercase suffixes for further description. Principal soil-mineral horizons and suffixes are presented in Appendix C.

16. Locate each soil order on the world map as you give a general description of it.

Utilize Table 18-6, for an integrated overview of the ten major soil orders and their descriptions, characteristics, areal distribution, former equivalent name and Canadian equivalent, and pronunciation. Figure 18-9 presents the worldwide distribution of all eleven soil orders. Small locator maps are also presented for six of the orders as follows: Figure 18-10 (Oxisols), Figure 18-15 (Mollisols), Figure 18-19 (Alfisols), Figure 18-13 (Aridisols), Figure 18-20 (Ultisols), Figure 18-21 (Spodosols), and Figure 18-24 (Vertisols). These smaller maps were prepared from Figure 18-9 so that students will be able to relate the individual soil order depicted back to the worldwide pattern.

17. How was slash-and-burn shifting cultivation, as practiced in the past, a form of crop and soil rotation and conservation of soil properties?

Earlier slash-and-burn shifting cultivation practices were adapted to equatorial and tropical soil conditions and formed a unique style of crop rotation. The scenario went like this: people in the tropics cut down (slashed) and burned the rain forest in small tracts, cultivated the land with stick and hoe, and planted maize (corn), beans, and squash. After several years the soil lost fertility, and the people moved on to the next tract to repeat the process. After many years of movement from tract to tract, the group returned to the first patch to begin the cycle again. This practice protected the limited fertility of the soils somewhat, allowing periods of recovery to follow active production. However, this orderly native pattern of land rotation was halted by the invasion of foreign plantation interests, development by local governments, vastly increased population pressures, and conversion of vast new tracts to pasturage.

18. Describe the salinization process in arid and semiarid soils. What associated soil horizons develop?

A soil process that occurs in Aridisols and nearby soil orders is salinization. Salinization results from excessive POTET rates in the deserts and semiarid regions of the world. Salts dissolved in soil water are brought to surface horizons and deposited there as surface water evaporates. These deposits appear as subsurface salic horizons, which will damage and kill plants when they occur near the root zone. Obviously, salinization complicates farming in Aridisols. The introduction of irrigation water may either water log poorly drained soils or lead to salinization. Nonetheless, vegetation does grow where soils are better drained and lower in salt content.

19. Which of the soil orders are associated with Earth's most productive agricultural areas?

Mollisols (grassland soils) are some of Earth's most significant agricultural soils. There are seven recognized suborders, not all of which bear the same degree of fertility. The dominant diagnostic horizon is called the mollic epipedon, which is a dark, organic surface layer some 25 cm (10 in.) thick. As the Latin name implies, Mollisols are soft, even when dry, with granular or crumbly peds, loosely arranged when dry. These humus-rich organic soils are high in base cations (calcium, magnesium, and potassium) and have a high CEC.

20. What is the significance to plants of the 51 cm (20 in.) isohyet in the Midwest relative to soils, pH, and lime content?

In North America, the Great Plains straddle the 98th meridian, which is coincident with the 51 cm (20 in.) isohyet of annual precipitation–wetter to the east and drier to the west. The Mollisols here mark the historic division between the short- and tall-grass prairies. The relationship among Mollisols, Aridisols (to the west), and Alfisols (to the east) is shown in Figure 18-17. The illustration also presents some of the important graduated changes that denote these different soil regions, including the level of pH concentration and the depth of available lime.

21. Describe the podzolization process associated with northern coniferous forest soils. What characteristics are associated with the surface horizons? What strategies might enhance these soils?

The Spodosols (northern coniferous forest soils) and their four related suborders occur generally to the north and east of the Alfisols. Spodosols lack humus and clay in the A horizons. An eluviated albic horizon, sandy and leached of clays and irons, lies in the A horizon instead and overlies a spodic horizon of illuviated organic matter and iron and aluminum oxides. The surface horizon receives organic litter from base-poor, acid-rich trees, which contribute to acid accumulations in the soil. The low pH (acid) soil solution effectively removes clays, iron, and aluminum, which are passed to the upper diagnostic horizon. An ashen-gray color is common in these subarctic forest soils and is characteristic of a formation process called podzolization. The low base-cation content of Spodosols requires the addition of nitrogen, phosphate, and potash, and perhaps crop rotation as well, if agriculture is to be attempted. A soil amendment such as limestone can significantly increase crop production in these acidic soils.

Podzolic (Russian, podzol)(25 subgroups) Soils formed in association with the conditions of coniferous forests and sometimes heath. Leaching of overlying horizons occurs in response to moist, cool-to-cold climates. Iron, aluminum, and organic matter from L, F, and H horizons are redeposited in podzolic B horizon.

22. What former Inceptisols now form a new soil order? Describe these soils: location, nature, and formation processes. Why do you think they were separated into their own order?

Andisols formerly were considered under Inceptisols and Entisols, but in 1990 they were placed in this new order. Andisols are derived from volcanic ash and glass. Previous soil horizons frequently are found buried by ejecta from repeated eruptions. Volcanic soils are unique in their mineral content and in their recharge by eruptions.

Weathering and mineral transformations are important in this soil order. Volcanic glass weathers readily into allophane (a noncrystalline aluminum silicate clay mineral that acts as a colloid) and oxides of aluminum and iron. Andisols feature a high CEC, high water-holding ability, and develop moderate fertility, although phosphorus availability is an occasional problem.

23. Why has a selenium contamination problem arisen in western soils? Explain the impact of agricultural practices, and tell why you think this is or is not a serious problem.

(See Focus Study 18-1) About 95% of the irrigated acreage in the United States is west of the 98th meridian. This region is increasingly troubled by salinization and water-logging problems. But at least nine sites in the West, particularly in California's western San Joaquin Valley, are experiencing related contamination of a more serious nature–increasing selenium concentrations. The soils in these areas were derived from former marine sediments that formed shales in the adjoining Coast Ranges. As parent materials weathered, selenium-rich alluvium washed into the semiarid valley, forming the soils that needed only irrigation water to become productive. After 1960, large-scale irrigation efforts intensified, resulting in subtle initial increases in selenium concentrations in soil and water. In trace amounts, selenium is a dietary requirement for animals and humans, but in higher amounts it is toxic to both. Toxic effects were reported during the 1980s in some domestic animals grazing on grasses grown in selenium-rich soils in the Great Plains.

According to U.S. Fish and Wildlife Service scientists, the toxicity moves through the food chain and genetically damages and kills wildlife. For example, birth defects and death were widely reported in all varieties of birds that nested at selenium-contaminated Kesterson Wildlife Refuge; approximately 90% of the exposed birds perished or were injured. Such damage to wildlife presents a real warning to human populations at the top of the food chain. At the very least we must acknowledge that soil processes are complex. Certainly, much remains to be learned to avoid environmental tragedies such as the one at Kesterson–remember that there are nine such threatened sites in the West, and Kesterson was only the first catastrophe.

Overhead Transparencies

As an adopter you are provided with the following figures for overhead projector use.

- Figure 18-1: Soil sampling and mapping units
- Figure 8-4: Soil texture triangle
- Figure 18-5: Soil texture types
- Figure 18-9: Worldwide distribution of the 11 soil orders
- Mounted together: Figure 18-11, Laterization process in Oxisols, Figure 18-18 Calcification process in Aridisol/Mollisol soils, and Figure 18-21c Podzolization process, typical in cool and moist climatic regimes
- Appendix C, Figure 1: Principal soil regions of the Canadian System of Soil Classification

19

Ecosystem Essentials

Overview

The interaction of the atmosphere, hydrosphere, and lithosphere produces conditions within which the biosphere exists. Chapter 19 begins the process of synthesizing all these "spheres" from throughout *GEOSYSTEMS* into a complete spatial picture of Earth that culminates in Chapter 20. In this complex age, the spatial tools of the geographic approach are uniquely suited to unravel the web of human impact on Earth's systems. Many potential career opportunities are based on a degree in this field, which can lead to planning, GIS applications, environmental impact assessment, and location analysis vocations. The same type of geographic synthesis at a site or environmental work-up is the approach of landscape architecture.

The biosphere extends from the floor of the ocean to a height of about 8 km (5 mi) into the atmosphere. The biosphere is composed of myriad ecosystems from simple to complex, each operating within general spatial boundaries. Ecology is the study of the relationships between organisms and their environment and among the various ecosystems in the biosphere. Biogeography, essentially a spatial ecology, is the study of the distribution of plants and animals and the diverse spatial patterns they create across Earth.

We are the species with developed technology that allows us to lift off the surface of our home planet despite the protests of gravity, to view Earth from afar. At a moment when these accomplishments fill us with pride, we also find ourselves overwhelmed with the immensity of the home planet and the smallness of our everyday reality, as revealed to us by photographs of Earth–such as the one on the back cover of the *GEOSYSTEMS* textbook. A proper understanding of this chapter on global ecosystems is best fueled by an inner picture of our Earth stimulated by the one taken from space, for it is only from that perspective that the importance of even the smallest community, or the most finite ecosystem, will arise in proper significance. And, from such awareness, even the loss of a single species will be a headline and a topic of concern to us all.

The *Student Study Guide* presents 20 "Learning Objectives" to guide the student in reading the chapter.

New to the Third Edition

(Note: This section highlights major changes, new features, and additions in the third edition. This does not describe all the rewrite and recast of the text.)

1. A list of key learning concepts begins the chapter.

2. Description of plants as food, medicines, and chemical compounds, recognizing the role plants play in sustaining the health of our ecosystem.

3. Figure 19-2, illustrates a community of plants and animals near the edge of the forest.

4. Photos of specific niches of a variety of plants and animals; Figure 19-4 (a) Elephant heads (a wildflower), (b) a eastern yellow bellied racer, (c) a dragonfly,

(d) killdeer chicks, and (e) coral mushrooms.

5. A satellite image of net primary productivity is attached to Figure 19-9.

6. Limiting factors that inhibit biotic operations have been put into a bullet format for clarity.

7. News Report #1: "Earth's Magnetic Field-An Abiotic Factor" summarizes studies which have shown that birds, bees and sea turtles use magnetic fields for direction and navigation.

8. News Report #2: "Humans Dump Carbon into the Atmosphere". This report provides recent statistics concerning the amount of carbon dioxide which is released into the atmosphere each year, and the sources of this pollution.

9. Figure 19-14 illustrates the nitrogen cycle.

10. New illustrations of energy, nutrient and food pathways in the environment are shown in Figure 19-16.

11. A terrestrial food chain for the Merlin Hawk is illustrated in Figure 19-17a , and an aquatic food chain for the Mako shark is illustrated in Figure 19-17b.

12. The description of food chain efficiency, including the conversion of energy from grain to beef, has been reworded for clarity.

13. The concept of sustainable agriculture has been introduced as an ecologically compatible use of the environment.

14. News Report #3: "Experimental Prairies Confirm the Importance of Biodiversity." This report summarizes findings that biological diversity in an ecosystem results in greater stability, productivity and soil nutrient use within an ecosystem.

15. Focus Study 1, entitled "The Great Lakes" is a description of the environmental history of the Great Lakes Ecosystems. This focus study is a superb holistic evaluation of the human impact upon the stability of the Great Lakes Ecosystems. It examines human use of the lakes, the source of lake contaminants, and how the physical structure of the individual lakes determine the ecological impact of such activities.

16. Focus Study Figures 1, 2, and 3 illustrate average depths, surface levels and surrounding land use of the Great Lakes Region, respectively.

17. Discussion of fire ecology has been added to this chapter. The impact of urban sprawl into wildlands is addressed as a key factor in the 1991 Berkeley, CA fires.

18. A photo of Yellowstone National Park after the 1988 fire is presented as an example of succession recovery in Figure 19-26.

19. A new summary and review section ends the chapter.

Key Learning Concepts

1. *Define* ecology, biogeography, and the ecosystem concept.

2. *Describe* communities, habitats, and niches.

3. *Explain* photosynthesis and respiration and *derive* net photosynthesis and the world pattern of net primary productivity.

4. *List* abiotic ecosystem components and *relate* those components to ecosystem operations.

5. *Explain* trophic relationships in ecosystems.

6. *Define* succession and *outline* the stages of general ecological succession in both terrestrial and aquatic ecosystems.

Expanded Outline Discussion

The following headings (boldfaced) match some of the first, second, and third order headings in Chapter 19. The narrative under each heading contains information, sources, and anecdotal facts relating to portions of the chapter. Not all text headings are discussed.

ECOSYSTEM COMPONENTS AND CYCLES

Humans need to be drawn into a discussion of ecosystems and the components and cycles of the biosphere. This will add relevance for the learner and enhance an interest in the subject. Years ago, I gave an environmental slide talk to a service club. The fellow sitting next to me said that "business has been more difficult since they passed that ecology stuff." He seemed to think that ecology was an act of Congress or some legislature. I told him that ecology was not a law; rather, it is a fact of life, and further, that he is ecology–a mass of it. As Gilbert White and Mostafa Tolba remind us in the quote on page 586: "The time is ripe to step up and expand current efforts to understand the great interlocking systems of air, water, and minerals nourishing Earth." And, here we are equipped with the powerful perspective and tools of physical geography to accomplish this task!

Naked without modern conveniences, a human could survive foraging in some immediate proximity to the equator, where climates are winterless and food abundant. Survival is enhanced by high productivity levels in that environment. Yet humans occupy a wide variety of habitats worldwide, surviving in places unlike the tropical rain forests. We are ecological "generalists," detached from the environment through our wits and application of our abilities. We draw from Earth resources and subsidize our food, shelter, and clothing needs, spreading forth to every corner of Earth. Humans therefore occupy "artificial habitats," and function within a multitude of "artificial niches." Perhaps it is our detachment from most any limiting factors that creates the tremendous cost human societies are exacting from Earth.

This chapter begins with a description of basic communities and the position and operation of living species–habitat and niche–and a suggestion that the symbiosis we observe in nature is analogous to human-Earth interrelationships. Remember from Chapter 1 that the biosphere occupies that point of overlap among the atmosphere, hydrosphere, and lithosphere.

Communities

This section contains important definitions for community, habitat, niche, competitive exclusion, mutualism, and symbiosis.

Plants: The Essential Biotic Component

An important way to begin the discussion of plants is to describe photosynthesis and leaf activity. Note the changes in the equations of photosynthesis and respiration on p. 586. Photosynthesis is followed by respiration and the derivative net photosynthesis which produces net primary productivity for an entire community (Table 19-1). From this the net dry weight of organic material–the biomass–feeds the food chain. This edition includes a satellite image of vegetation density in Figure 19-9b, which complements Figure 19-9a, net primary productivity, and offers students with insight as to methods which will be used in the future to evaluate deforestation and crop yields.

Beginnings

Earth's "evolutionary atmosphere" is discussed in Chapter 2 as a time of great change and the beginning of the biosphere (See Table 2-1, "Earth's Past Atmospheres"). The surface environment was cooling and stabilizing and liquid water was accumulating in lower sections of the crust. Complex chemical reactions were underway, shielded by these new surface waters. Approximately 3.6 billion years ago, chemosynthetic bacteria began producing organic materials from inorganic elements in a non-oxygen atmosphere. The early cells did not have organized genetic material (chromosomes). The "living atmosphere" is designated as beginning during the slow transition from bacterial action to the first photosynthetic reactions within simple plants, about 3.3 billion years ago. Early living cells soon developed an ability to deal with the evolving supply of oxygen gas (O_2). These earliest plant forms have left their trace in the fossil record.

Land plants and land animals became common about 430 million years ago (end of the Silurian Period), with plants as sophisticated as a tree dating to about 400 million years ago (beginning of the Devonian Period). Swampy forests formed in the Carboniferous Period (280 to 345 million years ago) and, as we learned in Chapter 11, the continents were migrating toward the collision that formed Pangaea. The equatorial location of this continental mass is important, for it placed these lands within the influence of warm and wet equatorial climates (*Af rain forest* climates).

Conifers (gymnosperms, or cone bearing trees in which seeds are not enclosed in an ovary structure) date from the Permian Period (280 million years ago to 225 million years ago), and were followed by the flowering plants (angiosperms, or those plants with seeds enclosed in an ovary or fruit) which date from the middle Cretaceous Period (110 m.y. ago). Angiosperms quickly succeeded and are the dominant type of plant on Earth today. The animal kingdom at this time was dominated by a diverse assemblage of reptiles from very small to the largest animals which ever lived on Earth, filling every conceivable niche and habitat.

Approximately two billion years ago, simple splitting of cells in which genetic information is equally distributed in two daughter cells began, a process known as *mitosis*. The rise of oxygen levels in the atmosphere coincides with the appearance of more complex chromosomes and nuclear material within the cell. Approximately one billion years ago, *meiosis*, or the more complicated sexual reproduction, involving special reproductive cells, appeared in primitive plants and animals. Male and female identities to plant pollen grains (male) and eggs in an ovarian base (female) in flowers mark the dramatic arrival of sex in plants.

In thinking this through, imagine that at some time in the past every carbon atom in your body, or oxygen atom in that breath you just inhaled, cycled through a plant somewhere on the face of Earth! Life is remarkable in its ability to transmit identity and function across the span of Earth's history; some algae cells today are quite similar to the fossil remains from the beginning of the living atmosphere.

Abiotic Ecosystem Components

The chapter divides nonliving and living, abiotic and biotic, ecosystem components. Figure 19-10 should prove to be valuable in illustrating the general relationship among temperature, precipitation, and vegetation types and communities. This illustration is in your overhead packet along with Figure 19-11 which shows the zonation of plants with altitude. In reading about Alexander von Humboldt, we can only wonder what hiking about the Andes of Peru–the conditions, scientific arrangements, and accommodations–was like in the 1790s. The spatial distribution of plants with changing physical conditions by latitude and altitude demonstrates the interrelationships between the abiotic and living systems.

Light, Temperature, Water, and Climate

Sunflowers rotate their heads and actually track the Sun across the sky. Poinsettias flower when days are short, whereas wheat flowers when days are long and nights are short. It may seem surprising, but crops grown in arctic latitudes respond well to the long summer days. Cabbages and other vegetables grow to larger sizes than those grown at lower latitudes. Certain crops actually mature more rapidly during the longer days: beans grow faster in Alaska than in Texas. Of course, growing seasons are much shorter and the threat of frost is more prevalent.

Photosynthetic rates are only slightly affected by temperature; a range of temperatures between 10°C to 35°C is optimum for plant productivity. However, photosynthesis is sharply reduced at high air and soil temperatures, and temperatures do enhance seed germination and fruit ripening. The duration of temperature exposure seems to be more important than minimum or maximum extremes. As you move poleward, the diurnal (daily) and seasonal variability grows in magnitude and becomes an increasingly important factor. Also, mountainous topography, both in slope and elevation, is inhibiting to plant growth due to colder temperatures and exposure variability. Plants in colder temperature regions have lower photosynthetic rates and are able to function at temperatures which approach freezing. Remember though, that below 4°C (39°F), water begins to expand as it cools and upon freezing can break cell walls and damage plants. As temperatures increase, photosynthetic reaction rates increase accordingly, although leaf temperatures over 43°C (109°F) are deadly to most plants. Temperature is an important influence on physiological plant processes, including respiration. Remember, increases in respiration reduce net productivity. Respiration releases carbon dioxide to the environment, causing one of the impacts of an enhanced global warming due to the "greenhouse effect," an increase in the atmosphere's carbon dioxide load from plant respiration.

The discussion in Chapter 9 defining the principal components of a water budget made clear the important role of water in the environment. Potential evapotranspiration represents an accurate analog of potential plant growth if adequate water is available. Farmers must manage water carefully both in quantity and timing to maximize net productivity in their crops. Plants use water to carry sugars, dissolved nutrients, and wastes throughout plant tissues. Typically, plants are about 70 percent water by weight.

Gaseous and Sedimentary Cycles

Figure 19-13 and the discussion of the carbon budget of the atmosphere and ocean can be tied back to Chapters 3, 5, and 10, and to global climate change. Table 5-2 shows the increasing percentage of CO_2 in the lower atmosphere from past to present, and gives estimates for the future. CO_2 is presently increasing in concentration at the rate of 0.4% per year. This increase appears sufficient to override any natural climatic tendencies toward a cooling trend, as well as to produce possible unwanted global warming in the next century.

All of life releases carbon dioxide as part of cellular respiration; this includes the work of the decomposers, combustion of fossil fuels, and releases from outgassing thermal features. Plants use this carbon dioxide in photosynthesis to form carbohydrates. Concentrations of carbon dioxide in the atmosphere vary seasonally and at differing elevations above the field as a function of plant operations. As with light and temperatures, there are optimum concentrations for CO_2; in fact, increases in CO_2 appear to increase the light saturation tolerance level of most plants and the efficiency of light utilization. These improvements in response to light brought on by higher CO_2 levels increase net productivity. The present level of CO_2 in the atmosphere (0.036 percent), as discussed in Chapter 3, is approximately one-fourth of the level needed for maximum photosynthesis. At present, carbon dioxide enhancement is in commercial use in greenhouses in some countries to increase production of high-priced vegetables and nursery stock. Yet increased outside atmospheric temperatures associated with heightened levels of greenhouse gases will also increase respiration rates, which would offset most of the new net plant productivity increase. It is estimated that a 1C° rise in temperature can alter rates of respiration by as much as 10 to 30 percent.

If society's use of the remaining 4.4 trillion tons of fossil fuel reserves over the next 100 years doubles, then atmospheric carbon dioxide levels will jump by 500 ppm. When added to the present 360 ppm, it will result in a carbon dioxide level of 860 ppm. This is a staggering amount when compared with past carbon dioxide levels.

Relative to biogeochemical cycles see: Samuel S. Butcher, *et al.* eds., *Global Biogeochemical Cycles*, San Diego: Academic Press, Inc., 1992. The edited volume is a thorough discussion of all aspects of these cycles: beginnings and evolution, modeling the cycles, analysis of systems, effects of tectonics, discussion of specific systems of soils, oceans, atmosphere, and specific cycles of carbon, nitrogen, sulfur, phosphorus, trace metals, and human impacts on all cycles.

Biotic Ecosystem Operations

This section builds up to Figure 19-21 and Odum's community metabolism study of Silver Springs, Florida, which I have redrawn here in a vertical arrangement. The human position as top carnivore can be pointed out as the ecological relationships of these food chains are discussed.

Concentration in Food Chains

Ever since Rachel Carson's book, *The Silent Spring*, appeared in 1962, there has been growing concern about the biological amplification of chemicals from pesticides, herbicides, and toxic wastes in the food chain. And even though DDT has been banned for almost 20 years, it has been legal for export since March 1981 following a presidential lifting of the ban on exporting banned products. Of course, such long-lasting chemicals end up back in the United States with imported vegetables and other food products. We really are all in this together.

A fascinating and enlightening exercise is to go back and read the newspaper and news magazine articles of 1960-62 to see the criticisms by corporate interests heaped upon Rachel Carson's research and writing. A CBS television documentary "The Silent Spring of Rachel Carson" is still available. You find that the same scripts the chemical and petroleum industries used against her are being used today–playing on the theme of uncertainty and lower-threshold risk assessment myths. Imagine how an improved public learning curve would impact these repeated old arguments against acting for public trust and safety!

Not only that, but a by-product of the application of a chemical in the first place simplifies the ecosystem and reduces population diversity, which in itself introduces a level of instability in the particular ecosystem. See: William M. Stigliani, *et al.*, "Chemical Time Bombs–Predicting the Unpredictable," *Environment*, 33, no. 4, May 1991: 4-9+. Note the graph on page 7 showing Herman Daly's calculation of U.S. per capita gross national product (today assessed as the gross domestic product) adjusted for the costs of environmental damages!

An additional disturbing fact is the inadequate testing given many of the chemicals in use. A government survey in 1985 showed that 62% of agricultural chemicals had never been tested for cancer-causing potential in humans, 60-70% were not tested for possible implication in birth defects, and 93% not been tested as a possible cause of genetic mutations in humans. The case can be made that a vast experiment is underway in the environment–one that is testing the basic resilience of communities and ecosystems. I am baffled constantly by the fact that with possible genetic and health effects on babies and the unborn that we do not see more unity among all political factions on these issues of environmental protection!

STABILITY AND SUCCESSION

Ecosystem Stability, Agricultural Ecosystems, Climate Change

The stability of many ecosystems is under stress due to increasing temperatures. The key question for scientists is to determine how fast plants can adapt to new conditions in given habitats, or migrate to remain within their specific habitat conditions. A serious question is being asked about the possibility of plant migrations (Figure 19-25). A new volume addresses many aspects of this question of climate change and plant adaptations and migrations–an induced succession. Peters, R. L. and T. E. Lovejoy, eds, *Global Warming and Biological Diversity*. New Haven: Yale University Press, 1992. (ISBN: 0-300-05056-9; phone 1-203-432-0940). Many different species and succession forecasts are presented with maps and discussion. See: Stuart L. Pimm and Andrew M. Sugden, "Tropical Diversity and Global Change," *Science*, 263, no. 5149, February 18, 1994: 933-34. Three maps are presented showing tree turnover rates, rates that may have profound consequences on the number of species in the ecosystem.

See: Edward O. Wilson, *The Diversity of Life*, Cambridge, MA: The Belknap Press of Harvard University, 1992, for insight into the great extinctions, an evolving biodiversity through times past, and the human impact that threatens biodiversity. His book includes a useful glossary and extensive references.

Note the focus study in Chapter 21 on the 1992 Earth Summit and the discussion of biodiversity as an issue and subject of a treaty. Also see *Global Biodiversity Strategy–Guidelines for Action to Save, Study, and Use Earth's Biotic Wealth Sustainably and Equitably*, co-authored by the World Resources Institute, The World Conservation Union, and UNEP, 1992, and available from these organizations.

The January/February 1994 issue of *Nature Conservancy*, dedicated the entire issue to biodiversity. Articles and reports covered a 1993 review, definition of biodiversity, protections strategies, management strategies, and issues of sustainable developments and preserving biodiversity. Available from the Nature Conservancy, 703-841-5300.

Agricultural Biotechnology; Prospects for the Third World, edited by John Farrington, Overseas Development Institute, 1989, describes recent biotechnology advances used to increase food and cashcrop yields in the third world. This source also identifies the role of the U.S.D.A. and major American and European MNC's in biotechnology research.

Ecological Succession; Terrestrial Succession.

A typical secondary succession of principal plants in the southeastern United States, as determined by Odum, is shown in Figure 19-28. The key in succession is the competitive struggle for light, water, nutrients, space, time, reproduction, and survival. An operative principle applies to this struggle: the *principle of competition* or competitive exclusion. When two or more species compete for limited food and resources, the more efficient organism will succeed and the other will fail. Two species in a stable community will occupy a similar niche in different ways that tend to reduce their mutual competition. Species in a community tend to niche-differentiating, complementing one another rather than directly competing, although they continue to interact. The principle of individuality explained in the text holds that each species succeeds in its own way.

Glossary Review for Chapter 19 (in alphabetical order)

biodiveristy
biogeochemical cycles
biogeography
biomass
carnivore
chlorophyll
community
consumers
decomposers
ecological succession
ecology
ecosystems
fire ecology
food chain
food web
habitat
herbivore
life zone
limiting factor
net primary productivity
niche
omnivore
photosynthesis
pioneer community
primary succession
producers
respiration
secondary succession
stomata
vascular plants

Annotated Chapter Review Questions

1. What is the relationship between the biosphere and an ecosystem? Define ecosystem and give some examples.

The interaction of the atmosphere, hydrosphere, and lithosphere produces conditions within which the biosphere exists. This sphere of life and organic activity extends from the floor of the ocean to a height of about 8 km (5 mi) in the atmosphere. The biosphere is composed of myriad ecosystems from simple to complex, each operating within general spatial boundaries. An ecosystem is a self-regulating association of living plants and animals and their nonliving physical environment. Earth itself is an ecosystem within the natural boundary of the atmosphere. Various smaller ecosystems–for example, forests, seas, mountain tops, deserts, beaches, islands, lakes, ponds–make up the larger whole.

2. What does biogeography include? Describe its relationship to ecology.

Biogeography, essentially a spatial ecology, is the study of the distribution of plants and animals and the diverse spatial patterns they create across Earth. Ecology is the study of the relationships between organisms and their environment and among the various ecosystems in the biosphere.

3. Briefly summarize what ecosystem operations imply about the complexity of life.

An ecosystem is a complex of many variables, all functioning independently yet in concert, with complicated flows of energy and matter.
"Life devours itself: everything that eats is itself eaten; everything that can be eaten is eaten; every chemical that is made by life can be broken down by life; all the sunlight that can be used is used. . . . The web of life has so many threads that a few can be broken without mak-

ing it all unravel, and if this were not so, life could not have survived the normal accidents of weather and time, but still the snapping of each thread makes the whole web shudder, and weakens it. . . . . You can never do just one thing: the effects of what you do in the world will always spread out like ripples in a pond." (Friends of the Earth and Amory Lovins, The United Nations Stockholm Conference: *Only One Earth*. London: Earth Island Limited, 1972, p. 20.)

An ecosystem is composed of both biotic and abiotic components. Nearly all depend upon an input of solar energy; the few limited ecosystems that exist in dark caves or on the ocean floor depend upon chemical reactions. Ecosystems are divided into subsystems, with the biotic portion composed of producers, consumers, and decomposers. Gaseous and sedimentary nutrient cycles comprise the abiotic flows in an ecosystem.

4. Define a community within an ecosystem.

A convenient biotic subdivision within an ecosystem is a community, which is formed by interacting populations of living animals and plants in an area. An ecosystem is the interaction of a community with the abiotic physical components of its environment. Many communities are included in an ecosystem. For example, in a forest ecosystem, a specific community may exist on the forest floor, whereas another functions in the canopy of leaves high above. Similarly, within a lake ecosystem, the plants and animals that flourish in the bottom sediments form one community, whereas those near the surface of the lake form another. A community is identified in several ways–by the physical appearance of the community, the number of species and the abundance of each, and the trophic (feeding) structure of the community.

5. What do the concepts of habitat and niche involve? Relate them to some specific plant and animal communities.

Within a community, two concepts are important: habitat and niche. Habitat is the specific physical location of an organism, the place in which it resides or is biologically suited to live. In terms of physical and natural factors, most species have specific habitat parameters (with definite limits) and a specific regimen of sustaining nutrients. Niche refers to the function, or occupation, of a life-form within a given community; it is the way an organism obtains and sustains its living. An individual species must satisfy several aspects in its niche; among these are a habitat niche, a trophic (food) niche, and a reproductive niche.

6. Describe symbiotic and parasitic relationships in nature. Draw an analogy between these relationships and human societies on our planet. Explain.

Some species have symbiotic relationships, or arrangements that mutually benefit and sustain each organism. For example, lichen (pronounced "liken") is made up of algae and fungus. The algae are the producers and food source, and the fungus provides structure and support. Their mutually beneficial relationship allows the two to occupy a niche in which neither could survive alone. Lichen developed from an earlier parasitic relationship in which the fungi broke into algae cells directly. Today the two organisms have evolved into a supportive harmony. Some scientists are questioning whether or not our human society and Earth constitute a global form of a symbiotic relationship.

7. Define a vascular plant. How many species are there on Earth?

Vascular plants have internal fluid and material flowing through their tissues through specialized conducting systems. At present there are almost 250,000 species of them. This great diversity and complexity compounds the difficulty of classification for convenient study.

8. How do plants function to link the Sun's energy to living organisms? What is formed within the light-responsive cells of plants?

The largest concentration of light-sensitive cells rests below the upper layers of the leaf. These are called chloroplast bodies, and within each resides a green, light-sensitive pigment called chlorophyll. Within this pigment, light stimulates photochemistry. Photosynthesis unites carbon dioxide and oxygen (derived from water in the plant) under the influence of certain wavelengths of visible light, subsequently releasing oxygen and producing energy-rich organic material.

9. Compare photosynthesis and respiration and the derivation of net primary photosynthesis. What is the importance of knowing the net primary productivity of an ecosystem and how much biomass an ecosytem has accumulated?

The difference between photosynthetic production and respiration loss is called net photosynthesis. The amount varies, depending on controlling environmental factors such as light, water, temperature, soil fertility, landforms, and the plant's site, elevation, and competition from other plants and animals. The net photosynthesis for an entire community is its net primary productivity. This is the amount of useful chemical energy (biomass) that the community "fixes" in the ecosystem. Biomass is the net dry weight of organic material; it is biomass that feeds the food chain. Net productivity is generally regarded as the most important aspect of any type of community, greater biomass supports greater biodiversity (more trophic levels) which results in greater stability in an ecosystem.

10. Briefly describe the global pattern of net primary productivity?

The net photosynthesis for an entire community is its net primary productivity. Primary productivity is mapped in terms of fixed carbon per square meter per year. See Figure 19-8 for a portrayal of net primary productivity. Table 19-1 lists various ecosystems, their net primary productivity per year, and an estimate of net total biomass (primary production) worldwide. Net primary productivity is estimated at 170 billion tons of dry organic matter per year.

11. What are the principal abiotic components in terrestrial ecosystems?

The pattern of solar energy receipt is crucial in both terrestrial and aquatic ecosystems. Solar energy enters an ecosystem by way of photosynthesis, with heat dissipated from the system at many points. The duration of Sun exposure is the photoperiod. Air and soil temperatures determine the rates at which chemical reactions proceed. Operations of the hydrologic cycle and water availability depend on rates of precipitation/evaporation and their seasonal distribution. Water quality is important–its mineral content, salinity, and levels of pollution and toxicity.

12. Describe what Humboldt found that led him to propose the life-zone concept. What are life zones? Explain the interaction among altitude, latitude, and the types of communities that develop.

Alexander von Humboldt (1769-1859) described a distinct relationship between altitude and plant communities, his life zone concept. As he climbed in the Andean mountains, he noticed that the experience was similar to that of traveling away from the equator toward higher latitudes. Each life zone possesses its own temperature, precipitation, and insolation relationships and therefore its own biotic communities. See Figure 19-11 for specific vegetation examples.

13. What are biogeochemical cycles? Describe several of the essential cycles.

Several important cycles of chemical elements operate. Oxygen, carbon, and nitrogen form gaseous cycles involving atmospheric phases. Other elements form sedimentary cycles principally involved in mineral and solid phases (such as phosphorus, calcium, and sulfur). Some elements combine the two cycles. The various processes are called biogeochemical cycles, indicating the occurrence of chemical reactions in both living (biotic) and nonliving (abiotic) spheres. The chemical elements themselves recycle over and over again in life processes. The carbon and oxygen cycles are illustrated together because they are so closely intertwined through, for example, photosynthesis and respiration (Figure 19-13).

The key link in the nitrogen cycle is provided by nitrogen-fixing bacteria, which live principally in the soil and are associated with the roots of certain plants–for example, legumes such as clover, alfalfa, soybeans, peas, beans, and peanuts. The bacteria reside in nodule colonies on the legume roots and fix the nitrogen from the air in the form of nitrates (NO_3) and ammonia (NH_3). The nitrogen in the organic wastes of these organisms is freed by a different type of bacteria that denitrifies wastes, recycling nitrogen back to the atmosphere.

14. What is a limiting factor? How does it function to control the spatial distribution of plant and animal species?

The term limiting factor identifies the one physical or chemical component that most inhibits biotic operations, through its lack or excess. A few examples include the low temperatures at high elevations, the lack of water in a desert, the excess water in a bog, the amount of iron in ocean surface environments, the phosphorus content of soils in the eastern United States or at elevations above 6100 m (20,000 ft), where there is a general lack of active chlorophyll. In most ecosystems, precipitation is the limiting factor, although variation in temperatures and soil characteristics certainly affect vegetation patterns. Each organism possesses a range of tolerance for each variable in its environment.

15. What role is played in an ecosystem by producers and consumers?

Producers and consumers are part of the food chain within ecosystems. Energy flows from producers, which manufacture their own food, to consumers, who feed on producers or consumers who are at lower trophic levels. Because producers are always plants, the primary consumer is called an herbivore, or plant eater. A carnivore is a secondary consumer and primarily eats meat (primary consumers). A consumer who eats both producers (plants) and consumers (meat) is called an omnivore. A tertiary consumer eats primary and secondary consumers, and is referred to as the "top carnivore" in the food chain.

16. Describe the relationship among producers, consumers, and decomposers in an ecosystem. What is the trophic nature of an ecosystem? What is the place of humans in a trophic system?

From producers, which manufacture their own food, energy flows through the system along a circuit called the food chain, reaching consumers and eventually decomposers. Ecosystems are generally structured as a food web, a complex network of interconnected food chains, in which consumers participate in several different food chains. Primary consumers feed on producers, which are always plants, so the primary consumer is called an herbivore, or plant eater. A carnivore is a secondary consumer and primarily eats meat for sustenance. A tertiary consumer eats secondary consumers and is referred to as the *top carnivore* in the food chain. A consumer that feeds on both producers (plants) and consumers (meat) is called an omnivore–a role occupied by humans, among others. The decomposers are the final link in the chain; they are the microorganisms–bacteria, fungi, insects, worms, and others–that digest and recycle the organic debris and waste in the environment.

17. What are biomass and population pyramids? Describe how these models help explain the nature of food chains and communities of plants and animals.

Figure 19-20 demonstrates the summertime distribution of various populations in specific grassland and temperate forest ecosystems. Such a stepped population pyramid is characteristic of summer conditions in those ecosystems. You can see the decreasing number of organisms at each successive trophic level. The base of the temperate forest pyramid is narrow, however, because most of the producers are large, highly productive trees and shrubs, which are outnumbered by the consumers they can support in the chain.

18. Follow the flow of energy and biomass through the Silver Springs, Florida, ecosystem. Describe the pathways in Figure 19-16 and 19-21a.

See the figures. A basic approach to ecological studies is an analysis of a community's metabolism, the way in which it uses energy and produces food for continued operation.

19. What is meant by ecosystem stability?

In a given ecosystem, a community moves toward maximum biomass and relative stability. However, inertial stability, the tendency for birth and death rates to balance and the composition of species to remain stable, does not necessarily foster resilience, or the ability to recover from change.

20. How does ecological succession proceed? What are the relationships that exist between existing communities and new, invading communities?

Ecological succession occurs when different communities of plants and animals (usually more complex) replace older communities (usually simpler). Each temporary community of species modifies the physical environment in a manner suitable for the establishment of a later set of species. Changes apparently move toward a more stable and mature condition, which is optimum for a specific environment. This end product in an area is called the ecological climax, with plants and animals forming a climax community–a stable, self-sustaining, and symbiotically functioning community with balanced birth, growth, and death.

Succession often requires an initiating disturbance, such as strong winds or a storm, or a practice such as prolonged overgrazing. Early communities may actually inhibit the growth of other species, but when existing organisms are disturbed or removed, new communities can emerge. At such times of transition, the interrelationships among species produce elements of chance, and species having an adaptive edge will succeed in the competitive struggle for light, water, nutrients, space, time, reproduction, and survival. When two or more species compete for limited food and resources, the more efficient organism will succeed and the other will fail in the principle of competition, or competitive exclusion. The principle of individuality holds that each species succeeds in its own way, according to its own requirements, relating to both the physical environment and the complex interactions with other species. Thus, the succession of plant and animal communities is an involved process with many interactive variables.

21. Discuss the concept of fire ecology in the context of the Yellowstone National Park fires of 1988. What were the findings of the government task force?

Over the past 50 years, fire ecology has been the subject of much scientific research and experimentation. Today, fire is recognized as a natural component of most ecosystems and not the enemy of nature it once was popularly considered to be. In fact, in many forests, undergrowth is purposely burned in controlled "cool fires" to remove fuel that could enable a catastrophic and destructive "hot fire." Ironically, when fire suppression and prevention strategies are rigidly followed, they can lead to abundant undergrowth accumulation, which allows the potential total destruction of a forest by a major fire. Fire ecology imitates nature by recognizing fire as a dynamic ingredient in community succession. In its final report on the fire in Yellowstone National Park, a government interagency task force concluded that: "an attempt to exclude fire from these lands leads to major unnatural changes in vegetation . . . as well as creating fuel accumulation that can lead to uncontrollable, sometimes very damaging wildfire." Thus, participating federal land managers and others reaffirmed their stand that fire ecology is a fundamentally sound concept.

22. Summarize the process of succession in a body of water. What is meant by cultural eutrophication?

A lake experiences successional stages as it fills with nutrients and sediment and as aquatic plants take root and grow, capturing more sediment and adding organic debris to the system (Figure 19-31). This gradual enrichment through various stages in water bodies is known as eutrophication. The progressive stages in lake succession are named oligotrophic (low nutrients), mesotrophic (medium nutrients), and eutrophic (high nutrients). Each stage is marked by an increase in primary productivity and resultant decreases in water transparency so that photosynthesis becomes concentrated near the surface. Energy flow shifts from production to respiration in the eutrophic stage, with oxygen demand exceeding oxygen availability. As society dumps sewage and pollution in waterways, the nutrient load is enhanced beyond the cleansing ability of natural biological processes, thus producing cultural eutrophication.

Overhead Transparencies

As an adopter you are provided with the following figures for overhead projector use.

- Figure 19-2: Abiotic and biotic components of ecosystems
- Figure 19-10: How temperature and precipitation affect ecosystems
- Figure 19-11: Vertical and latitudinal zonation of plant communities
- Figure 19-13: Carbon and oxygen cycles, simplified
- Figure 19-15: Limiting factors affect the distribution of plants and animals
- Figure 19-18: An Antarctic food web
- F.S. 19-1, Figure 1 and 2: The Great Lakes Basin and profile of lake elevations

20

Terrestrial Biomes

Overview

Chapter 20 synthesizes many of the elements of *GEOSYSTEMS*, bringing them together to create a regional portrait of the biosphere. To facilitate this process, I prepared Table 20-1 to portray aspects of the atmosphere, hydrosphere, and lithosphere that merge to produce the major terrestrial ecosystems. Beyond the specific description of ten major biomes, this chapter is meant as an overview of *GEOSYSTEMS*. The table is presented in columns covering vegetation characteristics, soil classes, Köppen climate designation, annual precipitation range, temperature patterns, and water balance characteristics.

Focus Study 20-1 presents the biosphere reserve effort to establish protected "islands" of natural biomes. The principal goal is to hold off the record number of extinctions now taking place. Even the most detached individual must recognize that humankind has been an agency of change on Earth–the creators of cultural, albeit artificial, landscapes. The spatial implications of these dynamic trends is of particular importance to physical geographers, for we have the potential for spatial analysis and synthesis. Our discipline is at the heart of geographic information system model construction. The biosphere is quite resilient and adaptable whereas many specific biomes and communities are greatly threatened by further destructive impacts. The irony is that some plants and animals in these biomes contain cures and clues to human disease, potential new food sources, and mechanisms to recycle the excessive levels of carbon dioxide now entering the atmosphere.

The *Student Study Guide* presents 18 "Learning Objectives" to guide the student in reading the chapter. The *Applied Physical Geography* lab manual has one exercise with five steps that involve aspects of this chapter.

New to the Third Edition

(Note: This section highlights major changes, new features, and additions in the third edition. This does not describe all the rewrite and recast of the text.)

1. A list of key learning concepts begins the chapter.

2. The discussion of life form designations is newly organized.

3. The discussion of aquatic ecosystems has been moved to News Report #1: "Aquatic Ecosystems and the LME Concept". This report describes the importance of examining oceanic regions as Large Marine Ecosystems (LMEs). This organizational concept encourages planners to consider complete ecosystems, rather than individual species that may be affected by human activity.

4. Examples of plant and animal life are described for each specific biome.

5. Figure 20-12 illustrates of the maturational stages of Giant Sequoias in the Sierra Nevada. Described as an example of a Needleleaf Forest and Montane Forest biome.

6. An aerial view of the rain forest has been added to Figure 20-5 to illustrate the rain forest canopy. Additionally, a new photo of the Zimbabwe rain forest has been added to Figure 20-6, and a new photo of the burning rain forest has been included in Figure 20-7.

7. News Report #2: "Drilling for Oil in the Rain Forest," describes the amount of oil exploration that has destroyed the Amazonian Rain Forest since 1972.

8. Discussion of the Arctic National Wildlife Refuge (ANWR) has been moved to News Report #3: "ANWR Faces Threats". The history of the ANWR is described, including recent Congressional laws which may allow development of Alaska's Arctic Coast for petroleum exploration to begin.

9. Updated statistics from the United Naitons Environmental Program regarding Known and Estimated Species on Earth, Table 1, Focus Study 1.

10. New photographs to illustrate several biomes: Figure 20-2, the Australian realm, Figure 20-17, the Sonoran desert, and Figure 20-20, Logan Pass, Montanta, an alpine landscape.

11. A new summary and review section ends the chapter.

Key Learning Concepts

1. *Define* the concept of biogeographical realms of plants and animals and *define* ecotone, terrestrial ecosystem, and biome.

2. *Define* six formation classes and the life-form designations and *explain* their relationship to plant communities.

3. *Describe* ten major terrestrial biomes and *locate* them on a world map.

4. *Relate* human impacts, real and potential, to several of the biomes.

Expanded Outline Discussion

The following headings (boldfaced) match some of the first, second, and third order headings in Chapter 20. The narrative under each heading contains information, sources, and anecdotal facts relating to portions of the chapter. Not all text headings are discussed.

BIOGEOGRAPHICAL REALMS

The two maps in Figure 20-1 form the basis of regionalization in biogeography, for they group plants in Botanical Geographic Realms (top) and animals in Zoogeographic Realms (bottom) as to the geographic regions in which they evolved. We are able to assemble a map as complex as Figure 20-4, and other maps of much greater resolution, because of the progress science has made in developing remote sensing techniques and data gathering efforts. The uniqueness of the Australian realm is good to highlight, for not only is it isolated by deep water with no ice-age land bridge, but it also drifted early on from the rest of Pangaea.

Aquatic Ecosystems

Important aspects of salt marshes and mangroves are discussed in Chapter 16 along with additional material on coastal planning and zoning, barrier islands, and a GIS-style analysis of Marco Island, Florida. The brief section on aquatic ecosystems in this chapter should certainly refer back to the coastal chapter. For more on large marine ecosystems (LMEs), consult the book from AAAS, Sherman, Kenneth, Lewis M. Alexander, and Barry D. Golds, Eds. *Large Marine Ecosystems–Patterns, Processes, and Yields*, Washington: American Association for the Advancement of Science, No.

90-30S, 1990. LMEs are the aquatic analogies of biosphere reserves.

Terrestrial Ecosystems

There are several ways to classify terrestrial ecosystems. I describe formation classes and life-form designations in this section. Botanists and biogeographers also may use leaf shape and size, the amount of ground coverage provided by the plant, seasonality of the vegetation, the texture of the leaves, and overall size and height structure of the formation. The book edited by J. A. Taylor, *Biogeography–Recent Advances and Future Directions*, Totowa, NJ: Barnes & Noble, 1984, contains an entire chapter overviewing "vegetation analysis."

One method of assessment of terrestrial ecosystems is a consideration of how plants differ in moisture availability requirements. These variations result in unique characteristics associated with each water-balance regime. These characteristics are important for further understanding of plant communities, formations, and distributions. Plants are classified as follows:

Plant type, moisture, and examples

- Hydrophytes (excessive moisture, seaweed, mangroves, bulrushes.
- Mesophytes (alternating deficit and surplus) common field and forest plants.
- Tropophytes (seasonal vegetation) maple, birch, larch, ash, etc.
- Xerophytes (water deficit and drought) cacti, sagebrush, junipers.

EARTH'S MAJOR BIOMES

I raise the question in the text as to what actually remains of the natural biomes. A map of human-induced artificial landscapes remains to be completed, for we do not have a satisfactory national land inventory in the United States, and there is political and corporate resistance to such an inventory on a global scale. I can only wonder if such fears are real when we see the U.S. *Landsat* budget threatened and a serious suggestion made by the previous administration to shut it down. If geographic analysis on a large scale is to be completed, improvement in remote-sensing capabilities will be of critical importance, such as construction and launch of the various components of an Earth Observation System, or EOS, that will introduce a new era in the monitoring of Earth's systems.

A large, stable terrestrial ecosystem is known as a biome. A *biome* is characterized by specific plant and animal communities and their interrelationship with the physical environment. Each biome is usually named for its *dominant vegetation*. We can generalize Earth's wide-ranging plant species into six broad biomes: *forest*, *savanna*, *grassland*, *shrubland*, *desert*, and *tundra*. We further define these general biomes into more specific vegetation units called *formation classes*. These units refer to the structure and appearance of dominant plants in a terrestrial ecosystem, for example, equatorial rain forest, northern needleleaf forest, Mediterranean shrubland, arctic tundra. Each formation includes numerous plant communities, and each community includes innumerable plant habitats. Within those habitats, Earth's diversity is expressed in 250,000 plant species. Despite this intricate complexity we can generalize Earth's numerous formation classes into 10 global terrestrial biome regions as portrayed in Figure 20-4 and detailed in Table 20-1.

Table 20-1 is meant to synthesize many elements throughout this text, and is organized along the basis of Earth's ten major biomes. The biome map (Figure 20-4) and Table 20-1 should work in concert as the student proceeds through the remainder of the chapter. Figure 20-4 is included in your overhead transparency packet.

Greater description of ecotones is included in this edition. A good example of ecotone research can be found in "Landscape Analysis of the Forest-Tundra Ecotone in Rocky Mountain National Park, Colorado," Wm. L. Baker and Peter J. Weisberg, *The Professional Geographer*, November 1995, pp 361-374.

"The Living Earth" composite image appears inside the front cover of the text. Artist Erik Bruhwiler and scientists at the Jet Propulsion Laboratory worked with hundreds of thousands of satellite images to produce a cloudless view of Earth in the true-colors of a local summer day. Compare the map in Figure 20-4 with this remarkable composite image and see what correlations you can make.

For detailed coverage of North America, you may want to obtain a copy of *North American Terrestrial Vegetation* edited by Barbour, Michael G., and William D. Billings, Cambridge: Cambridge University Press, 1988. From their detailed map of vegetation formations (facing the title page), you will be able to see the degree of generalization used in preparing Figure 20-4.

Four additional references are added to the suggested readings section by Ehrlich; FAO; and two from the World Resources Institute.

There are many great resources concerning biomes and habitat conservation:

Timothy Beatley, *Habitat Conservation Planningl Endangered Species and Urban Growth*, University of Texas Press, 1994.

John Richards, and Richard Tucker,eds. *World Deforestation in the Twentieth Century*, Duke University Press, 1988.

Douglas Ian Stewart, *After the Trees; Living on the TransAmazon Highway*, University of Texas Press, 1994.

Susan Hecht, and Alexander Cockburn, *Fate of the Rainforest*, Harper and Row, 1990.

Glossary Review for Chapter 20 (in alphabetical order)

arctic and alpine tundra
biogeographical realms
biome
boreal forest
chaparral
cold desert and semi desert
desertification
ecotone
equatorial and tropical rain forest
formation classes
Mediterranean shrubland
midlatitude broadleaf and mixed forest
midlatitude grasslands
taiga
needleleaf forest and montane forest
temperate rain forest
terrestrial ecosystem
tropical savanna
tropical seasonal forest and scrub
warm desert and semi desert

Annotated Chapter Review Questions

1. Reread the two opening quotations in this chapter. What clues do you have to the path ahead for Earth's forests? Is our future direction controllable? Explain.

Statement of personal analysis and opinion given the quotes from 1876 and 1992.

2. What is a biogeographical realm? How is the world subdivided according to plant and animal types?

Biogeographical realms of plants and animals are geographic regions where groups of species evolved. From these centers, species migrate worldwide according to their niche requirements, reproductive success, competition, and climatic and topographic barriers. Recognition that such distinct regions of flora and fauna exist was an early beginning of biogeography as a discipline. The map in Figure 20-1 illustrates the botanical (plant) and zoological (animal) regions forming these biogeographical realms. Each realm contains many distinct ecosystems that distinguish it from other realms.

3. Describe a transition zone between two ecosystems. How wide is an ecotone?

The transition zone between two ecosytstems is called an ecotone. Boundaries between natural systems are "zones of shared traits," therefore they are zones of mixed identity and composition, rather than rigidly defined boundaries. A tropical savanna is a good example of an ecotone. Situated between tropical forests and tropical steppes or deserts, tropical savannas are a mixture of trees and grasses. The savanna biome includes treeless tracts of grasslands, and in very dry savannas, grasses grow discontinuously in clumps, with bare ground between them.

4. Define biome. What is the basis of the designation?

A large, stable terrestrial ecosystem is known as a biome. A biome is characterized by specific plant and animal communities and their interrelationship with the physical environment. Each biome is usually named for its dominant vegetation. We further define these general biomes into more specific vegetation units called formation classes. These units refer to the structure and appearance of dominant plants in a terrestrial ecosystem, for example, equatorial rain forest, northern needleleaf forest, Mediterranean shrubland, arctic tundra.

Plant distributions are responsive to environmental conditions and reflect variation in climatic and other abiotic factors. Therefore, it is important to cross reference this discussion with Figure 10-7 and the Köppen classifications.

5. Distinguish between formation classes and life-form designations as a basis for spatial classification.

A community, or association of related species, is formed by interacting populations of plants and animals in an area. Large vegetation units, the floristic component of a terrestrial ecosystem characterized by a dominant plant community, are called plant formation classes. Each formation includes numerous plant communities, and each community includes innumerable plant habitats. Within those habitats, Earth's diversity is expressed in approximately 250,000 plant species.

More specific systems are used for the structural classification of plants. Such *life-form* designations are based on the outward physical properties of individual plants or the general form and structure of a vegetation cover. These physical life-forms, portrayed in Figure 20-3, include *trees* (larger woody main trunk, perennial, usually exceeding 3 m or 10 ft); *lianas* (woody climbers and vines); *shrubs* (smaller woody plants; branching stems at ground); *herbs* (small plants without woody stems above ground); *bryophytes* (mosses, liverworts); *epiphytes* (plants growing above the ground on other plants, using them for support); and *thallophytes*, which lack true leaves, stems, or roots (bacteria, fungi, algae, lichens).

6. Using the integrative chart in Table 20-1 and the world map in Figure 20-4, select any two biomes and study the correlations of vegetation characteristics, soil, moisture, and climate with their spatial distribution. Then contrast the two using each characteristic.

A specific assignment to correlate the integrative table and the world biome map. Also, a valuable correlation can be made to "The Living Earth" composite illustration inside the front cover of the text (see question #19 later).

7. Describe the equatorial and tropical rain forests. Why is the rain forest floor somewhat clear of plant growth? Why are logging activities for specific species so difficult there?

Biomass in a rain forest is concentrated high up in the canopy, that dense mass of overhead leaves with a vertical distribution of life that is dependent on a competitive struggle for sunlight. The canopy is composed of a rich variety of plants and animals. Lianas (vines) branch from tree to tree, binding them together with cords that can reach 20 cm (8 in.) in diameter. Epiphytes flourish there too: such plants as small orchids, bromeliads, and ferns that live completely above ground, supported physically but not nutritionally by the structures of other plants. The floor of the rain forest and the floor of the ocean are roughly parallel in that both are dark or dimly lit, relatively barren, and a place of fewer life-forms–although the rain-forest floor is much livelier than the sea floor. Logging is difficult because individual species are widely scattered; a species may occur only once or twice per square kilometer.

8. What are the issues surrounding deforestation of the rain forest? What is the impact of these losses on the rest of the biosphere? What new threat to the rain forest has emerged?

Burning is more common than logging in deforestation because of the scattered distribution of specific types of trees mentioned earlier. Fires are used to clear land for agriculture, which is intended to feed the domestic population as well as to produce cash exports of beef, rubber, coffee, and other commodities. Every year, approximately 6.1 million hectares (15.25 million acres) are thus destroyed, and more than 4 million hectares (10 million acres) are selectively logged. This total is a loss of 0.6% of tropical rain forest each year, at which rate they will be removed completely in fewer than 180 years if the destruction continues unabated.

The United Nations Food and Agricultural Organization (FAO) estimates that every year approximately 16.9 million hectares (41.7 million acres) are destroyed, and more than 5 million hectares (12.3 million acres) are selectively logged. This total–averaged for the period 1980-1991 in 76 countries that contain 97% of all rain forest–represents a 0.9% loss of equatorial and tropical rain forest worldwide each year (up from the previous average of 0.6%). If this destruction continues unabated, these forests will be completely removed by about 2050 A.D.! By continent, forest losses are estimated at more than 50% in Africa, over 40% in Asia, and 40% in Central and South America. Brazil, Colombia, Mexico, and Indonesia lead the list of lesser-developed countries that are removing their forests at record rates, although these statistics are disputed by the countries in question.

Another threat to the rain forest biome and indigenous peoples emerged in 1991: exploration for and development of oil reserves. U. S. oil corporations are going ahead in Yasuni National Park, Ecuador (near the equator at 77° W) with road building and drilling. One estimate of the ultimate petroleum reserve there is 1.5 billion barrels, or enough to satisfy about three months of the U.S. demand. Similar projects are being considered in Peru.

The present human assault on Earth's rain forests has put this diverse fauna and the varied flora at risk. It also jeopardizes an important recycling system for atmospheric carbon dioxide, sources of new forms of food, and potential sources of valuable pharmaceuticals and medicines.

9. What do caatinga, chaco, brigalow, and dornveld refer to? Explain.

Local names are applied to the tropical seasonal forest and scrub on the margins of the rain forest: the caatinga of northeast Brazil, chaco area of Paraguay and northern Argentina, the brigalow scrub of Australia, and the dornveld of southern Africa.

10. Describe the role of fire or fire ecology in the tropical savanna biome and the midlatitude broadleaf and mixed forest biome.

Savannas covered more than 40% of Earth's land surface before human intervention but were especially modified by human-caused fire. Elephant grasses averaging 5 m (16 ft) high and forests are assumed to have once penetrated much farther into the dry regions, for they are known to survive there when protected from fire.

11. Why does the northern needleleaf forest biome not exist in the Southern Hemisphere? Where is this biome located in the Northern Hemisphere, and what is its relationship to climate type?

Stretching from the east coast of Canada and the Maritimes westward to Alaska and continuing from Siberia across the entire extent of the Russia to the European Plain is the northern needleleaf forest, also called the taiga (a Russian word) or boreal forest. The Southern Hemisphere, lacking D climates except in mountainous locales, has no biome designated as such. However, montane forests of needleleaf trees exist worldwide at high elevation.

12. In which biome do we find Earth's tallest trees? Which biome is dominated by small, stunted plants, lichens, and mosses?

The temperate rain forest biome is recognized by its lush forests at middle and high latitudes, occurring only along narrow margins of the Pacific Northwest in North America, with some similar types in southern China, small portions of southern Japan, New Zealand, and a few areas of Chile. The tallest trees in the world, the coastal redwoods (*Sequoia sempervirens*), are found in this biome. Their distribution is shown on the map in Figure 19-15a. These trees can exceed 1500 years of age and typically range in height from 60 to 90 m (200 to 300 ft), with some exceeding 100 m (330 ft). Virgin stands of other representative trees–Douglas fir, spruce, cedar, and hemlock–have been reduced to a few remaining valleys in Oregon and Washington.

Tundra vegetation is characterized by low, ground-level herbaceous plants and some woody plants. Representative plant species are sedges, mosses, arctic meadow grass, snow lichen, and dwarf willow. The arctic tundra is found in the extreme northern area of North America and Russia, bordering on the Arctic Ocean and generally north of the 10°C (50°F) isotherm for the warmest month.

13. What type of vegetation predominates in the Mediterranean dry summer climates? Describe the adaptation necessary for these plants to survive.

The dominant shrub formations that occupy these regions are short, stunted, and tough in their ability to withstand hot-summer drought. The vegetation is called sclerophyllous (from *sclero* for "hard" and *phyllos* for "leaf"); it averages a meter or two in height and has deep, well-developed roots, leathery leaves, and uneven low branches. Plant ecologists think that this biome is well adapted to frequent fires, for many of its characteristically deep-rooted plants have the ability to resprout from their roots after a fire.

14. What is the significance of the 98th meridian in terms of North American grasslands? What types of eastern inventions were necessary for humans to cope with the grasslands?

In North America, tall grass prairies once rose to heights of 2 m (6.5 ft) and extended westward to about the 98th meridian, with shortgrass prairies farther west. The 98th meridian is roughly the location of the 50 cm (20 in.) isohyet, with wetter conditions to the east and drier to the west. Examples of the problems encountered are the great distances, groundwater too deep for hand-dug wells, lack of energy for pumping, lack of fencing materials, the existence of densely matted sod causing plowing difficulties, and mobile and adaptable native populations. All these conditions represented spatial problems for analysis and later resolution, i.e., railroads, oil-drilling techniques for water wells, wind energy, barbed wire, John Deere's self-scouring steel plow, and, for the unfortunate Native Americans, the Colt six-shooter.

15. Describe some of the unique adaptations found in a desert biome.

Much as a group of humans in the desert might behave with short supplies, plant communities also compete for water and site advantage. Some desert plants, called ephemerals, wait years for a rainfall event, at which time their seeds germinate quickly, develop, flower, and produce new seeds, which then rest again until the next rainfall event. The seeds of some xerophytic species open only when fractured by the tumbling, churning action of flash floods cascading down a desert arroyo, and of course such an event produces the moisture that a germinating seed needs.

Desert plants employ other strategies such as long, deep tap roots (e.g., the mesquite), succulence (i.e., thick, fleshy, water-holding tissue such as that of cacti), spreading root systems to maximize water availability, waxy coatings and fine hairs on leaves to retard water loss, leafless conditions during dry periods (e.g., palo verde and ocotillo), reflective surfaces to reduce leaf temperatures, and tissue that tastes bad to discourage herbivores.

The creosote bush (*Lorrea divaricata*) sends out a wide pattern of roots and contaminates the surrounding soil with toxins that prevent the germination of other creosote seeds, which are possible competitors for water. When a creosote bush dies, surrounding plants or germinating seeds work to occupy the abandoned site, but they must rely on infrequent rains to remove the toxins.

16. What is desertification? Explain its impact.

We are witnessing an unwanted expansion of the desert biome. This is due principally to poor agricultural practices (overgrazing and inappropriate agricultural activities), improper soil-moisture management, erosion and salinization, deforestation, and the ongoing climatic change. A process known as desertification is now a worldwide phenomenon along the margins of semi-arid and arid lands. The role of global climate change and the reduction of soil moisture in marginal lands is a subject of active study with distinct connections still elusive.

17. What physical weathering processes are specifically related to the tundra biome? What typeso f plants and animals are found there?

Winter is governed by intensely cold continental polar air masses and stable high-pressure conditions. A growing season of sorts lasts only from 60 to 80 days, and even then, frosts can occur at any time. Vegetation is fragile in this flat, treeless world; soils are poorly developed periglacial surfaces, which are continually underlain by permafrost. In the summer months only the surface horizons thaw, thus producing a mucky surface of poor drainage. Roots can penetrate only to the depth of thawed ground, usually about a meter. The surface is shaped by freeze-thaw cycles that create permafrost and frozen ground phenomena and gelifluction (solifluction) processes discussed in Chapter 17. Vegetation is described in Question 12 above.

18. What is the relationship between island biogeography and biosphere reserves? Describe a biosphere reserve. What are the goals?

Setting up formal natural reserves called biosphere reserves at continental sites involves principles of island biogeography. Island communities are special places for study because of their spatial isolation and the relatively small number of species present. They resemble natural experiments because the impact of individual factors, such as civilization, can be more easily assessed on islands than they can over larger continental areas. According to the equilibrium theory of island biogeography, developed by biogeographers R. H. MacArthur and E. O. Wilson, the number of species should increase with the size of the island, decrease with increasing distance from the nearest continent, and remain about the same over time, even though composition may vary. These considerations are important to establishing the optimum dimensions for biosphere reserves. The race is on between setting aside tracts of land in reserves and the permanent loss of remaining natural biomes.

The goal of biosphere reserves is to preserve species diversity. The intent is to establish a core in which genetic material is protected from outside disturbances. Ultimately, the goal is to establish at least one reserve for each of the 194 distinctive biogeographical communities presently identified. This way species diversity can be ensured, allowing us to further research and catalogue species which could prove to be medically useful, or successfully adapt to climatic change.

19. Compare the map in Figure20-4 with the composite satellite image inside the front cover of this text. What correlations can you make between the local summertime portrait of Earth's biosphere and the biomes identified on the map?

Comparison by student of these two illustrations (map and composite image), biome by biome.

Overhead Transparencies

As an adopter you are provided with the following figures for overhead projector use.

- Figure 20-4: The ten major global terrestrial biomes
- Figure 20-5: The three levels of a rain forest canopy
- Figure 20-16: Desertification estimates (by the United Nations)

21

The Human Denominator

Overview

We come to the end of our journey through the pages of *GEOSYSTEMS* and an introduction to physical geography. A final chapter seems appropriate, given the dynamic trends that are occurring in Earth's physical systems. Although I realize that there is a risk in such an inclusion since it has not been the practice in our field.

From the 1st to the 12th of June 1992, in Rio de Janeiro, a first-ever global conference called the *United Nations Conference on Environment and Development (UNCED)* took place. The purpose was to address with substance the many fundamental issues that relate to achieving sustainable world development. The historic conference and a summary of the five principal accomplishments are featured in Focus Study 21-1.

The agenda for these precedent-setting meetings was available in July 1991 and includes many topics relevant to physical and cultural/human geography. The first agenda item includes: protection of the atmosphere (climate change, depletion of the ozone layer, and transboundary air pollution); land resources (combating deforestation, soil loss, desertification, and drought); fresh water resources, and oceans, seas, and coastal areas; and the rational use and development of living resources. I think you can see that many aspects of physical geography were at the heart of this conference. The rest of the agenda relates to other aspects of human activity including toxics, biotechnology, species diversity, international traffic of wastes and toxic products, quality of life and the human condition, living and working conditions, and human population growth and per capita impact. Attendance in Rio exceeded 40,000! Of all the nations that attended, the United States held out declaring nonattendance or nonparticipation until just a month before the conference began.

See: United Nations. *Drafts–Agenda 21, Rio Declaration, Forest Principles.* New York: United Nations, E.92.I.16, 92-1-100482-9, June 1992; and, United Nations. *The Global Partnership for Environment and Development–A Guide to Agenda 21.* Geneva: United Nations Conference on Environment and Development, April 1992.

We can use the occasion of this conference to highlight specific applied topics within physical geography and find useful the many reports and documents that are now available. (See "Sustainable Future for Planet Earth," an editorial in *Science* Vol. 253, No. 5016, July 12, 1991: 117; and "Earth Summit" by Bruce Babbitt in *World Monitor* Vol. 5, No. 1, January 26, 1992): 26-31.

The *Student Study Guide* presents 20 "Learning Objectives" to guide the student.

New to the Third Edition

(Note: This section highlights major changes, new features, and additions in the third edition. This does not describe all the rewrite and recast of the text.)

1. A list of key learning concepts begins the chapter.

2. News Report #1: "Oil Spills: Global and Local," provides students with new statistics of recent oil spills as well as identification of common sources of oil waste, such as automobiles.

3. The Gaia Hypothesis has been highlighted in News Report #2: "Gaia Hypothesis Triggers Debate".

4. The January 2, 1989 issue of Time magazine named Earth as the "Planet of the Year." News Report #3 "Time Magazine and the 'Planet of the Year'" elaborates upon the reasons for this recognition and what the edition attempted to achieve for environmental awareness.

5. A detailed history of the Clean Air Act, its objectives and successes.

6. A new quotation by E.O. Wilson from the *Diversity of Life* to summarize the ethics of the environment.

7. The joy of Earth Day celebrations is captured in Figure 21-1.

8. Figure 21-2 illustrates wind power possibilities as an alternative to fossil fuels.

9. A new summary and review section ends the chapter.

Key Learning Concepts

1. *Determine* an answer for Carl Sagan's question, "Who speaks for Earth?"

2. *Analyze* the monetary and health benefits of the U.S. Clean Air Act.

3. *Analyze* the section "The Oily Bird" and *relate* this to energy consumption patterns in the United States and Canada.

4. *Explain* the essential elements of the five Earth Summit agreements and *relate* them to physical geography and *GEOSYSTEMS.*

5. *Appraise* your place in the biosphere and *realize* your physical identity as an Earthling.

Expanded Outline Discussion

The following headings (boldfaced) match some of the first, second, and third order headings in Chapter 21. The narrative under each heading contains information, sources, and anecdotal facts relating to portions of the chapter. Not all text headings are discussed.

THE CLEAN AIR ACT BRINGS A WINDFALL

Students may have differing insights to the environmental legislation, partly due to conflicting viewpoints around the appropriateness and fairness of the Clean Air Acts. I try to encourage students to see the causal factors behind global atmospheric problems such as the urban heat island, urban plume, acid rain, ozone depletion and global warming. The significant causal force they recognize is pollution from automobiles. This exercise encourages students to recognize spatial relationships between Earth systems, and most students have good ideas concerning the appropriateness of policies which could reduce pollution from automobiles.

AN OILY DUCK, The Larger Picture

The estimate for domestic reserves of oil is still a valid issue at this time and should provide a strong basis for the discussion of conservation and energy efficiency (see: *GEOSYSTEMS,* p. 598). This also relates directly to reducing our input to the atmosphere of excessive carbon dioxide from the combustion of fossil fuels–the United States is the largest producer, as shown in Figure 10-32. The October 4, 1991, issue of *Science* (254, no. 5028, p. 29) reported results of U.N. meetings in Nairobi: "United States negotiators staunchly refused to consider making a commitment to reduce carbon dioxide emissions, even as the rest of the industrialized world solidified its stand for emission cuts." The U.S. has modified its position following the election and now has taken a much stronger stance in agreeing to freeze carbon dioxide emissions.

So, beginning with the oil-contaminated cormorant in Figure 21-3 you should be able to weave a fabric of complex interrelated systems that produced this devastation. The bird becomes a metaphor for much larger issues.

THE NEED FOR INTERNATIONAL COOPERATION

See my earlier comments on the first-ever global conference, the UNCED. The dynamics of global atmospheric and oceanic circulation necessitates international cooperation relative to the environment. Many of the contemporary changes occurring within Earth's physical systems are truly global in stature. The UNCED will no doubt be the first of many broad-agenda world conferences specifically dealing with a sustainable future.

Gaia Hypothesis

As I say in the text, "The debate is vigorous regarding the true applicability of this hypothesis to nature, although it remains philosophically intriguing in its portrayal of the relationship between humans and Earth. The hypothesis could form the basis of open discussion and may stimulate individual speculation about the fate of humans and the fate of Earth.

The writings and lectures of American biologist Lynn Margulis speak strongly about aspects of the interrelationships between humans and Earth. The idea that there may be a synergistic, perhaps symbiotic relationship, between the abiotic and biotic spheres (Figure 1-7) is certainly intriguing. Although, the thought of humans being eliminated by evolving changes in Earth's systems remains a disturbing possibility to the senses.

There are many resources concerning the environment. Yet, two perspectives written by geographers are The Human Mosaic, by Andrew Goudie, available through MIT Press, 3rd edition, 1990. And *Environmental Problems; Nature, Economy, and State*, by R.J. Johnson, Belhaven Press, 1989.

WHO SPEAKS FOR EARTH?

We end again with "Planet of the Year" from *Time* magazine and Carl Sagan's eloquent question and concluding answer as to "Who speaks for Earth?" We all do!

The best to all of us: "May we all perceive our spatial importance within Earth's ecosystems and do our part to maintain a life-supporting Earth far into the future."

Annotated Chapter Review Questions

1. What part do you think technology, politics, and future thinking should play in science courses?

This is an important question, especially for an author arguing with the powers that be about whether or not such a chapter as this should be included in a physical geography text. The answer for me was found in the 30-year-old quote from Marston Bates "that we are part of the system of nature." Admittedly, we have to be sensitive to the core content of physical geography, yet I think it is important that students can identify their dependent role in the biosphere and impact on natural systems. Many of our students take this class as their only science course, and this represents their only exposure to the spatial implications of the ongoing experiment Earth's cultures are conducting on this planet. This is especially important because of the specific global impact of Americans and Canadians. Gilbert White has been a leader all these years in the application of physical geography to the real world (see his quote at the beginning of Chapter 19).

2. Given the assessment of monetary benefit from the U.S. Clean Air Act, what is your opinion as to any continued opposition to its regulations?

Personal analysis and response.

3. According to the discussion in the chapter, what worldwide-scale factors led to the *Exxon Valdez* accident? Describe the complexity of that event from a global perspective. In your analysis, examine both supply-side and demand-side issues, as well as environmental and strategic factors.

On 24 March 1989, in Prince William Sound off the southern coast of Alaska, in clear weather and calm seas, a single-hulled supertanker operated by Exxon Corporation, an international energy corporation, struck a reef that was outside the normal shipping lane and spilled 41.64 million liters (11 million gallons) of oil. It took only 12 hours for the *Exxon Valdez* to spill its contents, yet a reasonable cleanup will take years and billions of dollars. Because contingency emergency plans were not in place, and promised equipment was unavailable, response by the oil industry took 10 to 12 hours to activate, about the same time that it took the ship to empty.

Many factors influence our demand for oil. Improvement in automobile efficiency began in 1975 due to federal regulations. During the 1980's, there was a rollback of auto efficiency standards, a reduction in gasoline prices, large reductions in funding for rapid transit development, and the continuing slow demise of America's railroad network. The demand for fossil fuels was also affected by the slowing of domestic conservation programs, elimination of research for energy alternatives, such as solar and wind power, and even the political delay of a law requiring small appliances to be more energy-efficient.

The immediate effect on wildlife was contamination and death, but the issues involved are bigger than these damaged ecosystems. A great many factors influence our demand for oil. Well over half of our imported oil goes for transportation. The demand for fossil fuels in the 1980s also was affected by the slowing of domestic conservation programs, the elimination of research for energy alternatives such as solar and wind, and even delays of a law requiring small appliances to be more energy efficient. Conservation plans again were politically blocked in the Department of Energy in 1990 and early 1991.

The death toll for animals was massive: at least 3000 sea otters killed (or about 20% of the resident otters), 300,000 birds, and uncounted fish, shellfish, plants, and aquatic microorganisms. Sublethal effects, namely mutations, now are appearing in fish. This latter side effect of the spill is serious because salmon fishing is the main economy in Prince William Sound, not oil. Conflicting scientific reports emerged in 1993 as to longterm damage in the region–scientific studies commissioned by Exxon disagreed with scientific studies prepared outside the industry.

4. What is meant by the Gaia hypothesis? Describe several concepts from this text that might pertain to this hypothesis.

Just as the abiotic spheres affect the biosphere, so do living processes affect abiotic functions. All of these interactive effects in concert influence Earth's overall ecosystem. In essence, the planetary ecosystem sets the physical limits for life, which in turn evolves and helps to shape the planet. Thus, Earth can be viewed as one vast, self-regulating organism. The hypothesis contends that life processes control and shape inorganic physical and chemical processes, with the biosphere so interactive that a very small mass can affect a very large mass. Thus, Lovelock and Margulis think that the material environment and the evolution of species are tightly joined; as the species evolve through natural selection, they in turn affect their environment. The present composition of the atmosphere is given as proof of this coevolution of living and nonliving systems (see Table 2-1). "It follows that, if the world is made unfit by what we do, there is the probability of a change in regime to one that will be better for life but not necessarily better for us."

5. Relate the content of the various chapters in this text to the integrative Earth systems science concept. Which chapters in this text help you better understand Earth-human relations and human impacts?

Personal analysis and response by the student. The discussion introducing global wind circulation in Chapter 6 and the AVHRR images of the atmospheric effects of Mount Pinatubo's eruption might be a good place to start.

An important corollary to international environmental efforts is the linkage of academic disciplines. A positive step in that direction is the Earth systems science approach illustrated in this text. Exciting progress toward an integrated understanding of Earth's formation and the operation of its physical systems is happening right now, driven by insights drawn from our remote-sensing capabilities. Never before has society been able to monitor Earth's physical geography so thoroughly. Geographic traditions of spatial analysis and our holistic approach to planetary systems are important aspects of the emerging new science. The GIS revolution is indicative of this evolving role for geography.

6. Explain the potential spatial impact of nuclear warfare on the environment. What is the nuclear winter hypothesis, and what are its potential implications for the environment?

From 1982 to the present, scientific publications have described the nuclear winter hypothesis. This hypothesis indicates that a "nuclear winter" might follow the detonation of only modest numbers of nuclear weapons. As you have learned, the atmosphere is very dynamic. The soot and ash from the many urban fires following a nuclear exchange would enter the atmosphere and raise Earth's albedo, absorbing insolation in the stratosphere and upper troposphere, reradiating it to space, and thus cooling Earth's surface to below freezing even during midlatitude summer months. As the nuclear winter hypothesis developed, others suggested a milder, although still significant, "nuclear autumn" as a consequence.

Also, in 1975, the National Academy of Sciences published its study on the impact of modern nuclear war on stratospheric ozone. This was the beginning of public awareness that modern technological warfare is not an environmentally sound activity. This NAS study states that a nuclear war could lead to the reduction of 40% to 70% of the stratospheric ozone, a catastrophic change for food chains, plants, animals, and humans. The Association of American Geographers issued a strongly worded statement in 1986 regarding nuclear war and its effects (see suggested reading).

7. This chapter states that we already know many of the solutions to the problems we face. Why do you think these solutions are not being implemented at a faster pace?

A question for personal reflection and conclusions.

8. Who speaks for Earth?

Carl Sagan answered his question "who speaks for Earth" with this perspective:

> **We have begun to contemplate our origins: starstuff pondering the**

stars; organized assemblages of ten billion billion billion atoms considering the evolution of atoms; tracing the long journey by which, here at least, consciousness arose. Our loyalties are to the species and the planet. We speak for Earth. Our obligation to survive is owed not just to ourselves but also to that Cosmos, ancient and vast, from which we spring.

(Carl Sagan, *Cosmos, New* York: Random House, 1980, p. 345. Reprinted by permission.)

May we all perceive our spatial importance within Earth's ecosystems and do our part to maintain a life-supporting Earth far into the future. And may we as geography educators realize our role in the lives of the thousands of students that pass through our classrooms.

Please feel free to communicate criticism, questions, ideas, opinions, and your thoughts on *GEOSYSTEMS.* I will do everything I can to respond, revise, correct, and update future editions of this text. Thanks ahead of time for any feedback you might extend to me. The best to you and to geographic education.

Overhead Transparencies

As an adopter you are provided with the following figures for overhead projector use.

- Figure 21-5 and 21-4: Recent worldwide oil spills; 1989 *Exxon Valdez* oil spill map

---